Smart Nanomaterials Technology

Series Editors

Azamal Husen, Wolaita Sodo University, Wolaita, Ethiopia

Mohammad Jawaid, Department of Chemical and Petroleum Engineering, United Arab Emirates University, Al Ain, United Arab Emirates

Nanotechnology is a rapidly growing scientific field and has attracted a great interest over the last few years because of its abundant applications in different fields like biology, physics and chemistry. This science deals with the production of minute particles called nanomaterials having dimensions between 1 and 100 nm which may serve as building blocks for various physical and biological systems. On the other hand, there is the class of smart materials where the material that can stimuli by external factors and results a new kind of functional properties. The combination of these two classes forms a new class of smart nanomaterials, which produces unique functional material properties and a great opportunity to larger span of application. Smart nanomaterials have been employed by researchers to use it effectively in agricultural production, soil improvement, disease management, energy and environment, medical science, pharmaceuticals, engineering, food, animal husbandry and forestry sectors.

This book series in Smart Nanomaterials Technology (SNT) publishes cutting edge research monographs and edited volumes by top researchers and professionals across the globe. This book series aims to comprehensively cover topics in the fabrication, synthesis and application of these materials for applications in the following fields:

- Energy Systems—Renewable energy, energy storage (supercapacitors and electrochemical cells), hydrogen storage, photocatalytic water splitting for hydrogen production
- Biomedical—controlled release of drugs, treatment of various diseases, biosensors
- Agricultural—agricultural production, soil improvement, disease management, animal feed, egg, milk and meat production/processing
- Forestry—wood preservation, protection, disease management
- Environment—wastewater treatment, separation of hazardous contaminants from wastewater, indoor air filters

Review Process

The proposal for each volume is reviewed by the main editor and/or the advisory board. The chapters in each volume are individually reviewed single blind by expert reviewers (at least two reviews per chapter) and the main editor.

Ethics Statement for this series can be found in the Springer standard guidelines here - https://www.springer.com/us/authors-editors/journal-author/journal-author-helpdesk/before-you-start/before-you-start/1330#c14214.

Arafat Abdel Hamed Abdel Latef
Editor

Chitosan Nanoparticles for Plant Resilience Under Abiotic and Biotic Stressors

Springer

Editor
Arafat Abdel Hamed Abdel Latef
Department of Botany and Microbiology,
Faculty of Science
Qena University
Qena, Egypt

ISSN 3004-8273 ISSN 3004-8281 (electronic)
Smart Nanomaterials Technology
ISBN 978-981-95-6421-7 ISBN 978-981-95-6422-4 (eBook)
https://doi.org/10.1007/978-981-95-6422-4

This Springer imprint is published by the registered company Springer Nature Singapore Pte Ltd.
The registered company address is: 152 Beach Road, #21-01/04 Gateway East, Singapore 189721, Singapore

Preface

The global agricultural sector is at a pivotal crossroads. With the world's population expected to approach 10 billion by 2050, ensuring food security has become an urgent priority. This challenge is further intensified by the growing impacts of climate change, which bring more frequent and severe abiotic stresses while aggravating persistent and adaptive biotic stressors. Although traditional farming methods, heavily reliant on chemical fertilizers and pesticides, have offered temporary solutions, they have come at a steep cost to environmental health, soil biodiversity, and sustainability in the long term. The situation calls for innovative, eco-friendly, and highly effective approaches to enhance crop resilience and guarantee sustainable agricultural productivity.

It is within this context that the emergence of nanotechnology in agriculture offers a beacon of hope. Among the myriad of nanomaterials, chitosan nanoparticles (CNPs) have distinguished themselves as a uniquely promising tool. Derived from chitin, the second most abundant natural biopolymer on Earth, chitosan is inherently biodegradable, biocompatible, non-toxic, and possesses innate antimicrobial and eliciting properties. When engineered into nanoparticles, these intrinsic qualities are profoundly enhanced, granting CNPs superior ability to penetrate plant tissues, act as smart delivery systems for agrochemicals, and effectively modulate plant physiological and molecular responses to stress.

This edited volume, *Chitosan Nanoparticles for Plant Resilience Under Abiotic and Biotic Stressors*, is conceived to provide a comprehensive and in-depth exploration of this dynamic field. The book is systematically structured to guide the reader from fundamental concepts to cutting-edge applications and future perspectives.

We begin by establishing a solid foundation. The opening chapters provide an overview of chitosan nanoparticles themselves, detailing their synthesis, properties, and characterization, followed by a thorough examination of the devastating agricultural impacts of both abiotic and biotic stresses. This sets the stage for the core of the book: a detailed, evidence-based dissection of the role of CNPs in mitigating these challenges.

The subsequent sections are dedicated to specific stress scenarios. A series of chapters meticulously documents the efficacy of CNPs in ameliorating abiotic stresses, including salinity, drought, temperature extremes, and heavy metal toxicity. Parallel to this, another suite of chapters delves into the application of CNPs against biotic threats, showcasing their potential in suppressing viral, bacterial, and fungal diseases, as well as their role in controlling insect pests.

Finally, we look toward the horizon. The concluding chapters address the emerging trends and future prospects of CNP research, highlighting novel applications and integration with other technologies. We also confront the practical challenges in application, addressing concerns regarding large-scale synthesis, cost-effectiveness, regulatory frameworks, and environmental fate, to present a balanced and realistic view of the path from laboratory breakthrough to field implementation.

This book brings together the expertise of leading scientists and researchers from around the world. It is designed to be a valuable resource for a wide-ranging audience, including plant scientists, nanotechnologists, agricultural engineers, environmental researchers, graduate students, and industry professionals eager to harness the potential of this innovative technology. The overarching objective is that this compilation not only educates and informs but also inspires further advancements, collaboration, and responsible development. By doing so, it contributes to building a more resilient and sustainable agricultural system for future generations.

The editor extends heartfelt gratitude to all contributing authors for their exceptional chapters and to the publishing team for their unwavering support in bringing this important project to fruition.

Qena, Egypt Arafat Abdel Hamed Abdel Latef

Contents

About the Editor

Arafat Abdel Hamed Abdel Latef is a professor of plant physiology at the Department of Botany and Microbiology, Faculty of Science, Qena University (formerly South Valley University), Egypt. He received his B.Sc. in 1998, M.Sc. in 2003, and Ph.D. in 2005 from South Valley University, Egypt. Professor Arafat has published 98 research/review articles in high-quality international and local journals. He has edited two books for Academic Press (Elsevier), two books for CRC Press, Taylor & Francis Group, LLC, and four books for Springer-Verlag, Heidelberg He has also published 32 book chapters in international volumes, including John Wiley and Sons Inc., Taylor & Francis Group, Academic Press (Elsevier), and Springer-Verlag, Heidelberg He has received several awards, such as the Springer Nature Editor of Distinction Award 2025, the Egypt State Award of Excellence in Agricultural Sciences 2021, the South Valley University Award of Scientific Excellence 2019, the Top Peer Reviewer Award 2019 Powered by Web of Science and Publons, and the South Valley University Award of Scientific Publication for the years from 2010 to 2024.

Chitosan Nanoparticles: An Overview

Muhammad Nauman Khan, Nasir Assad, Tooba, Barkat Ullah, Abdul Razzaq, Tabassum Yaseen, Maryam Bibi, Syed Mukaram Shah, Barkat Ali, and Muhammad Adnan

Abstract Chitosan nanoparticles (CNPs) have gained increasing attention as sustainable, biocompatible, and multifunctional nanomaterials with diverse applications in agriculture, medicine, food technology, and environmental remediation. Chitosan, derived from chitin through deacetylation, possesses inherent antimicrobial and biodegradable properties that are further enhanced at the nanoscale, enabling better solubility, surface activity, and bioavailability. This chapter provides a comprehensive overview of the synthesis of CNPs through methods such as ionic gelation, reverse micelles, emulsification, and spray drying, and shows how the processing parameters influence particle size, stability, and functional performance.

M. N. Khan (✉) · B. Ullah
Department of Botany, Islamia College Peshawar, Peshawar, Pakistan
e-mail: nomiflora@uop.edu.pk

M. N. Khan
Biology Laboratory, University Public School, University of Peshawar, Peshawar, Pakistan

N. Assad · Tooba
Institute of Chemistry, University of Sargodha, Sargodha, Pakistan

A. Razzaq
Qarshi Herb Research Center, Qarshi Industries (Pvt.) Ltd. Hattar, Haripur, Pakistan

Qarshi University, Lahore, Punjab, Pakistan

T. Yaseen
Department of Botany, Bacha Khan University, Charsadda, Khyber Pakhtunkhwa, Pakistan

M. Bibi
Department of Botany, University of Malakand, Lower Dir, Chakdara, Khyber Pakhtunkhwa, Pakistan

S. M. Shah
Center of Plant Biodiversity, University of Peshawar, Peshawar, Pakistan

B. Ali
College of Horticulture, Sichuan Agricultural University, Wenjiang, Chengdu, China

M. Adnan
Department of Chemistry, Bacha Khan University, Charsadda, Khyber Pakhtunkhwa, Pakistan

A. A. H. Abdel Latef (ed.), *Chitosan Nanoparticles for Plant Resilience Under Abiotic and Biotic Stressors*, Smart Nanomaterials Technology,
https://doi.org/10.1007/978-981-95-6422-4_1

The physicochemical properties of CNPs—including particle size distribution, zeta potential, morphology, porosity, and chemical composition—are studied in relation to their mechanisms of action, such as membrane disruption, endocytosis, controlled release, and elicitation of host defense responses. Key applications are being investigated across sectors: as smart agrochemical carriers and elicitors of plant immunity, targeted drug and gene delivery systems, antimicrobial food packaging, wastewater treatment agents, aquaculture enhancers, and cosmetic additives. Despite their potential, challenges remain in terms of large-scale production, raw material variability, regulatory approval, and long-term ecological safety. The chapter concludes with an outlook for the future prospects, focusing on green synthesis, standardized characterization, interdisciplinary optimization, and integration with new technologies. Overall, CNPs represent a promising frontier in sustainable nanotechnology, bridging the gap between natural biopolymers and advanced applications to address pressing global challenges in health, agriculture, and the environment.

Keywords Biomedical applications · Chitosan nanoparticles · Green synthesis · Physicochemical properties · Sustainable nanotechnology

1 Introduction

Nanotechnology has emerged as a revolutionary discipline that stands between the fields of physics, chemistry, biology, and engineering synthesizing materials with at least one dimension at the nanoscale [1, 8, 35, 39]. At such small dimensions, matter exhibits novel properties that are not observed in the bulk, including quantum confinement effects, high catalytic activity, altered optical behavior, and different modes of interaction with biological systems [9, 22]. These qualities have propelled the application of nanotechnology in many fields, ranging from personalized medicine to green agriculture and environmental remediation [3, 29]. In the context of such dynamic development, biopolymer-based nanostructures present a unique niche as they represent the combination of nanoscale functionality with the sustainability and functionality of renewable natural resources [9, 10, 26, 32, 61, 66]. Of these, chitosan nanoparticles (CNPs) are of particular interest due to their unusual biocompatibility, biodegradability, and potential for extreme chemical modification, allowing these nanoparticles to be used in an enormous number of high-value applications [19]. As illustrated in Fig. 1, CNPs demonstrate multifunctional applications in agriculture, biomedical sciences, antimicrobial activity, and photocatalysis.

Chitosan is derived from chitin, a structural polysaccharide with multiple industrial uses that are not yet fully utilized. The molecular structure of chitin consists of β-(1 → 4)-linked N-acetyl-D-glucosamine (chitin), which forms a crystalline structure that makes chitin poorly soluble and imparts physical properties to the material [58]. Chitosan is widespread in the shells of marine crustaceans, in the cuticle of insects and in the cell walls of fungi, making it the most widespread polysaccharide in nature after cellulose [74]. In the industrial production of chitosan, crustacean exoskeletal

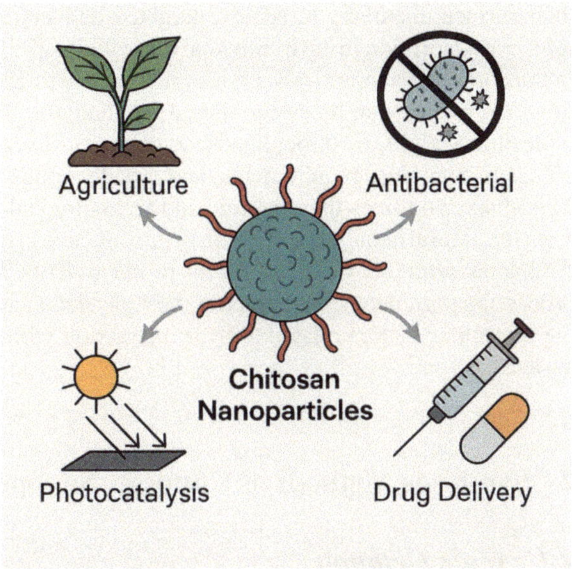

Fig. 1 Central schematic representation of chitosan nanoparticles (CNPs) and their major applications

waste from seafood processing is usually processed and subjected to either alkaline deacetylation or enzymatic catalysis for deacetylation [27]. This produces a polymer in which the degree of deacetylation controls the number of amine groups per polymer, the solubility characteristics, and the reactivity pattern [21]. One of the definitive characteristics of chitosan is that it protonates under acidic conditions, resulting in a net positive charge in the polymer chain [18]. Its unusual cationic character as a polysaccharide forms the basis for its ability to bind with anionic biomolecules and surfaces, which accounts for its antimicrobial weapon system, film-forming biopolymer, and adaptability [54].

Although chitosan is very useful in its bulk form, especially in wound dressings, food coatings, and wastewater treatment reagents, certain physico-chemical limitations restrict its use in situations requiring stepwise release control, rapid bioavailability, or penetration of biological barriers [28]. Chitosan's low solubility at physiological pH, a modest surface area, and poor dissolution rate limit its suitability for various activities [7]. Such limitations can be methodologically addressed by designing the polymer at the nanoscale, thus changing its functionality to the most fundamental level. CNPs form a more homogeneous dispersion in aqueous fluids, interact better with the target tissues or molecules and can be transported much more effectively through mucosal and cellular barriers [2].

The nanoscale modification of chitosan has several application-defining benefits [67]. The vast surface-to-volume ratio expands the contact points of the molecules and promotes binding interactions and catalytic capacities. The size of the particles and their aspect can be refined during synthesis to achieve optimal transport, adhesion, or diffusion characteristics, depending on the intended application [30]. CNPs can also be further functionalized by conjugating or adsorbing functional ligands to their surface chemistry to achieve targeted delivery in therapeutic applications, or selective adsorption in environmental remediation [37]. Active ingredients entrapped in the cores of nanoparticles are less susceptible to chemical or enzymatic degradation, and release can be easily altered by changing the cross-link density, polymer molecular weight, or through surface coatings. These capabilities result in higher efficacy, lower dose requirements, and broader applications than bulk chitosan [75]. This chapter outlines the scientific and technological basis by tracing the evolution from the natural abundance of chitin to the selective activity of chitosan nanoparticles. It explains synthesis strategies for the production of CNPs, provides insights into the processing parameters that can alter their physicochemical properties, and discusses the molecular factors related to their interaction with biological and environmental systems.

2 Synthesis Methods of Chitosan Nanoparticles

2.1 Ionic Gelation

Ionic gelation is also the most commonly described technique for the preparation of CNPs, which became widely known in the 1990s as a fast, solvent-free method suitable for biomolecules [46]. The method is based on the fact that chitosan is a polycationic material at an acidic pH, which causes the primary amino groups to become protonated (instead of being free of charge) and form an electrostatic bond with multivalent polyanions, mostly sodium tripolyphosphate (TPP). The procedure is initiated by dissolving chitosan in weak organic acids like acetic, citric, or lactic acid, which ensures protonation and complete retraction of the molecules. When the TPP solution is added in a controlled manner with vigorous stirring, immediate ionic cross-linking occurs and nucleation and growth of the NPs take place within milliseconds (ms). The ratio of nucleation and growth strongly depends on the flexibility of the polymer chains, the ionic strength, the pH, and the stoichiometric ratio between TPP and chitosan (Guide). These parameters make it possible to change the particle size without restriction between <50 and >500 nm. For example, increasing the TPP/chitosan ratio leads to increased cross-linking density, but can lead to aggregation as soon as the zeta potential falls below the stability threshold. On the other hand, lower concentrations of crosslinkers lead to softer NPs, characterized by slow release kinetics, which is favorable for prolonged drug delivery. The molecular weight of the chitosan is also an important factor; low molecular weight chitosan can be used to produce smaller, more uniform particles, whereas high molecular weight types produce larger particles with strong mucosal adhesion. The mild conditions of this method allow the direct encapsulation of thermostable compounds such as enzymes, growth factors, nucleic acids, and antimicrobial agents of plant origin without loss of bioactivity. It is also environmentally friendly and economically feasible, which makes it interesting for the agricultural field of nanocarriers and the preservation of foodstuffs. Ionic gelation often results in batch-to-batch instability and low long-term stability in saline media and solutions, although stabilizers and co-crosslinking systems continue to be developed.

2.2 *Reverse Micelle Technique*

The reverse micelle approach to CNPs production allows extreme control of NP uniformity by sequestering the chitosan crosslinking molecules in nanosized water-filled cores suspended in the non-polar continuous phase. The method was developed based on microemulsion chemistry, which was later transferred to biopolymer nanoparticles to achieve precise size control. The interaction of surfactants with the water/oil interface causes them to assemble into nanoscopic water invasion structures supported by surfactants such as sodium bis(2-ethylhexyl) sulfosuccinate (AOT) or cetyltrimethylammonium bromide (CTAB). The agents such as glutaraldehyde or TPP crosslink chitosan dissolved in these aqueous pools. The size of the resulting NP size is controlled via the size of the micellar cores, which depends on a molar ratio of water to surfactant (W/S) [15].

Since the growth of the particles is physically limited by the edges of the micelles, it is possible to produce consistently monodisperse NPs with a size of less than 100 nm (nm), which is a significant advantage in the field of intravenous administration, since in this case the limitation of the size distribution has a direct impact on biodistribution and excretion rates. Once the cross-linking reaction is complete polar solvents like ethanol are added and the surfactants form a micelle that can be separated by shaking to release the NPs, which are purified by centrifugation and subsequent washing to remove traces of surfactants and organic compounds. Although direct control of size and morphology is not possible with any other method, the scalability of this technique is desirable due to the large amounts of surfactants, the complicated solvent recovery and purification requirements to meet the safety standards required in biomedicine. However, the production of reverse micelles is still under development for advanced nanomedicine, where precise physicochemical modification is necessary to accomplish specific receptor-directed (target) binding, sustained delivery and diagnostic imaging.

2.3 *Emulsification Method*

The emulsification method is a very versatile technique for formulating CNPs that can encapsulate hydrophilic and hydrophobic active ingredients, and it can be used as a preferred technique in the pharmaceutical, nutraceutical, and cosmetic industries. It usually consists of an emulsification mechanism (a water-in-oil (W/O) type emulsion) in which an aqueous chitosan solution is dispersed in an immiscible organic solvent such as liquid paraffin or cyclohexane, and stabilized by emulsifiers such as Span 80 or Tween 80. Typically, homogenization, rotor–stator mixing, or ultrasonication are used to further diminish the size of the droplets to nanometers (nm) [24]. Cross-linking is then initiated, either by a chemical method using glutaraldehyde, genipin, or epichlorohydrin, or an ionically-based technique with the use of TPP, diffused into the droplets to solidify the NPs' structure. The recovery of NPs is done by removing the solvent by rotary evaporation, followed by several drying processes to eliminate unavoidable reagent residues. The encapsulation of lipophilic drugs can be achieved by emulsification either into the hydrophobic phase or into the particle structure, which can increase the bioavailability of pharmaceutically difficult-to-utilize drugs. The size of the particles is controlled by parameters like type and amount of emulsifier, degree of homogenization, ratio between aqueous and oil phase, and amount of crosslinker, which are usually between 100 and 800 nm [52]. Co-encapsulation strategies are also possible, in which several active ingredients with different solubilities are introduced into a particle system to be released synergistically with each other. However, there are other problems that complicate the method, such as the possible toxicity of residual solvents or crosslinkers, environmental problems with the disposal of the organic phase, and the energy requirements of high-shear emulsification on an industrial scale [69].

2.4 Spray Drying Technique

Spray drying is an established synthesis process that has been adapted for the production of NPs, especially when long stability and solid forms of powders are required [51]. In the procedure, chitosan solution or suspension with the active ingredient is atomized into fine droplets using a nozzle or rotary atomizer and placed in a hot air chamber. The droplets solidify quickly through rapid evaporation of the solvent, leading to dry NPs that are collected through a cyclone separator. The final morphology, size, and moisture content of the particles are determined by the process parameters, i.e., inlet and outlet air temperature, atomization pressure, feed rate, nozzle setup, and chitosan concentration. The particle sizes created by spray drying are between 200 nm and a few micrometers; with more precise optimization, it is even possible to achieve sizes below 300 nm. Spray drying also offers unrivaled scale, low unit manufacturing costs and ease of downstream processing, in cases where fine powders are needed, such as seed coating, inhalation products, or as an ingredient in dry foods. Nevertheless, heat-sensitive materials can degrade at high input temperatures, so alternative modified spray drying systems like freeze or vacuum-spray drying or protection in protective matrices are required to maintain bioactivity. This also increases the functionality of the resulting NPs as chitosan can be flexibly mixed with other polymers prior to spray drying (e.g., alginate, gelatin).

2.5 *Effect of Synthesis Methods on Particle Size, Stability, and Functionality*

The synthetic method used to produce CNPs selects not only the size and shape of the particles and their congruence, but also their colloidal stability, surface chemistry, loading efficiency, and resulting biological or functional performance. These interdependent parameters are based on changes in the reaction environment, type of cross-linking, energy levels, solvent systems, and post-processing methods. The most significant parameter is particle size, as it affects biodistribution, cell uptake, and diffusion properties. One method is to physically trap particle growth, as in the inverted micelle approach, which usually produces ultra-small NPs ($\ll$100 nm) with narrow size distributions, which favor intravenous delivery of drugs and deep penetration of tissue. Conversely, the spray drying technique employs the use of droplet generation before evaporation of the solvent tends to yield larger and more polydisperse particles (>200 nm) that can be beneficial in agriculture as they ensure long release and minimal leaching. Ionic gelation usually produces NPs in the 50–500 nm size range, and the size can be tuned by manipulating the polymer-to-cross-linker ratio, pH, and ionic strength of the medium. Small sizes are normally achieved with a lower chitosan concentration and higher stirring speeds, while higher viscosity or slower mixing favors aggregation and the formation of larger particles [23].

The level of protonation of the amino groups of chitosan and the character of the crosslinker also determine the stability, which is characteristic of the zeta potential. Electrostatic repulsion is caused by high positive surface charges (>+30 mV) resulting from ionic gelation with TPP under mildly acidic conditions, leading to decreased particle aggregation in aqueous media. In contrast, NPs synthesized through emulsification might have a lower surface charge due to the masking of the amino groups by the addition of a surfactant or by hydrophobic surface stabilizers, which in this case require additional stabilizing measures in the form of a coating with polyanions (e.g., alginate) or polyethylene glycol (PEG) conjugation. Spray-dried particles are in a solid state and are subject to higher physical stability during storage, so additional care in the re-dispersion methods avoids irreversible aggregation during rehydration. The synthesis route is closely linked to functionality as it can determine not only the structural parameters but also the surface accessibility of the functional groups, porosity, and loading.

Highly hydrated porous NPs with exposed amino groups can be achieved by ionic gelation, making them ideal for use in drug delivery systems to create mucoadhesive or antimicrobial properties on the surface. Particles synthesized using reverse micelles tend to have smooth surfaces and are less porous, which can be of interest for controlled release formulations where diffusion over time is required. Emulsification offers multifunctional systems in the form of co-encapsulation of hydrophobic and hydrophilic agents in the different areas of the same particle, as used in the combined coating of foods with antimicrobial and antioxidant actives. Spray drying results in compact, less-permeable particles, which is advantageous for the protection of labile compounds during storage or transportation under harsh environmental conditions.

The bioactivity is also indirectly influenced by the release profile of the encapsulated compounds, which is controlled by the synthesis method. The smaller, high surface area particles that can be produced by ionic gelation or reverse micelle techniques tend to have a more explosive release without secondary coating. Spray-dried or higher viscosity emulsions may contain a larger number of smaller particles that are released slowly by diffusion which again is advantageous when long-lasting action is required in agriculture or treatment. Last, the architecture of the particles may differ due to method-specific variations, which may affect surface functionalization with targeted ligands, imaging agents reactive components and responsive moieties. For example, ionic gelation involves mild aqueous conditions that allow direct post-synthesis conjugation of sensitive biomolecules, as opposed to reverse micelle or emulsification approaches that may require post-synthesis washing steps to eliminate the surfactants before functionalization. Overall, the interaction between synthesis conditions and NP properties emphasizes why it is important to align both methods and applications. The technology that is successful in the domain of engineering, monodisperse, nanoscale drug-carrying vehicles for targeted therapy, can be less suitable in the modest application area where agriculture, mechanical stability and long-term drug release are key factors. New hybrid and green synthesis technologies are attempting to fill these (technological) gaps by integrating the precision of laboratory-based manufacturing with the sustainability and high throughput of larger-scale applications to ultimately increase the applicability and efficiency of chitosan nanoparticles in a variety of industries. Table 1 provides a comparative summary of CNPs and their synthesis methods, highlighting the impact on particle size, stability, functionality, and limitations.

3 Physicochemical Properties of CNPs

3.1 Particle Size and Size Distribution

Particle size determines the physicochemical reactivity, mobility, and biological fate of CNPs and influences everything from the kinetics of cell internalization to transport into the environment. The nanosize typically achieved for CNPs (20–500 nm) leads to a significant increase in surface energy, which can enhance interactions with biological membranes or catalytic surfaces. Ultra-small NPs (<50 nm) produced by methods such as reverse micelle or controlled ionic gelation are especially effective in traversing biological barriers like the intestinal epithelium or plant leaf cuticle. Conversely, larger particles (>300 nm) fabricated by spray drying or high-viscosity emulsions are advantageous in controlled-release agricultural formulations as they reduce premature leaching into the environment [11]. Size distribution uniformity, quantified through the polydispersity index (PDI), is essential for reproducible performance. Monodisperse populations (PDI < 0.2) ensure consistent pharmacokinetics in biomedical applications, while broader distributions may be acceptable—or even

Table 1 Comparison of CNP synthesis methods and their effects

Method	Particle size and control	Stability and surface charge	Functional advantages	Limitations	References
Ionic gelation	Tunable between <50 and >500 nm by adjusting TPP/ chitosan ratio, pH, stirring, and Molecular weight (MW)	High positive zeta potential (>+30 mV) promotes colloidal stability; MW influences size uniformity and adhesiveness	Mild, solvent-free; preserves bioactivity; suitable for enzyme or drug encapsulation	Batch variability; unstable in high-salt media; less uniform size distribution	[70]
Reverse micelle	Produces ultra-small (<100 nm), highly monodisperse NPs by confining growth in micellar cores	Smooth, less porous surface; structural uniformity	Excellent for intravenous drug delivery due to narrow size distribution	Complex, uses surfactants/ organic solvents; difficult to scale; requires extensive purification	[71]
Emulsification	Broader size range (100–800 nm), controllable via homogenization and emulsifier parameters	Surface charge may be masked by surfactants; it often requires post-coating for stability	Capable of co-encapsulating hydrophilic and hydrophobic agents; versatile for pharma/ cosmetics	Uses organic solvents; energy-intensive; potential toxicity from residual solvents or crosslinkers	[6]
Spray drying	Yields larger particles (≥200 nm) with the potential to fine-tune below 300 nm	Dry powders with good storability; re-dispersal needs care to prevent aggregation	Industrially scalable; low cost; good for powders like seed coatings or inhalables	Heat may degrade sensitive molecules; may require freeze- or vacuum-drying variants to preserve activity	[44]

advantageous—in environmental contexts where staged release is desired. Particle size is a function of numerous synthesis parameters: Polymer concentration, degree of deacetylation, cross-linker stoichiometry, ionic strength, stirring speed, and even reagent addition order. For example, high shear mixing during ionic gelation limits particle growth, while low-shear conditions promote aggregation. Post-synthesis processing, such as ultrasonic treatment, can further reduce particle size, but may also destabilize sensitive payloads. In addition to dynamic light scattering (DLS), high-resolution imaging using transmission electron microscope (TEM) or cryo-TEM is critical for distinguishing between individual NPs and soft aggregates, which may appear as larger hydrodynamic diameters when measured in aqueous solution. This distinction is particularly important in regulatory evaluations, where the actual size of the NPs needs to be verified.

3.2 Zeta Potential and Stability

The zeta potential reflects the electrokinetic potential at the slip plane of NPs in suspension and serves as an indicator of colloidal stability. In CNPs, the positive surface charge results primarily from protonated amino groups, the extent of which depends on the pH of the environment and the degree of deacetylation. At acidic pH, protonation is maximal and leads to high positive zeta potentials ($> +40$ mV), which stabilize dispersions via strong electrostatic repulsion. At near-neutral or basic pH, partial deprotonation reduces surface charge and increases the risk of aggregation unless steric stabilizers are introduced. The ionic environment plays a dual role: while electrolytes can shield repulsive forces and induce flocculation, divalent cations such as Ca^{2+} or Mg^{2+} may also serve as secondary cross-linkers and alter mechanical stability. The Zeta potential also influences functional performance: when administering drugs, a positive charge enhances binding to negatively charged mucosal and cell surfaces and thus absorption; in antimicrobial coatings, the electrostatic attraction to bacterial membranes promotes the destruction of the cell wall. However, an excessively high surface charge can lead to rapid excretion in the bloodstream due to opsonization. The analytical determination of the zeta potential is usually carried out by electrophoretic light scattering, but the particle size and the viscosity of the dispersion medium must be taken into account in the interpretation. For agricultural sprays, stability tests should simulate ionic strengths and pH conditions in the field to ensure that laboratory stability is transferable to real-world conditions.

3.3 Morphology and Surface Characteristics

The morphological features of CNPs—shape, texture, porosity, and internal structure—are central to their functional behavior. The spherical morphology prevalent in ionic gelation and reverse micelle systems facilitates predictable flow and uniform coating behavior, while non-spherical or irregular particles may offer unique release kinetics by altering surface area exposure. Porosity is especially important for high loading capacity: mesoporous or microporous CNPs enable sustained diffusion of encapsulated compounds, while dense particles with low porosity provide excellent protective encapsulation for sensitive active. Nanoscale surface roughness can influence the adsorption profiles of proteins in biological fluids or the adhesion of particles to plant surfaces. The hydrophilicity or hydrophobicity, which is determined by the ratio of exposed amino, hydroxyl, and acetyl groups, affects the dispersion behavior: Highly hydrophilic CNPs can be easily suspended in aqueous media, whereas hydrophobically modified surfaces can interact better with lipid membranes or waxy plant cuticles. Advanced characterization techniques provide insights beyond shape observation. Atomic force microscopy (AFM) can quantify surface roughness at the nanometer scale, nitrogen adsorption–desorption isotherms provide information on porosity and surface area, and small-angle X-ray (SAXS) can provide data on internal nanostructuring. These properties can be highly influenced by post-synthesis modifications such as polymer grafting, layer-by-layer assembly, or coating with biopolymers like alginate or gelatin [56].

3.4 Chemical Composition and Functional Groups

The chemical composition of CNPs is based on the polysaccharide backbone of chitosan, which consists of β-(1 → 4)-linked D-glucosamine and N-acetyl-D-glucosamine units. The abundance of amino groups not only determines the charge but also serves as reactive sites for cross-linking, conjugation, or coordination with metal ions. The most significant factors affecting these properties are the degree of deacetylation (DD). The more deacetylated the substance, the higher the density of free amino groups, the better the solubility in acidic media and the higher the antimicrobial activity [20]. The structural integrity after synthesis can be confirmed by analyzing the functional groups of the synthesis product by Fourier transform infrared spectroscopy (FTIR) or solid-state nuclear magnetic resonance spectroscopy (NMR), and the chemical changes in the product can also be observed. The appearance of shifts in the amide I or amide II bands could be an indication of cross-linker binding, whereas the shift in the hydroxyl stretching regions could be caused by hydrogen bonding to the encapsulated molecules. Additional surface-sensitive information about the chemical state of the sample can be obtained by X-ray photoelectron spectroscopy (XPS), but this method usually does not have the sensitivity to confirm the success of functionalization with ligands or polymer coatings.

The adaptability of the CNP surface is supported by the chemical reactivity. Targeted peptides, antibodies or fluorescent probes can be added through covalent modification, and ionic interactions are suitable for binding polyanionic bioactives like DNA or heparin. In the environmental field, functional groups may filter out specific pollutants, like heavy metals or dyes, through functionalization making CNPs efficient adsorbents [28]. By perfecting control over these physicochemical properties, researchers can customize CNPs to meet specific performance requirements, e.g., to maximize antimicrobial efficacy in food packaging or achieve precise biodistribution in nanomedicine. Since all properties interact optimization must be done at the system level and include incorporating synthesis engineering, surface engineering, and the targeted application environment.

4 Mechanisms of Action

CNPs unfold their functionality in a cascade of linked physicochemical and biological events that begin as soon as they come into contact with a biological interface. The essential feature of their action lies at the interface of the highly electrostatically attractive charge between the positively charged amino groups of chitosan and the negatively charged assemblies on cellular or microbial surfaces, e.g., phosphate chemicals in phospholipids, carboxyl sites in proteins, and anionic polysaccharides (e.g., teichoic acids or lipopolysaccharides). The interaction is not only adhesive, but can also actively destabilize the structural integrity of the membrane, by removing the stabilizing divalent cations, Mg^{2+} and Ca^{2+}. In microorganisms, the result of this destabilization may be membrane thinning, the formation of nano-sized pores, and the loss of vital intracellular contents. Functionalization of NPs in plants may involve binding to the anionic areas of the leaf cuticle or root epidermal cells, which can increase retention, which can be used to deliver bioactive molecules locally or systemically to target sites. In mammalian systems, these interactions can be tuned through careful control of particle size, charge density, and surface functionalization so that the NPs promote cell penetration without causing significant cytotoxicity [45].

After surface adhesion, CNPs can facilitate the movement of drugs across restrictive biological barriers. In epithelial tissue, their cationic nature allows them to interact with tight junction complexes and transiently loosen them by modulating the conformation of proteins such as claudins and occludins. This temporarily increases paracellular permeability, enabling hydrophilic or high-molecular-weight molecules to bypass transcellular pathways. In addition, many CNPs are internalized via energy-dependent processes such as clathrin-mediated endocytosis, caveolae-mediated endocytosis, or macropinocytosis, with the preferred pathway largely dependent on the size of the NP, the surface chemistry, and the specific cell type. Smaller particles, often smaller than 200 nm, are more likely to be taken up by receptor-mediated pathways, whereas larger particles are more likely to be ingested non-specifically. In the endosomal compartment, the buffering capacity of the chitosan amino groups contributes to the so-called proton-sponge effect, in which osmotic swelling and possible rupture of the endosomal membrane release the NPs and their charge into the cytosol. In plants, penetration can also occur through stomatal openings or microchannels in the cuticle, with subsequent distribution through the vascular system, enabling agrochemical payloads to reach distal tissues [62].

The antimicrobial and bioactive responses elicited by CNPs are based on several converging mechanisms. Their ability to bind microbial cell surfaces can impair the synthesis of structural biomolecules and thus weaken the cell walls of bacteria or the cell membranes. In Gram-negative bacteria, penetration into the outer membrane disrupts the lipopolysaccharide layer and increases permeability, while in Gram-positive bacteria, the dense peptidoglycan matrix is weakened, making the cells more vulnerable to osmotic stress. After internalization, CNPs can bind nucleic acids through electrostatic interactions and hydrogen bonding, inhibiting transcription and translation processes. They can also chelate essential trace metals such as iron, magnesium, or zinc, thereby inactivating metal-dependent enzymes that are critical for microbial survival. In some systems, NPs may directly or indirectly generate reactive oxygen species that cause oxidative stress, which damages proteins, lipids, and nucleic acids [49]. In addition to their direct antimicrobial effect, CNPs can also trigger positive bioactive effects in host organisms. In plants, they act as elicitors of innate defense mechanisms by stimulating the biosynthesis of phytoalexins, chitinases, and other proteins associated with pathogenesis. This primes the plant for systemic acquired resistance, allowing it to respond more effectively to subsequent pathogen attacks. In mammalian applications, CNPs can modulate immune function by influencing cytokine release profiles, promoting macrophage activation, and supporting tissue regeneration processes. These effects can be enhanced or tailored through chemical modification or the addition of therapeutic agents, enabling a synergistic interplay between the inherent bioactivity of the NP and the delivered payload [65]. The versatility of CNPs, which enables the material to perform multiple functions simultaneously (e.g., to adhere, promote penetration, transfer substances or active ingredients, and have inherent antimicrobial or immunomodulatory properties), makes them highly suitable for applications in agriculture, medicine, food preservation, and environmental remediation in terms of sustainability and high performance. Their physical, chemical, and biological activity is more likely to lead to resistance to microbial populations and makes them versatile when it comes to responding to various challenges in different fields. An overview of the diverse physicochemical and biological mechanisms through which CNPs act is presented in Fig. 2.

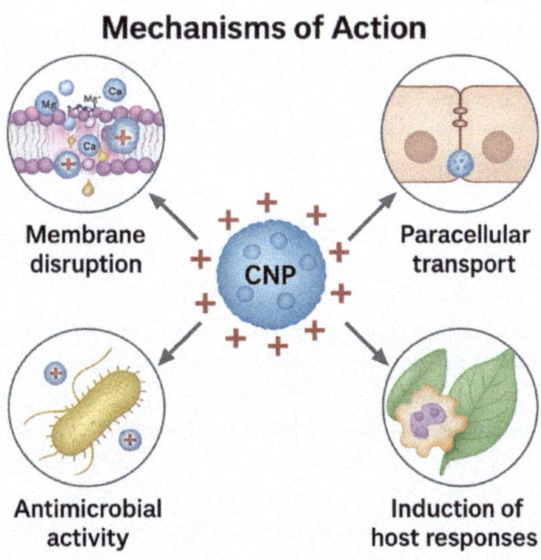

Fig. 2 Schematic representation of the mechanisms of action of chitosan nanoparticles (CNPs). The illustration shows electrostatic surface interactions with microbial membranes, tight junction modulation, endocytosis with proton-sponge effect, nucleic acid and enzyme interference, reactive oxygen species (ROS) generation, plant uptake and defence elicitation, and immune modulation in mammalian systems

5 Applications of Chitosan Nanoparticles

CNPs are highly versatile due to their outstanding combination of biological compatibility, chemical reactivity, and a wide range of adaptability to diversity, which is why they are among the most actively studied biopolymer nanomaterials today. By scaling down the materials to the nanoscale, these natural properties of biocompatibility, biodegradability, and antimicrobial activity of chitosan are maintained while increasing reactivity, surface-to-volume ratio, and solubility over an extended pH range. This nano-engineering makes it possible to precisely control the interaction of materials with a biological system, environmental contaminants, and active compounds, providing the background for multifunctional applications. The versatility of CNPs extends to agriculture, biomedicine, food preservation, environmental remediation, aquaculture, veterinary medicine, cosmetics, and various undeveloped technologies, and each of these areas can be optimized via specific physical, chemical, and functional changes in particle size, surface, area and encapsulation efficiency [16].

CNPs have also proven to be a renewable solution to replace typical agrochemicals in agriculture, eliminating the trade-off between maintaining crop health and environmentally friendly practices. They have inherent antimicrobial potential that can be used to suppress pathogenic fungi, bacteria, and viruses ultimately minimizing the prevalence of diseases like powdery mildew, bacterial blight, and viral leaf mosaic. Importantly, they also trigger plant immunity by activating systemic acquired resistance and promoting the production of compounds involved in plant defense, thus improving a plant's resistance without genetic engineering. CNPs are also used as transporters for pesticides, fungicides, and herbicides, which are released in a controlled manner and delivered to the site of infection or feeding by the pest. A dual-targeted approach can reduce the amount of runoff chemicals and decrease the amount of available active ingredient required to impact the environment. In terms of plant nutrition, CNP-based nano-fertilizers encapsulate micronutrients like zinc, copper, iron, and manganese, in microcapsules, releasing them more slowly and thus improving their uptake. They are used for seed preparation to improve the uniformity of germination and the vitality of seedlings and to increase resistance to abiotic stress factors such as salinity, drought, and extreme temperatures. Further developed systems can be smart CNP systems that dispense agrochemicals when triggered by environmental conditions—pH, moisture, or enzyme activity—which is a step toward precision agriculture [59].

The multifunctionality of CNPs has made them a core material for the creation of targeted drug delivery systems in biomedicine. Due to their mucoadhesive properties, they also provide enhanced bioavailability of orally, nasally, or ocularly administered therapeutics, resulting in prolonged mucosal retention. CNPs safeguard the bioactive material contained in the capsules, such as proteins, peptides and nucleic acids, from acids and other harsh physiological conditions and deliver them safely and intact to the target tissue. Targeting is site-specific, and facilitated by the surface modification of different ligands such as antibodies, peptides, or aptamers, an important feature in oncology where drug delivery should only target tumors, and not healthy tissue [36]. In cancer treatment, CNP-based delivery systems are effective in increasing the intracellular uptake of drugs through endocytosis, releasing the drugs over a longer period of time, and overcoming the effects of multidrug resistance [73]. The potential of CNPs also lies in the area of gene therapy with the formation of a stable polyplex between the CNP and the DNA, small interfering RNA (siRNA) or mRNA into which the genetic profile can be incorporated and adjusted. In wound healing and regenerative medicine, CNPs incorporated into hydrogels or scaffolds stimulate fibroblast proliferation, angiogenesis, and collagen deposition while preventing infections. Ophthalmic applications benefit from the ability of CNPs to cross the corneal barrier and deliver anti-inflammatory, antimicrobial, or anti-glaucoma agents with improved efficacy. In respiratory medicine, CNP-based inhalable formulations enable the delivery of antibiotics or anti-inflammatory drugs, deep into the lungs, which is promising for the treatment of chronic respiratory diseases.

The food industry utilizes CNPs both as active antimicrobial agents and as carriers for functional ingredients to improve the safety, quality, and shelf life of products [53]. In edible coatings, CNPs form semi-permeable barriers that reduce moisture loss, slow oxidative degradation, and control the respiration rate in fresh produce. These coatings inhibit microbial spoilage and pathogen contamination in fruits, vegetables, dairy, and meat products, reducing post-harvest losses. Active packaging materials with CNPs are also used, because they provide antimicrobial protection and can include freshness indicators that make spoilage visible. They can include natural preservatives, essential oils and antioxidants that can slowly leach out over time, maintaining sensory and nutritional values. CNPs are used in fermented foods to maintain the activity of probiotics during storage and through the gastrointestinal tract, and to increase the functional value of the product. CNPs are also used to stabilize emulsions in beverages, sauces, and dairy ice cream systems to prevent phase separation.

In environmental remediation, CNPs have a cationic surface that allows them to create a powerful electrostatic interaction with anionic pollutants, making them very useful in the purification of wastewater [34]. They have high adsorption capacities, especially for radium such as lead, cadmium, mercury, and chromium and in many cases outperform conventional adsorbents as they have high reactivity due to their large surface area and have functional groups that can be modified. In industrial wastewater, Industrial effluents can remove organic pollutants, such as pesticides, dyes and pharmaceutical residues can be removed by CNPs. By incorporating magnetic NPs, easily recyclable composites can be formed with the help of applied magnets, which increases their reusability. In microbial disinfection, CNPs disrupt the bacterial cell membrane and stop the growth of a biofilm, contributing to safer drinking water. They are also utilized in the recovery of nutrients through nutrient recovery systems, where phosphates and ammonium ions are collected for use in agriculture. In combination with photocatalytic materials, CNPs also increase the solar-based degradation of recalcitrant organic pollutants, providing environmentally friendly prospects for water purification. In addition to water treatment, they have made their antimicrobial coatings available for air filtration to limit the spread of airborne diseases in healthcare facilities and other common areas.

Aquaculture has recently tackled the issue of extending the lifespan of farmed animals and increasing their productivity through the use of CNPs [55]. They contain vaccines and immunostimulants so that the immune responses of fish and shellfish can be boosted and the injection rate decreased. The beneficial effect of their use in fish feed is that they enhance the bioavailability of nutrients and lower the burden of pathogens in aquaculture system facilities. In water management, CNP treatment prevents excessive proliferation of bacteria and algae to ensure sufficient discharge of aquatic life. Likewise, NP-based drug delivery systems in veterinary medicine can enable sustained release of antimicrobials, antiparasitics, or hormone therapies, with greater tolerance in dosing accuracy and lower costs for animal production.

CNPs are used in the beauty industry in cosmetics and personal care products for their film-forming, moisturizing and antimicrobial characteristics. In skin care formulas, they help to increase the efficacy of the active ingredients contained in the formulas, e.g., vitamins, peptides, and plant extracts, and act as protection against environmental stressors. CNPs can also protect cosmetic formulations from microbial contamination without the use of preservatives. When used in hair care, they add strength to fibers, decrease breakage, and provide sun protection, while when used in oral care products, they reduce bacterial colonization, improve enamel remineralization of enamel, and freshen breath [40].

The current applications reveal that the potential of CNPs lies not only in biological or ecological systems. They have been tested as binders in lithium-ion battery electrodes, where their mechanical flexibility and ionic conductivity are beneficial for battery performance and lifetime. The CNP-based bio-inks allow the construction of mechanically and degradation-tailorable, cell-loaded structures using the 3D bioprinting process. The textile industry uses CNP coatings to coat fabrics to give them antimicrobial activity, UV resistance, and hydrophobicity, extending their life and minimizing laundering. CNPs are even used in heritage conservation to prevent microbial degradation of paper, textiles, and artwork, demonstrating their prevalence in specific but important areas [41].

The most interesting aspect of these applications is that they enable multifunctional applications; a single CNP formulation could therefore fulfill multiple functions. An agricultural CNP platform could be used for nutrient delivery, pest control and enhancing plant immunity, and a biomedical drug formulation with an imaging modality could have a theranostic utility by combining imaging and drug functions. The flexibility of CNPs to meet different functional demands, by changing the size, surface charge, degree of deacetylation, and addition of effective molecules means that they can be molded for very specific demands. CNPs are expected to move beyond experimental demonstration to large-scale commercial applications, with the still-developing green synthesis pathways, in combination with scalable manufacturing technologies [28]. It would be important to pay close consideration to regulatory frameworks, safety assessments, and life cycle analyses during this development so that they can responsibly be introduced into different industrial sectors. Finally, CNPs offer a new level of future and solutions to difficult global complications in food security, health care, environmental safety and more, as they are multifunctional, biodegradable, and compatible (Table 2).

Table 2 Applications of CNPs across various fields

Field	Key functions of CNPs	Examples/benefits	References
Agriculture	Antimicrobial activity; elicitation of plant immunity; controlled agrochemical delivery; nano-fertilizers	Disease suppression (powdery mildew, bacterial blight, viral mosaics); seed priming; stress tolerance (salinity, drought, heat); smart release triggered by pH/moisture	[48]
Biomedicine	Drug/gene delivery; mucoadhesion; cancer therapy; wound healing; regenerative medicine	Enhanced oral/nasal/ocular bioavailability; tumor-targeted therapy; gene therapy (DNA/siRNA delivery); tissue scaffolds/hydrogels; ophthalmic and respiratory formulations	[71]
Food industry	Antimicrobial coatings; active packaging; delivery of functional ingredients	Edible coatings for fruits/vegetables; preservation of dairy/meat; encapsulation of essential oils, antioxidants, probiotics; emulsion stabilization	[42]
Environmental remediation	Adsorption of pollutants; photocatalytic activity; antimicrobial water/air disinfection	Removal of heavy metals (Pb, Cd, Hg, Cr); organic pollutant degradation (dyes, pesticides, pharmaceuticals); antimicrobial water treatment; nutrient recovery (phosphates, ammonium)	[47]
Aquaculture and veterinary	Vaccine/nutrient carriers; immunostimulants; controlled drug release	Boosted immunity in fish/shellfish; improved feed bioavailability; control of bacterial/algal overgrowth; veterinary antimicrobials, antiparasitics	[63]
Cosmetics and personal care	Film-forming; moisturizing; antimicrobial	Improved delivery of vitamins, peptides, plant extracts; protection against microbial contamination; hair fiber strengthening; oral care (enamel demineralization, fresh breath)	[40]
Advanced/industrial uses	Functional coatings; energy storage; bioprinting; heritage preservation	Battery electrode binders; 3D bio-inks; antimicrobial/UV-resistant textiles; conservation of paper, textiles, artworks	[31]

6 Challenges and Limitations

The increasing attention paid to CNPs points to their enormous potential as sustainable nanomaterials with wide-ranging applications in medicine, agriculture, food technology, and environmental remediation. Their biocompatibility, biodegradability, and functional versatility set them apart from many synthetic nanomaterials, where toxicity and long-term persistence can be a problem. Despite their promising properties, the successful commercialization of CNPs is fraught with scientific, technical, and regulatory risks as well as environmental considerations. If left unaddressed, these drawbacks have the potential to hinder the economic and societal contribution of CNP-based technologies. A considered assessment of these barriers not only puts the current state of the field into perspective but also provides insight into the strategic direction of future developments [59]. Priorities include increasing CNP production while maintaining the quality and price of the product. Existing synthesis processes such as ionic gelation, reverse micelle systems, emulsification, and spray drying are optimized under high-quality laboratory conditions. The properties of NPs lose reproducibility when the nanoscale is transferred to the industrial scale. The uniformity of mixing, control of ionic interactions, and prevention of aggregation become exponentially more complicated as the batch volume increases. An example of this is ionic gelation, where the stoichiometry of chitosan and the cross-linkers such as tripolyphosphate must be very carefully controlled; any variation at larger volumes may lead to a wide particle size distribution and low encapsulation efficiency. This inconsistency is not tolerated in sensitive areas such as drug delivery, where the efficacy and safety of drug therapy depend on the exact physicochemical characteristics [64, 72].

Economic factors also pose a challenge. Although chitin is abundant as a by-product of seafood and fungi, its conversion into high-quality chitosan requires labor-intensive refining steps such as deacetylation, purification, and fractionation. Costs are further increased by pharmaceutical-grade solvents, stabilizers, and sterilization, as well as stringent regulatory quality controls such as freeze-drying and endotoxin testing. These costs limit the competitiveness of CNPs over conventional polymers or carriers, unless their added value, such as controlled release and multifunctional properties, clearly justifies the costs [4, 33]. A key challenge for scalability is to maintain consistency in physicochemical properties. Critical factors such as size, zeta potential, morphology, porosity, and surface functionalization strongly influence the performance of CNPs. In biomedicine, size governs biodistribution and cell uptake, while zeta potential affects protein binding and immune interactions. In agriculture, these properties determine adhesion, wash-off resistance, and release kinetics. However, even small variations can drastically alter functionality, and batch-to-batch reproducibility is difficult due to sensitivity to reaction conditions such as temperature, pH, ion concentration, and degree of chitosan deacetylation [17].

Another challenge arises from the different chitosan raw materials. Chitosan derived from crustacean shells, squid pens, or fungal biomass differs in molecular weight, chain length, branching, and purity, all of which affect solubility, viscosity, and binding ability in NP synthesis. Even within a single source, seasonal and geographic factors contribute to compositional variability, which undermines uniformity. In the absence of standardized quality-control systems, CNPs from different laboratories, even batches within the same laboratory, have inconsistent properties, complicating both research outcomes and industrial implementation [50]. Another limitation is the absence of standardized characterization methods for CNPs. Different techniques—such as dynamic light scattering, nanoparticle tracking analysis, or electron microscopy—often provide contradictory results, while reporting practices vary between hydrodynamic diameter and morphology in the dry state. Without harmonized protocols or reference standards, cross-study comparisons remain difficult, and regulators cannot reliably assess product safety. This lack of standardization remains a major obstacle to both academic progress and commercial acceptance [13, 43].

Another serious limitation is legal and safety issues. Even though chitosan is largely considered safe as a bulk material, its shape change at the nanoscale offers a new dynamic that requires intensive risk evaluation of CNPs. NPs can penetrate biological barriers, interact with organelles, and cause reactions not seen with bigger particles. NP-specific risks, such as cytotoxicity, immunogenicity, genotoxicity, and biodistribution, should therefore be considered in the safety evaluations. For example, negatively charged cell membranes can interact strongly with positively charged CNPs, which can lead to destabilization of cell membranes, hemolysis, or even unwanted inflammatory reactions. There is little data on long-term toxicity, and little data is available on long-term exposure, accumulation, and reproductive and developmental toxicity [38]. To make matters worse, there is a wide range of applications. Intravenously delivered CNPs in biomedicine should be investigated to determine whether they interact with the reticuloendothelial system, whether they can accumulate in the liver and spleen, and whether they can cross the blood–brain barrier. Oral formulations need to be tested with regard to gastrointestinal stability and interactions with the gastric microbiota. In farming, CNPs applied through foliar sprays or even in the soil may unintentionally endanger pollinators or other beneficial insects, or microorganisms in the soil. The migration of NPs into food, especially when exposed to heat, acidity and prolonged storage, is a problem with food packaging. The risk assessment also depends on the specific application, which further complicates the regulatory evaluation [12].

There is no effective international system for regulating nanomaterials. Whereas some authorities have published guidance documents on the safety assessment of nanoparticles, these tend to apply existing procedures for bulk materials and may not effectively address nano-specific risks. There are no common rules, which creates uncertainty for regulators. This not only slows down the uncertainty to the regulators. This not only slows the approval of products but also discourages investment in the private sector, as companies do not know what regulatory requirements they may face, and may have to bear significant compliance costs. Moreover, the existing differences between national regulations present international barriers to trade, preventing CNP-based products from spreading worldwide [5].

Environmental safety remains a key concern, as the fate and behavior of CNPs in ecosystems is not fully understood. In aquatic systems, interactions with salts, organic matter, and colloids can alter aggregation, bioavailability, photosynthesis in algae, microbial diversity, and even gill function in fish. On land, CNPs may affect soil nutrient cycling, enzyme activity, and beneficial microbes such as fungi and bacteria, potentially reducing soil fertility [60]. In terms of agriculture, the unintended impact on non-target organisms would be highly critical. Both bees and butterflies as pollinators and ladybugs as predators of pests could be exposed to CNP residues. Although chitosan is non-toxic, the ecological outcomes of chronic exposure to nanoparticles are not yet clear. In addition, the properties of nanoparticles (which are relevant for toxicity) may change upon exposure to the environment, e.g., through photodegradation, metabolization by microorganisms, or binding to soil particles. Such transformations and their ecological implications are daunting and need to be modeled in risk assessment [68].

There is a substantial gap in standardized ecotoxicological tests. The various organisms, concentrations, and exposure durations used in current studies vary, with exposure taking place in an artificial laboratory environment that does not match real-world conditions. The lack of a harmonized testing framework makes it impossible to compare test results or determine universal safety levels. There is a need to develop robust and plausible ecotoxicological models to enable the safe use of CNPs in the environment [14]. In addition to here than the scientific and regulatory hurdles, there are other obstacles, such as the public perception of the people. NPs can be considered a potential term of concern as safety problems have been associated with the use of engineered nanomaterials like silver nanoparticles (AgNPs) in the past. Although CNPs are natural and renewable, the negative perception of nanotechnology could delay integration, particularly in sensitive areas such as food and agriculture. Open information on benefits, risks, and safety resources is therefore of utmost importance to build trust among consumers, stakeholders, and policy makers [57].

The barriers to the expansion of CNP production lie in process engineering, automation, cost-effective bio-valorization of chitin, and strict quality control to ensure batch consistency. The regulatory framework will increasingly require nanospecific guidelines based on comprehensive toxicological and ecotoxicological data, including long-term exposure studies. These challenges are not insurmountable obstacles, but rather opportunities for innovation. Advances such as continuous-flow synthesis can improve reproducibility and reduce costs, while standardization of chitosan quality could benefit numerous industries. Regulatory hurdles can encourage international cooperation and investment, while safer CNPs with predictable degradation will reduce ecological risks. Ultimately, overcoming these barriers will require multidisciplinary collaboration between scientists, engineers, policymakers, and industry to ensure that CNPs are developed in a technically sound, socially acceptable, and sustainable manner.

7 Future Prospects

Chitosan nanoparticles are poised to reshape sustainable nanotechnology with applications in biomedicine, agriculture, food safety, energy, and environmental engineering. In the coming decades, the transition from laboratory research to commercially viable, regulatory-approved products will depend on scalable green synthesis methods, reproducible quality, and interdisciplinary collaboration. Their multifunctionality, ranging from drug and gene delivery, wound healing, and plant protection to wastewater treatment, active food packaging, and energy storage, illustrates their adaptability and potential impact.

Future progress will require optimization of synthesis through eco-friendly techniques, the use of renewable raw materials, and advanced process control supported by computational modeling and machine learning. Integration into smart agricultural systems, diagnostic tools, and sustainable industrial materials will further expand their role. Success will also depend on standardized characterization protocols, strong regulatory frameworks, and transparent communication to build public confidence. With these advances, CNPs can become indispensable tools for food security, public health, environmental remediation, and the green industry, embodying innovation, sustainability, and responsible technological development.

8 Conclusion

CNPs offer a unique convergence of natural sustainability and innovation at nanoscale innovation, making them one of the most versatile nanomaterials being researched today. Their advantages stem from their inherent biocompatibility, biodegradability, and antimicrobial properties, which are further enhanced by nanoscale development. As discussed in this chapter, CNPs exhibit a high surface-to-volume ratio, tunable size, and modifiable surface chemistry, enabling improved solubility, stability, and targeted functionality. These features support a wide range of applications: In agriculture, CNPs act as smart nanocarriers for fertilizers and pesticides, while triggering plant immune responses to protect against pathogens; in biomedicine, they enhance drug and gene delivery through mucoadhesion, controlled release, and endocytosis, and also play a role in wound healing, cancer therapy, and regenerative medicine. In food preservation, CNP-based coatings extend shelf-life, prevent microbial spoilage, and protect nutritional quality. Their strong adsorption capacity and positive surface charge make them highly effective in wastewater treatment, pollutant removal, and antimicrobial filtration. In addition, their use in aquaculture, cosmetics, textiles, energy storage, and even cultural heritage preservation reflects their multifunctionality and adaptability. Nevertheless, there are still problems in scaling laboratory findings into industrial practice. The variability of raw material sources, lack of standardization, and limited safety data require urgent attention. Progress depends on the development of environmentally friendly and cost-effective synthesis methods, harmonized characterization protocols, and multidisciplinary collaboration. If these hurdles are overcome, CNPs can evolve from promising research results to commercially viable, regulatory-approved, and socially responsible solutions. Their future impact will rely not only on scientific innovation but also on sustainable and ethical integration into global systems.

References

1. Abdel Latef AAH, Zayed EM, Omar AA (2025) Bioremediation and Nanotechnology for Climate Change Mitigation. Springer Nature, Singapore Pte Ltd. pp 1–652. https://doi.org/10.1007/978-981-96-3069-1
2. Aibani N, Rai R, Patel P, Cuddihy G, Wasan EK (2021) Chitosan nanoparticles at the biological interface: implications for drug delivery. Pharmaceutics 13(10):1686
3. Ajmal M, Ullah R, Muhammad Z, Khan MN, Kakar HA, Kaplan A, Okla MK, Saleh IA, Kamal A, Abdullah A (2023) Kinetin capped zinc oxide nanoparticles improve plant growth and ameliorate resistivity to polyethylene glycol (PEG)-induced drought stress in *Vigna radiata* (L.) R. Wilczek (Mung Bean). Molecules 28(13):5059
4. Ali G, Sharma M, Salama E-S, Ling Z, Li X (2024) Applications of chitin and chitosan as natural biopolymer: potential sources, pretreatments, and degradation pathways. Biomass Convers Biorefin 14(4):4567–4581
5. Allan J, Belz S, Hoeveler A, Hugas M, Okuda H, Patri A, Rauscher H, Silva P, Slikker W, Sokull-Kluettgen B (2021) Regulatory landscape of nanotechnology and nanoplastics from a global perspective. Regul Toxicol Pharmacol 122:104885
6. Amor IB, Hemmami H, Laouini SE, Abdelaziz AG, Barhoum A (2024) Influence of chitosan source and degree of deacetylation on antibacterial activity and adsorption of AZO dye from water. Biomass Convers Biorefin 14(14):16245–16255
7. Aranaz I, Alcántara AR, Civera MC, Arias C, Elorza B, Heras Caballero A, Acosta N (2021) Chitosan: an overview of its properties and applications. Polymers 13(19):3256
8. Assad N, Abbas A, ur Rehman MF, Naeem-ul-Hassan M (2024) Photo-catalytic and biological applications of phyto-functionalized zinc oxide nanoparticles synthesized using a polar extract of *Equisetum diffusum D.* RSC Adv 14(31):22344–22358
9. Assad N, Naeem-ul-Hassan M, Ajaz Hussain M, Abbas A, Sher M, Muhammad G, Assad Y, Farid-ul-Haq M (2025) Diffused sunlight assisted green synthesis of silver nanoparticles using *Cotoneaster nummularia* polar extract for antimicrobial and wound healing applications. Nat Prod Res 39(8):2203–2217
10. Assad N, Naeem-ul-Hassan M, Sher M, Alatawi FS, Alatawi MS, Omran AM, Jame RM, Adnan M, Khan MN, Ali B (2023) *Aconitum lycoctonum* L.(Ranunculaceae) mediated biogenic synthesis of silver nanoparticles as potential antioxidant, anti-inflammatory, antimicrobial and antidiabetic agents. BMC Chem 17(1):128
11. Augustine R, Hasan A, Primavera R, Wilson RJ, Thakor AS, Kevadiya BD (2020) Cellular uptake and retention of nanoparticles: insights on particle properties and interaction with cellular components. Mater Today Commun 25:101692
12. Bashir SM, Ahmed Rather G, Patrício A, Haq Z, Sheikh AA, Shah MZuH, Singh H, Khan AA, Imtiyaz S, Ahmad SB (2022) Chitosan nanoparticles: a versatile platform for biomedical applications. Materials 15(19):6521
13. Bellich B, D'Agostino I, Semeraro S, Gamini A, Cesàro A (2016) The good, the bad and the ugly of chitosans. Mar Drugs 14(5):99
14. Breitholtz M, Rudén C, Hansson SO, Bengtsson B-E (2006) Ten challenges for improved ecotoxicological testing in environmental risk assessment. Ecotoxicol Environ Saf 63(2):324–335

15. Chaurasiya RS, Hebbar HU (2017) Reverse micelles for nanoparticle synthesis and biomolecule separation. In: Nanoscience in food and agriculture, vol 4. Springer, pp 181–211
16. Cruz-Segundo CM, Vasquez-Garcia SR, Flores-Ramirez N, Sanchez-Orozco R, Abdel-Gawwad HA, Santiago AA, Vargas J (2025) Development and characterization of electrospun chitosan-chitin-PVA nanofiber: exploring structural integrity, crystallinity, and thermal stability for advanced applications. J Mater Sci Mater Eng 20(1):103
17. Dadashi H, Vandghanooni S, Karamnejad-Faragheh S, Karimian-Shaddel A, Eskandani M, Jahanban-Esfahlan R (2024) A rapid protocol for synthesis of chitosan nanoparticles with ideal physicochemical features. Heliyon 10(11)
18. de Alvarenga ES (2011) Characterization and properties of chitosan. Biotechnol Biopolym 91:48–53
19. Divya K, Jisha M (2018) Chitosan nanoparticles preparation and applications. Environ Chem Lett 16(1):101–112
20. El-Naggar NE-A, Shiha AM, Mahrous H, Mohammed AA (2024) A sustainable green-approach for biofabrication of chitosan nanoparticles, optimization, characterization, its anti-fungal activity against phytopathogenic *Fusarium culmorum* and antitumor activity. Sci Rep 14(1):11336
21. Gagné N (1993) Production of chitin and chitosan from crustacean waste and their use as a food processing aid. Master of Science, Thesis, Submitted to Department of Food Science and Agricultural Chemistry, McGill University. https://escholarship.mcgill.ca/concern/theses/m039k714w
22. Ghaffar S, Abbas A, Naeem-ul-Hassan M, Assad N, Sher M, Ullah S, Alhazmi HA, Najmi A, Zoghebi K, Al Bratty M (2023) Improved photocatalytic and antioxidant activity of olive fruit extract-mediated ZnO nanoparticles. Antioxidants 12(6):1201
23. Grande Tovar CD, Castro JI, Valencia Llano CH, Navia Porras DP, Delgado Ospina J, Valencia Zapata ME, Herminsul Mina Hernandez J, Chaur MN (2020) Synthesis, characterization, and histological evaluation of chitosan-*Ruta graveolens* essential oil films. Molecules 25(7):1688
24. Grenha A (2012) Chitosan nanoparticles: a survey of preparation methods. J Drug Target 20(4):291–300
25. Guide Q Chitosan nanoparticles preparation with ionic gelation method
26. Gula H, Khan HA, Nasreen Z, Assad N, Turab SA, Hanif M (2025) Exploring the antibacterial, anticoagulant, and hemolytic potential of green-synthesized Fe_2O_3 nanoparticles by cucurbita pepo pulp. IEEE Trans NanoBiosci 24(4):411–420
27. Hamed I, Özogul F, Regenstein JM (2016) Industrial applications of crustacean by-products (chitin, chitosan, and chitooligosaccharides): a review. Trends Food Sci Technol 48:40–50
28. Harugade A, Sherje AP, Pethe A (2023) Chitosan: a review on properties, biological activities and recent progress in biomedical applications. React Funct Polym 191:105634
29. Hashmi SS, Ibrahim M, Adnan M, Ullah A, Khan MN, Kamal A, Iqbal M, Kaplan A, Assad N, Elshikh MS (2024) Green synthesis of silver nanoparticles from *Olea europaea* L. extracted polysaccharides, characterization, and its assessment as an antimicrobial agent against multiple pathogenic microbes. Open Chem 22(1):20240016
30. Herdiana Y, Wathoni N, Shamsuddin S, Joni IM, Muchtaridi M (2021) Chitosan-based nanoparticles of targeted drug delivery system in breast cancer treatment. Polymers 13(11):1717
31. Huseyn PI, Ibrahimova MJ, Abbasov VM, Nasirov FA, Aslanbeyli AM (2024) Chitosan biopolymer in industrial wastewater treatment. Process Petrochem Oil Refin 25(1)
32. Hussain SB, Zahra S, Hussain S, Mahmood H, Haq Au, Shafique M, Gill AR, Ali A, Javed MA, Khan MN (2025) Characterization and comparative antibacterial activities of gold nanoparticles synthesized by *Tagetes patula* L.(Asteraceae) flower extract. Microsc Res Tech 88(6):1869–1880
33. Iber BT, Kasan NA, Torsabo D, Omuwa JW (2022) A review of various sources of chitin and chitosan in nature. J Renew Mater 10(4):1097
34. Issahaku I, Tetteh IK, Tetteh AY (2023) Chitosan and chitosan derivatives: Recent advancements in production and applications in environmental remediation. Environ Adv 11:100351

35. Jabbar A, Abbas A, Assad N, Naeem-ul-Hassan M, Alhazmi HA, Najmi A, Zoghebi K, Al Bratty M, Hanbashi A, Amin HM (2023) A highly selective Hg^{2+} colorimetric sensor and antimicrobial agent based on green synthesized silver nanoparticles using *Equisetum diffusum* extract. RSC Adv 13(41):28666–28675
36. Jafernik K, Ładniak A, Blicharska E, Czarnek K, Ekiert H, Wiącek AE, Szopa A (2023) Chitosan-based nanoparticles as effective drug delivery systems—a review. Molecules 28(4):1963
37. Jhaveri J, Raichura Z, Khan T, Momin M, Omri A (2021) Chitosan nanoparticles-insight into properties, functionalization and applications in drug delivery and theranostics. Molecules 26(2):272
38. Khan Z, Maqsood Q, Baradoke A, Ferreira LFR, Franco M, Schmidt JE, Hussain N (2024) Environmental and toxicological implications of chitosan nanostructures. In: Advances in chemical pollution, environmental management and protection, vol 10. Elsevier, pp 137–172
39. Khan ZuR, Assad N, Naeem-ul-Hassan M, Sher M, Alatawi FS, Alatawi MS, Omran AM, Jame RM, Adnan M, Khan MN (2023) *Aconitum lycoctonum* L.(Ranunculaceae) mediated biogenic synthesis of silver nanoparticles as potential antioxidant, anti-inflammatory, antimicrobial and antidiabetic agents. BMC Chem 17(1):128
40. Kulka K, Sionkowska A (2023) Chitosan based materials in cosmetic applications: a review. Molecules 28(4):1817
41. Kush P, Singh R, Kumar P (2025) Chitosan nanoparticles: fundamental chemistry, properties, and current research trends in biomedical field. Fundam Biomed Appl Chitosan Nanoparticles 1–51
42. Manigandan V, Karthik R, Ramachandran S, Rajagopal S (2018) Chitosan applications in food industry. In: Biopolymers for food design. Elsevier, pp 469–491
43. Marques C, Som C, Schmutz M, Borges O, Borchard G (2020) How the lack of chitosan characterization precludes implementation of the safe-by-design concept. Front Bioeng Biotechnol 8:165
44. Másson M (2024) The quantitative molecular weight-antimicrobial activity relationship for chitosan polymers, oligomers, and derivatives. Carbohyd Polym 337:122159
45. Moura D, Mano JF, Paiva MC, Alves NM (2016) Chitosan nanocomposites based on distinct inorganic fillers for biomedical applications. Sci Technol Adv Mater 17(1):626–643
46. Naskar S, Kuotsu K, Sharma S (2019) Chitosan-based nanoparticles as drug delivery systems: a review on two decades of research. J Drug Target 27(4):379–393
47. Pal P, Pal A, Nakashima K, Yadav BK (2021) Applications of chitosan in environmental remediation: a review. Chemosphere 266:128934
48. Pandey P, Verma MK, De N (2018) Chitosan in agricultural context—a review. Bull Environ Pharmacol Life Sci 7(4):87–96
49. Panigrahi A, Halder J, Rai VK, Dash P, Das C, Kar B, Sarangi MK, Ghosh G, Rath G (2025) Herbal bioactive loaded chitosan therapeutics: a promising strategy for wound healing. J Drug Deliv Sci Technol 107321
50. Pellis A, Guebitz GM, Nyanhongo GS (2022) Chitosan: sources, processing and modification techniques. Gels 8(7):393
51. Peltonen L, Valo H, Kolakovic R, Laaksonen T, Hirvonen J (2010) Electrospraying, spray drying and related techniques for production and formulation of drug nanoparticles. Expert Opin Drug Deliv 7(6):705–719
52. Perera U, Rajapakse N (2013) Chitosan nanoparticles: preparation, characterization, and applications. In: Seafood processing by-products: trends and applications. Springer, pp 371–387
53. Qu T, Wang X, Zhang F (2025) Antibacterial food packaging with chitosan and cellulose blends for food preservation. Polymers 17(13):1850
54. Rahangdale D, Joshi N, Kumar A (2020) Chitosan and its derivatives: a new versatile biopolymer for various applications. In: Functional chitosan: drug delivery and biomedical applications. Springer, pp 1–42

55. Rakhi R, Shahariar M, Alam M (2022) Nanoparticles (NPs) in aquaculture industry: current status and future perspectives. Ann Bangladesh Agric 26(1):113–125
56. Rampino A, Borgogna M, Blasi P, Bellich B, Cesàro A (2013) Chitosan nanoparticles: preparation, size evolution and stability. Int J Pharm 455(1–2):219–228
57. Rathore A, Mahesh G (2021) Public perception of nanotechnology: a contrast between developed and developing countries. Technol Soc 67:101751
58. Rinaudo M, Domard A (1989) Chitin and chitosan. Solution properties of chitosan London. Elsevier, Functional, pp 71–86
59. Riseh RS, Vatankhah M, Hassanisaadi M, Varma RS (2024) A review of chitosan nanoparticles: Nature's gift for transforming agriculture through smart and effective delivery mechanisms. Int J Biol Macromol 260:129522
60. Rizeq BR, Younes NN, Rasool K, Nasrallah GK (2019) Synthesis, bioapplications, and toxicity evaluation of chitosan-based nanoparticles. Int J Mol Sci 20(22):5776
61. Roheen T, Bibi M, Rehman MFu, Assad N, Gondal HY, Nadeem M, Atif FA, Batool F, Perveen N, Qadir R (2025) Enhanced antibacterial and anti-inflammatory efficiency of serratiopeptidase immobilized on CMC-silver nanoparticles. J Basic Microbiol e70060
62. Saleh HM (2025) Preparation and characterization of chitosan/starch/Ag@ TiO_2-NPs bionanocomposite. J Ind Chem Soc 102022
63. Şenel S, McClure SJ (2004) Potential applications of chitosan in veterinary medicine. Adv Drug Deliv Rev 56(10):1467–1480
64. Souza VG, Pires JR, Rodrigues C, Coelhoso IM, Fernando AL (2020) Chitosan composites in packaging industry-current trends and future challenges. Polymers 12(2):417
65. Wael K, Abdelgawad AM, Elsherbiny DA, El-Naggar ME (2025) Chitosan nanoparticles for antimicrobial applications. In: Fundamentals and biomedical applications of chitosan nanoparticles. Elsevier, pp 363–404
66. Wajid M, Uzair M, Muhammad G, Siddique F, Bashir M, Nadeem S, Ashraf A, Assad N, Mushtaq A, Rafay MZ (2024) Sulfamethoxazole-derived Schiff bases: synthesis, characterization, biological activities, molecular docking, DFT, and ADME studies. J Mol Struct 1312:138640
67. Wang J, Zhuang S (2022) Chitosan-based materials: preparation, modification and application. J Clean Prod 355:131825
68. Wang X, He M, Wang X, Liu S, Luo L, Zeng Q, Wu Y, Zeng Y, Yang Z, Sheng G (2024) Emerging nanochitosan for sustainable agriculture. Int J Mol Sci 25(22):12261
69. Wen C, Tang J, Cao L, Fan M, Lin X, Liu G, Liang L, Liu X, Zhang J, Li Y (2025) Strategic approaches for co-encapsulation of bioactive compounds: technological advances and mechanistic insight. Foods 14(12):2024
70. Wu K, Yan Z, Wu Z, Li J, Zhong W, Ding L, Zhong T, Jiang T (2024) Recent advances in the preparation, antibacterial mechanisms, and applications of chitosan. J Funct Biomater 15(11):318
71. Yadav H, Malviya R, Kaushik N (2024) Chitosan in biomedicine: a comprehensive review of recent developments. Carbohydr Polym Technol Appl 8:100551
72. Yanat M, Schroën K (2021) Preparation methods and applications of chitosan nanoparticles; with an outlook toward reinforcement of biodegradable packaging. React Funct Polym 161:104849
73. Yee Kuen C, Masarudin MJ (2022) Chitosan nanoparticle-based system: a new insight into the promising controlled release system for lung cancer treatment. Molecules 27(2):473
74. Younes I, Rinaudo M (2015) Chitin and chitosan preparation from marine sources. Structure, properties and applications. Mar Drugs 13(3):1133–1174
75. Zielińska A, Carreiró F, Oliveira AM, Neves A, Pires B, Venkatesh DN, Durazzo A, Lucarini M, Eder P, Silva AM (2020) Polymeric nanoparticles: production, characterization, toxicology and ecotoxicology. Molecules 25(16):3731

Overview of Abiotic Stress and Its Agricultural Impact

Mainak Barman, Souvik Das, Rimpa Kundu, Subhra Mukherjee, Satish Kumar Singh, and Arafat Abdel Hamed Abdel Latef

Abstract The global climate scenario is unpredictable and constantly changing. As a result, various abiotic stress factors are impacting cultivation practices and production worldwide. Changes in temperature and rainfall patterns, seawater inundation and heavy metal accumulation in cultivable soil are worsening existing burdens on agriculture. Human actions have complicated this situation through the overexploitation of natural resources, the indiscriminate use of fertilisers and poor irrigation practices. It is essential to identify the various abiotic stress factors that negatively affect global crop production. Understanding the critical morpho-physiological, biochemical and molecular mechanisms within plants, as well as their interactions in response to these stresses, is extremely important in addressing these challenges. Advances in conventional breeding and biotechnological tools have significantly strengthened efforts to combat these issues, which require further understanding to progress. This chapter provides a brief overview of the different abiotic stress factors impacting global crop production, their effects and the current status of plant breeding and biotechnology-based mitigation strategies.

Keywords Biochemical · Biotechnology · Climate change · Molecular · Plant breeding

M. Barman (✉) · S. Das · R. Kundu · S. Mukherjee
Department of Genetics and Plant Breeding, Bidhan Chandra Krishi Viswavidyalaya, Mohanpur, Nadia, West Bengal, India
e-mail: mainakbarman4@gmail.com

S. K. Singh
Department of Genetics and Plant Breeding, Dr. Rajendra Prasad Central Agricultural University, Pusa, Samastipur, Bihar, India

A. A. H. . . Abdel Latef
Department of Botany and Microbiology, Faculty of Science, Qena University, Qena, Egypt

A. A. H. Abdel Latef (ed.), *Chitosan Nanoparticles for Plant Resilience Under Abiotic and Biotic Stressors*, Smart Nanomaterials Technology,
https://doi.org/10.1007/978-981-95-6422-4_2

1 Introduction

Environmental conditions have always played a crucial role in the agriculture sector. Its productivity is highly sensitive to the environmental fluctuations [68]. Abiotic factors are essential elements in the environment that influence crop productivity, and these represent the most persistent and damaging among various factors detrimentally affecting crop yield [48, 70]. These stresses include many in the list, like drought, salinity, heat and cold extremes, low light, submergence, etc. They cause yield reductions often more than those caused by pests and diseases [56, 67, 78]. The distribution of arable land further exemplifies the severity of the problem. Only about one-tenth of the global agricultural land can be considered favourable for continued crop production. The residual is affected mostly by these stress factors, often in combination [92]. The situation poses a substantial problem, especially for regions with erratic rainfall and resource constraints. These severe hazards are casting a dark shadow for millions of people and jeopardizing future food security. The impact of various abiotic stresses is not limited only to crops. These intensely impact livestock, fisheries and related farming enterprises [37]. As the concerns of climate change rise, the impending risks are also expected to upsurge further. Consequently, it is essential to understand how plants sense stress signals and adjust to unfavourable climatic conditions.

Changes in temperature and rainfall patterns with unexpected flooding, along with the growing uncertainties of weather events, are projected to exaggerate prevailing burdens on agriculture. Also, irregular rainfalls and temperature fluctuations trigger various biotic stresses [4]. Human activities have made it more complex with overexploitation of natural resources, indiscriminate use of fertiliser and poor irrigation practices [63, 72]. If these concerns are left unaddressed, it can drastically hamper global food supplies, especially in circumstances where we need to produce almost double than today by the year 2050 to feed the ever-increasing global population [81]. Despite dwindling resources and harsher growing conditions, we need to confront these challenges with comprehensive strategies that can balance short-term adaptation with long-term sustainability. This calls for hunting for novel sources of abiotic-stress tolerance, along with understanding their detailed biochemical and molecular mechanism for making further identification easier. Conventional plant breeding approaches are widely adopted for developing stress-tolerant varieties in different crops, but they must now be aided with advanced biotechnological tools. This chapter has been crafted to deliver an overview of major abiotic stresses challenging today's agriculture and focus on their consequences in global food security. It also describes various physio-biochemical and molecular responses of plants struggling with different stress factors and highlights on status of plant breeding and biotechnology in fighting these dilemmas.

2 Major Types of Abiotic Stress

Water, carbon, minerals and light (energy) are the elementary sources that a plant requires at an optimum level to survive [16]. However, beyond the optimum level, the plant shows abnormal behaviours. Abiotic stress is defined as a phenomenon in the environment that affects the phenology of plants, as well as reduces the yield of the crops. As plants cannot move, various physiological, biochemical and molecular mechanisms are employed by plants to defend against these stresses. The main components of abiotic stress include drought, salinity, temperature (heat and cold), flooding and waterlogging, heavy metals and toxic elements, light stress and UV radiation and oxidative stress [14]. Figure 1 illustrates these major types of abiotic stresses affecting global crop production. It is imperative to have a basic understanding of these to apply different mitigation strategies.

2.1 *Drought Stress*

Drought is a natural, recurring feature of the climate that happens almost everywhere, from developed to developing countries. It is the most severe and important among all abiotic stresses with a major impact on the growth and development of plants [36, 91, 95]. It affects crop plants in terms of both qualitative and quantitative parameters. If a region experiences water deficiency for a long time due to an abnormal weather phenomenon, it may be referred to as drought [57, 86]. But it is also differently classified as agricultural, meteorological and physiological droughts. From an agronomic perspective, when plants do not receive available water for a certain period from the soil due to insufficient soil moisture, it is also referred to as drought stress [17]. In plants, water stress occurs due to a high rate of transpiration in the leaves and less uptake of water by the roots [77]. Drought stress may result from different complex

Fig. 1 Different types of abiotic stress affecting global agriculture

physiological mechanisms. Drought can occur for several reasons, such as low rainfall, high light intensity, salinity, temperature fluctuations and other factors [54, 77]. The primary causes of drought stress are irregular rainfall, increased temperature and atmospheric Carbon dioxide levels [30]. Sometimes soil has optimum moisture, but the plant can't uptake it due to various physiological factors. This type of drought can be defined as pseudo-drought or physiological drought [87]. Gebrechorkos et al. [33] conducted an assessment from 1901 to 2022 and found that AED (atmospheric evaporative demand) increased drought severity by 40% worldwide. It has been observed by probabilistic modelling that during severe droughts in the top-producing countries, yield loss may exceed 70% for soybeans and maize [52].

Various adaptive responses are taken by plants to survive against drought stress, like stomatal closure, osmotic adjustment and modification of the root architectures, etc. [23, 94]. Abscisic acid signalling pathways also play a vital role in alleviating drought stress. Also, many stress-responsive genes and proteins are involved in drought tolerance, such as ABA-responsive genes, dehydrin and LEA (late embryogenesis abundant proteins) [6, 22, 43].

2.2 *Salinity Stress*

Soil salinity is a long-term problem for farming, and it has arisen from mismanagement in agricultural practices. It is projected that by the year 2050, around 50% of the cultivated land will be salinized [12, 24]. About 831 Mha of land is affected by salt. It is around 7% of the total land in the world. Also, 33% of irrigated and 20% of the total cultivated land are majorly affected by salinity [65]. Soil salinity is described by a collection of soluble salts that are injurious to plants. Salt-affected soils are mainly of two types: saline and alkaline or sodic soils [18]. Saline soil mostly consists of different types of salt, like sulphate and chloride of magnesium (Mg), calcium (Ca), potassium (K) and sodium (Na). Saline soil has more than 4 dS/m of EC (electrical conductivity). EC above 8 dS/m is regarded as a highly saline soil, which is deleterious for crop plants [60]. Whereas alkali soils contain more than 15% exchangeable sodium (ESP), and sodium carbonate is the major constituent in these types of soils [19]. The pH of this type of soil is always greater than 8.2. ESP values of more than 70 represent that the soil is extremely alkali or sodic in nature and not suitable for cultivation except for highly tolerant varieties.

Salinity is widely seen in coastal areas and recurrent flood-affected regions, or due to seawater intrusions. A large area worldwide is facing salt stress, such as the Rann of Kutch, the Sundarbans and the Indo-Gangetic plains in India, Bangladesh, Indonesia, Central and South Asia, North Africa and many more [12]. The severe effects of this stress are shown by two distinct phases: osmotic and ionic toxicity. In the osmotic phase, plants experience physiological drought conditions due to the accumulation of salt in the root zone, which reduces the water uptake capacity of plants [30, 59]. In the ionic phase, an excessive amount of sodium and chloride ions accumulates in the plant tissue and disrupts the essential cellular processes such as protein synthesis, photosynthesis and enzyme activity [30, 54, 59]. Several nutrient uptakes are restricted by Na^+ toxicity and create a nutritional imbalance in the plant. It has been observed that susceptible wheat genotypes show a decrease in grains per spike and yield in saline soil [74]. In rice, salt-susceptible genotypes experienced 15.3% more grain yield reduction than the tolerant genotypes [89]. Besides systematic management practices, proper irrigation schedule, time of planting and choice of crop help to mitigate this stress. Also, reclamation methods such as the application of soluble calcium amendments and the leaching of soluble salts are effective measures for preventing salt stress.

2.3 Temperature Stress

Temperature stress is another critical and devastating abiotic factor. Global warming, deforestation and environmental pollution cause a rise in temperature worldwide, which severely affects crop growth and development. By the end of the twenty-first century, regarding global climate models, the mean optimum temperature is predicted to increase by 1–6 °C [5, 26]. When the temperature rises above 50 °C, plant parts that lack transpiration cooling mechanisms (e.g. twigs, fruits, trunks, etc.) can lead to cell and tissue death. Temperature stress can be assessed by the intensity of temperature (low or high), duration of exposure or rate of temperature changes. It is highly stage-specific in plants. It has been classified into two categories: heat stress (high temperature) and cold stress (chilling and freezing temperature). High temperature denotes long exposure to temperatures above 27 °C, specifically at both the reproductive and vegetative stages, and short exposure to temperatures above 37 °C during the vegetative stage [46]. On the other hand, cold stress can be classified into two categories: 0–15 °C is called chilling stress, and less than 0 °C is represented as freezing stress. Plants exposed to temperature stress show low germination, reduced photosynthesis, reduced pollen viability, disruption of translocation of photosynthate, protein denaturation, and loss of membrane stability and low yield [51, 61].

Plants involve different mechanisms to protect themselves from temperature stress, engaging antioxidants, proteins, metabolites, regulatory factors, various protectants and membrane lipids. The synthesis of heat shock protein (HSP), which functions as chaperones that prevent protein denaturation, is activated in response to heat stress. Plants employ various survival mechanisms, like heat avoidance through transpiration, leaf reflectance and insulation using bark, as well as heat tolerance by maintaining membrane stability, photosystem II stability and osmoregulation. Choukri et al. [20] reported a significant impact on the plant's phenology and reduction in yield, along with degraded seed nutritional quality in lentil exposed to a temperature of 43 °C at the reproductive stage. Reduction in wheat yield was observed with a $6.0 \pm 2.9\%$ °C increase in temperature [29, 96]. Application of heat-escaping strategies helps to mitigate heat stress, such as morning irrigation, mulching to retain soil moisture, seed priming, deep watering, physical barriers (shade cloth) and introducing heat-tolerant plant varieties.

Cold Stress also shows severe deleterious effects on plant phenology, growth and yield. The most common type of cold stress is chilling stress, which can significantly reduce yield in major crops [1, 41, 83]. Tropical plants are more sensitive to chilling stress as compared to other plants. Cold stress affects crop plants worldwide, including the Central Plains and the Midwest of the USA, South Korea, China, Japan, some parts of Australia and India, which are experiencing such temperature extremes. Reduction in germination rate, decrease in root growth, 'locked open' the stomata, pollen sterility, abnormal flower or fruit production and disruption in the normal rate of photosynthesis are the major concerning factors due to chilling stress in plants [62]. Whereas ice formation in the internal organs of plants (intercellular and xylem vessels), membrane disruption (frost plasmolysis), increases the production of reactive oxygen species (ROS), is the impact of freezing stress in plants [15]. Cold stress exposure causes various physiochemical disturbances in plants, including growth retardance. Plants respond to cold stress by signal transduction through the activation of transcription factors and cold-responsive genes. These genes help to sustain during cold stress by improving the tolerance ability of the plants. It was observed that overexpression of mitogen-activated protein kinase *OsMAPK5* in rice helps to improve the tolerance ability of rice seedlings under various abiotic stress, including cold stress [88, 90]. According to Yue et al. [93], 85% of the wheat cultivated area is impacted by cold stress every year globally during the booting stage in March and April [39]. Cold stress can be managed and mitigated by developing tolerant varieties through advanced breeding methods or advanced genome editing tools. Cold-tolerant varieties such as Himali, Himdhan and Meghalaya1 are also used in rice production. Various agronomic practices like adjusted sowing dates, mulching, wind machines, greenhouse farming, seed priming and sprinklers help to manage cold stress. Application of phytohormones can be beneficial in that case.

2.4 Flooding and Waterlogging

Flooding and waterlogging are two of the major stresses that affect the plant root zone severely and cause various negative impacts on yield. Flooding can be caused by excessive precipitation, an inappropriate drainage system, natural calamities, deforestation, urbanization on floodplains, and dam failures. The effect of flooding in agricultural land causes a waterlogging situation in plants, which degrades plant health. Around 12% of the global agricultural land is experiencing these issues [47, 84].

Low oxygen concentration in soil is known as hypoxia, whereas complete absence of oxygen is called anoxia [28]. Both are severe effects of flooding stress in plants, affecting respiration, carbohydrate, photosynthesis and the translocation of growth regulators. It can decrease the yield of crops grown in this type of soil due to the absence of a proper drainage system. Flooding slows the movement of air in water, so the diffusion of oxygen is also slowed. Deficiency of oxygen causes a change from aerobic to anaerobic respiration in the root. So, the consumption of organic compounds becomes less efficient, and the production of toxic products such as ethanol and lactate increases [66]. Flooding also affects the cytoplasmic pH, reduces phloem transport, decreases the availability of essential nutrients, decreases cellular energy and accumulates toxic substances. Waterlogging decreases the rate of transpiration and photosynthesis, causing stomata closure. Therefore, it reduce chlorophyll content, decreased leaf area and induced leaf senescence. Stress-tolerant varieties develop specialized adaptations like aerenchyma tissue formation, adventitious root development and metabolic adjustments [71].

2.5 *Heavy Metal and Toxic Element Stress*

Heavy metals are defined as the elements having a density greater than 5 g cm^{-3}. Some heavy metals, namely, cobalt (Co), copper (Cu), iron (Fe), manganese (Mn), molybdenum (Mo), nickel (Ni) and zinc (Zn) are essential for plants, whereas chromium (Cr) and antimony (Sb) for animals. These metal elements can directly affect growth, senescence and energy-generating processes. Their concentration in soil beyond permissible limits is toxic to plants. These either cause oxidative stress through free radicals and/or disrupt the functions of enzymes by replacing essential metals and nutrients. 10% of the land is estimated to be contaminated with heavy metals worldwide, which causes risk for crop production. Heavy metal toxicity disrupts various enzymatic activities, produces ROS, damages the cellular membrane and interrupts photosynthesis. For example, the second leaf of wheat seedlings showed a significant growth reduction due to the presence of high Cd^{2+} (50 μM) and Zn^{2+} (600 μM) concentrations [73].

2.6 *Light Stress*

Light stress can be classified based on its intensity into high light stress and low light stress. The plant requires solar energy that lies within the optimum range to perform the necessary activities through photosynthesis. However, beyond this range, it causes physiological damage. The plant initiates signal transduction from the chloroplast to all the other cells, and these signals activate several biochemical, physiological and molecular mechanisms to defend against the high-light intensity stress, including photo-inhibition and photo-damage. Production of anthocyanin, chloroplast and stomata movement, and ROS scavenging are the major protection mechanisms employed by plants against light stress [80]. Wheat and lentil showed significant differences for chlorophyll a and chlorophyll b under high-intensity invisible UV-light.

Low light intensity is a serious problem, particularly for rice, wheat, etc. Recurrent cloudy weather or continuous rainfall during the reproductive stage causes substantial yield loss and deteriorates the grain quality. 48% of rice production out of 58% of total rice cultivated area is contributed by the majorly India's eastern and north-eastern hilly regions. But in the wet season, light intensity drops to 500 from 1400 $\mu mol/m^2/s$ in the summer season due to cloudy weather. That low-light intensity causes stress by reducing grain filling and an increase in the number of chalky grains [32]. Low light also influences the abundance of proteins such as magnesium chelatase subunit H (ChlH), magnesium chelatase subunit I (ChlI), and protoporphyrinogen oxidase (PPOX), which are linked to chlorophyll synthesis [53].

2.7 UV Radiation and Oxidative Stress

Sunlight is the source of energy for all the photoautotrophs. However, Solar ultraviolet radiation (UVR) is a hidden stress in sunlight that affects the plants' morpho-physiological, biochemical and molecular mechanisms. Three types of bands are there: UV-C (100–280 nm), UV-B (280–315 nm) and UV-A (315–400 nm), but UV-B is more injurious to plants as its potentiality and energy are much higher than the others [25]. UV-B is known to cause severe damage in the photosynthesis process. The main concern is the production of cytotoxic and reactive compounds, which ultimately cause the oxidation of protein and lipid. Plant utilizes enzymatic (superoxide dismutase, catalase, peroxidase, etc.) as well as non-enzymatic (ascorbate, glutathione, jasmonic acid, flavonoids, etc.) antioxidants to protect themselves from this stress. Dai et al. [21] suggested that UV-B can be associated with the interruption of active oxygen metabolisms by the destruction of various defence systems (enzymatic and non-enzymatic) in rice. Jan et al. [44] showed that both drought and UV-radiation increase oxidative damage by ROS, reducing the growth and development in rice.

3 Physiological, Biochemical and Molecular Responses of Plants

Plants exhibit various physiological, biochemical and molecular responses when subjected to abiotic stresses. Table 1 displays the different responses of plants to various stress factors, including drought, heat and salinity.

Table 1 Physiological, biochemical and molecular responses of plants to different abiotic stress

Type of stress	Physiological response	Biochemical response	Molecular response	Reference(s)
Drought	– Wilting – Decrease transpiration, stomatal conduction and assimilation of CO_2 – Reduction of the area and the relative water content of the leaf – Lessen the biomass and growth – Root: shoot ratio increase	• Osmolytes acquisition in the cell • For scavenging ROS, the antioxidants like Superoxide Dismutase (SOD), Peroxidase (POD), Catalase (CAT), etc., enzymes increased • Reduce electrolyte leakage and improve membrane stability	• Dehydration Responsive Element Binding (DREB), ABA Responsive Element Binding Protein (AREB)/ ABA Binding Factor (ABF) transcription factor overexpression • Late Embryogenesis Abundant (LEA) proteins and dehydrins producing gene upregulated • For the rule of water transport, aquaporin expression • Ca^{2+} signalling	[13, 76]
Heat stress	– Increased leaf temperature and respiration rate – Quicker senescence – Affects membrane stability – Delayed heading – Increased sterility	• Increased Heat Shock Proteins (HSPs) production • Stimulate the activity of the antioxidant enzyme • Membrane lipid peroxidation for damage repair	• Heat Shock Factor genes (HSFs) gene upregulation • Signal transduction via Ca^{2+}, Mitogen-Activated Protein (MAP) kinase pathway • Activation of chaperones • Misfold protein denaturation	[38, 49, 51]

(continued)

Table 1 (continued)

Type of stress	Physiological response	Biochemical response	Molecular response	Reference(s)
Salinity	– Reduce water uptake due to osmotic stress – Leaf drying from the tip in older leaves and chlorosis – Imbalance in Na^+ and Cl^- ion – Stunting growth	• Compatible osmolytes acquisition, like proline, betaine, etc • Non-enzymatic antioxidants improved antioxidant defence • Vascular sequestration • Ionic exclusion for maintenance of a balanced Na^+/K^+ ratio • Maintenance of membrane integrity	• For homeostasis, Salt Overly Sensitive (SOS) pathway gene activation • NAC, DREB transcription factor involve in tolerance • ABA, ethylene-like hormonal regulation	[3, 76]
High temperature + dought combined	– Damaging of PSII and reducing the photosynthesis rate – Increasing the transpiration rate where water is available – In tuber crops, tuberization is a common phenomenon (In Potato, decreases in tuber size and number) – Leaf rolling – Inactivation of Rubisco and reduction of CO_2 fixation	• Reactive Oxygen Species (ROS) produced • Increased enzyme productivity • Synthesis of HSPs • Membrane modification for stability	• Heat shock transcription factors upregulated • Changes during transcription hamper hormonal regulation • Downregulation of Genes linked to the light-harvesting complex in the tolerant one • Conquest of programmed cell death pathways • Combined stress sometimes induces unique gene expression outlines not under single stress	[27, 49]

(continued)

Table 1 (continued)

Type of stress	Physiological response	Biochemical response	Molecular response	Reference(s)
Cold/low temperature	– Reduction in germination rate – Reduce shoot length and delay growth – Metabolic rate becomes slower – Cell injury – Viscosity increase transpiration	• More production of unsaturated lipids for membrane fluidity • During freezing, produce more antioxidants to prevent ROS production • For cryoprotection anti anti-freezing protein accumulation occurs	• Induce cold-tolerant genes • Upregulation of Genes responsible for freezing-tolerant proteins like dehydrins, LEA proteins • MAPKs, ICE1, Ca^{2+} transcription factor signalling • Cold memory generation by Epigenetic amendments in some plants	[31, 85]
Low light stress	– Leaf elongation – Increase in stem length – Increase in leaf area – Reduce the chlorophyll a/b ratio – Reduce the rate of photosynthesis – Delayed heading – Increase spikelet sterility – Yield hampered	• Reformed biosynthesis of chlorophyll (more chlorophyll b) • Enhanced production of light-harvesting proteins • Channelized photosynthate partitioning • Plastid movement to the upper surface of the leaf	• Upregulation of phytochrome-related genes • Increase the expression of shade-responsive genes • Enlarged the expression of light-harvesting complex II • Modification of auxin signalling	[32, 75]

(continued)

Table 1 (continued)

Type of stress	Physiological response	Biochemical response	Molecular response	Reference(s)
UV stress	– Leaf area reduction – Necrosis due to high UV light – Damage PSII and increase photorespiration – In Epidermal cells, DNA-damaging occurs	• UV-absorbing compounds (flavonoids, phenolics) production • Boost the enzyme activity for DNA repair	• Stimulation of UVR8 photoreceptor pathway • Enhanced expression of genes like flavonoid biosynthetic enzymes • Upregulation of DNA repair genes • Instigation of MAPK cascades	[34, 97]
Multiple/ combined abiotic stresses	– More impactful than single stress – Collective reduction in photosynthesis and yield – Phenology alters strongly – In crop plants, hampering grain filling, pollen viability and yield components	• Multiple stresses modify osmotic solute gathering differently from single stresses • Definite ROS generation, probably causes more oxidative damage • Revised different types of metabolite profiles, such as sugar alcohols, etc. act differently	• Activation of a unique transcriptome specific to combined stresses • Numerous genes are activated under combination, not in either stress alone • Crosstalk between hormones, ROS, Ca^{2+}, and Transcription factor (TFs) pathways is more complex	[27, 45, 69]

4 Breeding and Biotechnological Interventions in Abiotic Stress Management

4.1 Conventional Breeding for Stress Tolerance

Conventional breeding has been at the heart of crop improvement for ages. These focus on selecting genotypes that can endure stresses through natural genetic variation. Breeders often utilize high-quality germplasm from wild relatives, landraces and established varieties to enhance traits associated with tolerance in conventional breeding approaches [2]. Tolerant genotypes are crossed with commercially viable, high-yielding varieties in these methods. Then the best-performing genotype is chosen after multiple breeding cycles. However, abiotic stress tolerance is complex and hence, these conventional techniques frequently encounter difficulties. Several genes regulate tolerant traits, where each gene has a minor impact. This makes it more difficult to identify accountable genes and superior genotypes. The process is more laden by the lengthy duration and limited availability of diverse genetic resources in some crops. In the early phases of developing stress-tolerant varieties, conventional breeding is still decisive despite these obstacles. These are most useful when the genetic mechanisms underlying stress tolerance are not understood in detail [82]. Conventional breeding has resulted in many stress-tolerant crops. For example, drought-tolerant varieties in maize (ZM521, ZM523, ZM521-DT and Sorghum (M35-1,salt-tolerant varieties in wheat (Kharchia-65 and rice (Pokkali); submergence-tolerant varieties in rice (Swarna-Sub1/IR64-Sub1); heat-tolerant varieties in wheat (HD 2967, HD 3086); cold-tolerant variety in barley (Himalaya 292); low-light-tolerant variety in rice (Swarnaprabha); UV-B-tolerant variety in rice (Lalkain, Meghadambaru), etc.

4.2 Marker-Assisted Selection and QTL Mapping

Conventional plant breeding primarily depends upon the phenotypic traits in the field, which are strongly influenced by the environment and management practices. It makes the improvement of complex traits pertaining to tolerance/resistance slow and impulsive. The development of molecular markers has completely changed plant breeding by DNA-level selection. Marker-assisted selection (MAS) and quantitative trait loci (QTL) mapping allow breeders to find, monitor and combine advantageous genes with much greater speed and meticulousness. A detailed comparison of traditional breeding and MAS has been given in Table 2.

MAS is the practice of using molecular markers to distinguish plants that carry desirable alleles. These markers can be single-nucleotide polymorphisms or simple sequence repeats [10]. As they are closely linked to the desired traits, breeders can evaluate plants at any developmental stage rather than waiting for their complete expression. It has been widely used to introduce abiotic stress tolerance in many crops like rice, wheat, maize, groundnut, etc., in the last few years. Major advantages of MAS are as follows:

Table 2 Comparison between conventional breeding and marker-assisted selection

Aspect	Conventional breeding	Marker-assisted selection
Trait-carrier identification period	Multiple growing seasons	Stage-independent
Environment influence	High	Nil
Precision	Lesser	High
Multiple-trait combination	Tough and slow	Efficient
Field testing cost	High	Reduced for initial screening

(a) Selection can be done very early in the crop's life cycle.
(b) Selection is not affected by environmental variation.
(c) Selection process reduces the need for expensive multi-location field trials.
(d) Enables multiple tolerance genes into a single plant variety.

The majority of the traits contributing to various abiotic stress tolerance are complex and controlled by multiple genes with minor effects. These are again highly impacted by the environments. QTL mapping helps to recognize the chromosomal regions linked with these traits [81]. The technique includes the creation of a mapping population, genotyping of the population with a dense set of molecular markers, measurement of the phenotypic trait of desired tolerance and use of statistical models to link markers with trait variation and detect the positions of QTLs on the chromosomes. Markers that are closely linked with these QTLs can be used directly in MAS for effective selection in breeding populations once they have been found and verified. Breeders can use MAS to track and select for those regions in new breeding lines once QTL mapping reveals the location of a trait in the genome. This combination (Fig. 2) has allowed for more strategic crossings, increased precision and reduced breeding cycles. For example, in rice, the *Sub1* QTL (chromosome 9) was introduced into the high-yielding but flood-sensitive variety Swarna to produce Swarna-Sub1, which can survive up to two weeks of complete submergence; *Saltol* QTL (chromosome 1) is used in breeding for salt-tolerance in rice; in wheat, drought tolerance QTLs were identified on chromosomes 4A and 7D.

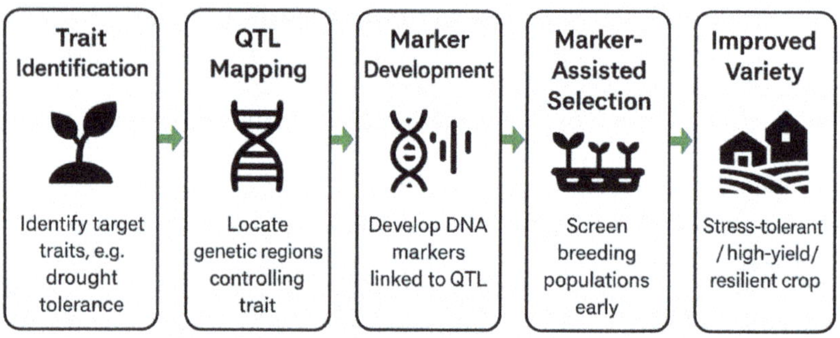

Fig. 2 Flowchart of quantitative trait loci (QTL) and marker-assisted selection (MAS) in climate-resilient crop breeding

4.3 Transgenic and Genome Editing Approaches

Crop improvement has attained noteworthy milestones with the progress of genetic engineering. Transgenic approaches have been broadly used to enhance crop performance against various abiotic stresses. In this, explicit genes of interest, usually from other species, are inserted into a plant genome to introduce novel traits [11, 55]. The introduction of the *AtDREB1A* gene from *Arabidopsis thaliana* into rice and wheat is a well-known example [50, 58, 64]. Many downstream stress-response genes are activated by the transcription factor dehydration-responsive element binding protein (DREB). The overexpression of the encoding gene in transgenic rice has resulted in increased drought tolerance. Another notable example is the transfer of antifreeze protein (AFP) genes from winter flounder fish into tomato plants to enhance frost resistance. There are many more in the list for various crops. Despite these success stories, there are still many regulatory barriers and public perception issues regarding genetically modified organisms. These include concerns over labelling, biosafety and the environment. As a result, Interest in genome editing technologies like CRISPR systems has increased in recent times, which can achieve similar outcomes without permanently adding foreign genes [79]. CRISPR/Cas9 is a specific gene-editing technique that employs a guide RNA to direct the Cas9 enzyme to a particular sequence of DNA, where it makes a clean cut in both strands [35]. When the plant engages to repair the break, it can cause small insertions or deletions in that zone, which may cause silencing, activating or fine-tuning genes associated with vital traits conferring various abiotic stress factors. The changes made by this technique use the crop's own DNA. Hence, the outcome is often considered 'non-transgenic', which makes them easier to accept. This approach accelerates crop improvement more quickly than conventional plant breeding methods, delivers far higher precision and allows researchers to develop stress-tolerant, high-yielding cultivars. For example, editing of transcription factor *OsMYB2* and OsDRAP1 has improved drought tolerance in rice by knocking out negative regulators and activating downstream drought-responsive genes [8, 40]. Aside from Cas9, the toolbox is expanded by CRISPR/Cas12 (Cpf1) and base-editing tools [7, 9, 42]. The base editors enable single-nucleotide modifications without producing double-strand breaks. In contrast, Cas12 cuts DNA another way. It produces staggered ends that are suitable for precise insertions or deletions. In contrast to conventional transgenic techniques, genome editing offers enhanced precision, expedited development and increased regulatory flexibility, as plants generated without foreign DNA often do not fall under strict GMO regulations.

5 Future Perspectives

The future of agriculture depends upon a comprehensive policy combining advanced genetic and plant breeding tactics, and erudite agronomic practices with solid governmental frameworks and socio-economic support systems. Development of various abiotic stress-tolerant varieties using precise approaches like CRISPR/Cas9 is becoming increasingly decisive with the increasing frequency and intensity of numerous abiotic stresses. However, their effectiveness is boosted with integrated field-level modernizations like precision irrigation, optimized soil fertility management and diversified crop rotations. Emerging technologies like remote sensing, high-resolution satellite imagery, drone-based spectral sensors, etc., are aiding in assessing stress and executing site-specific interventions to accelerate breeding processes and enrich site-specific decision-making. The technological revolution alone cannot meet the need for stress tolerance of the crops. Integrating agroecological principles and resource-efficient techniques is crucial to adopt with biotechnology and precision farming. In addition, equitable access to tolerant seeds, cheap inputs and knowledge transfer to smallholder farmers must be vigorously improved with the help of targeted policy support, institutional framework and community rendezvous.

6 Conclusion

Abiotic stressors are among the most significant threats to agricultural productivity worldwide. They possess wide-ranging negative effects on plant growth, yield and nutritional quality as well as on socio-economic outcomes for farmers. The collective effects of these factors impede food security, especially in those areas that are highly vulnerable to climatic fluctuations and constrained by resource limitations. These underpin the urgent need to develop effective mitigation strategies and tolerant genotypes to various stressors. However, plants have complex responses to various abiotic stresses. These involve numerous physiological, biochemical and molecular mechanisms. Progress in conventional breeding, molecular markers and biotechnological techniques has significantly boosted the ability to breed stress-tolerant varieties in different plant species. The complex nature of abiotic stress needs comprehensive and interdisciplinary techniques to tackle it. Integration of genetics, biotechnology, agronomy and cutting-edge technologies offers a promising approach. In addition, it is decisive to consider favourable socio-economic factors and policy frameworks to achieve the desired success from prevailing approaches.

References

1. Abdel Latef AA, Chaoxing H (2011) Arbuscular mycorrhizal influence on growth, photosynthetic pigments, osmotic adjustment and oxidative stress in tomato plants subjected to low temperature stress. Acta Physiol Plant 33:1217–1225
2. Acquaah G (2015) Conventional plant breeding principles and techniques. In: Advances in plant breeding strategies: breeding, biotechnology and molecular tools. Springer, Cham.

3. Adak S, Datta S, Bhattacharya S, Ghose TK, Lahiri Majumder A (2020) Diversity analysis of selected rice landraces from West Bengal and their linked molecular markers for salinity tolerance. Physiol Mol Biol Plants 26(4):669–682. https://doi.org/10.1007/s12298-020-00772-8
4. Adhikari B, Roy A, Reddy H, Roy D, Das C, Ghosh D, Das S, Mondal S, Nath R, Bhattacharyya PK, Jambulkar SK, Bhattacharyya S (2024) Identification and analysis of gamma-irradiation-induced Stemphylium blight tolerant lentil (Lens culinaris) mutant. Int J Rad Biol 100(12):1722–1730. https://doi.org/10.1080/09553002.2024.2409667
5. Akter N, Islam MR (2017) Heat stress effects and management in wheat. Agron Sustain Dev 37:5. https://doi.org/10.1007/s13593-017-0443-9
6. Aslam MM, Waseem M, Jakada BH, Okal EJ, Lei Z, Saqib HSA, Yuan W, Xu W, Zhang Q (2022) Mechanisms of abscisic Acid-Mediated drought stress responses in plants. Int J of Mol Sci 23(3):1084. https://doi.org/10.3390/ijms23031084
7. Badon IW, Oh Y, Kim HJ, Lee SH (2024) Recent application of CRISPR-Cas12 and OMEGA system for genome editing. Mol Therapy 32(1):32–43
8. Baldoni E, Genga A, Cominelli E (2015) Plant MYB transcription factors: their role in drought response mechanisms. Int J Mol Sci 16(7):15811–15851. https://doi.org/10.3390/ijms160715811
9. Bandyopadhyay A, Kancharla N, Javalkote VS, Dasgupta S, Brutnell TP (2020) CRISPR-Cas12a (Cpf1): a versatile tool in the plant genome editing tool box for agricultural advancement. Front Plant Sci 11:584151. https://doi.org/10.3389/fpls.2020.584151
10. Barman M, Kundu S (2019) Molecular markers and a new vista in plant breeding: a review. Int J Curr Microbiol Appl Sci 8(12):1921–1929. https://doi.org/10.20546/ijcmas.2019.812.230
11. Barman M, Kumar SS, Kumar A, Parveen R (2021) Cisgenics as new horizon in crop improvement. Res J Biotechnol 16:1
12. Basak N, Rai AK, Barman A, Mandal S, Sundha P, Bedwal S, Kumar S, Yadav RK, Sharma PC (2022) Salt affected soils: global perspectives. In: Environmental science and engineering. Springer, Cham. https://doi.org/10.1007/978-3-031-09270-1_6
13. Bashir SS, Hussain A, Hussain SJ, Wani OA, Zahid Nabi S, Dar NA, Mansoor S (2021) Plant drought stress tolerance: Understanding its physiological, biochemical and molecular mechanisms. Biotech Biotechnol Equip 35(1):1912–1925. https://doi.org/10.1080/13102818.2021.2020161
14. Baweja P, Kumar G (2021) Abiotic stress in plants: an overview. In: Plant stress biology: strategies and trends. Springer, Singapore
15. Benson EE, Bremner D (2004) Oxidative stress in the frozen plant: a free radical point of view. In: Life in the frozen state. CRC Press, USA
16. Bhatla SC, Kathpalia R (2023) Essential and functional mineral elements. Springer, Singapore
17. Bodner G, Nakhforoosh A, Kaul H (2015) Management of crop water under drought. Agron Sustain Develop 35(2):401–442
18. Chhabra R (2021) Nature and origin of salts, classification, area and distribution of salt-affected soils. Springer, Cham
19. Chorom M, Rengasamy P (1997) Carbonate chemistry, pH, and physical properties of an alkaline sodic soil as affected by various amendments. Soil Res 35(1):149. https://doi.org/10.1071/S96034
20. Choukri H, Hejjaoui K, El-Baouchi A, Haddad NE, Smouni A, Maalouf F, Thavarajah D, Kumar S (2020) Heat and drought stress impact on phenology, grain yield, and nutritional quality of lentil (Lens culinaris Medikus). Front Nut 7:596307. https://doi.org/10.3389/fnut.2020.596307
21. Dai Q, Yan B, Huang S, Liu X, Peng S, Miranda MLL, Chavez AQ, Vergara BS, Olszyk DM (1997) Response of oxidative stress defense systems in rice (Oryza sativa) leaves with supplemental UV-B radiation. Physiol Planta 101(2):301–308. https://doi.org/10.1111/j.1399-3054.1997.tb01000.x
22. Dar NA, Amin I, Wani W, Wani SA, Shikari AB, Wani SH, Masoodi KZ (2017) Abscisic acid: a key regulator of abiotic stress tolerance in plants. Plant Gene 11:106–111. https://doi.org/10.1016/j.plgene.2017.07.003

23. Dawood MF, Tahjib-Ul-Arif M, Shirazy BJ, Latef AAHA (2025) Unraveling the role of κ-carrageenan on the combined effect of drought and chromium stress in wheat (*Triticum aestivium* L.). Plant Physiol Biochem 223:J109781
24. Dawood MFA, Tahjib-Ul-Arif M, Sohag AAM, Abdel Latef AA (2024) Role of acetic acid and nitric oxide against salinity and lithium stress in canola (*Brassica napus* L.). Plants 13(1):51
25. Dawood MFA, Tahjib-Ul-Arif M, Sohag AA, Abdel Latef AA, Ragaey MM (2021) Mechanistic insight of allantoin in protecting tomato plants against ultraviolet C stress. Plants 10(1):11
26. De Costa W (2011) A review of the possible impacts of climate change on forests in the humid tropics. J Natl Sci Found Sri Lanka 39(4):281
27. Demirel U, Morris WL, Ducreux LJ, Yavuz C, Asim A, Tindas I, Hancock RD (2020) Physiological, biochemical, and transcriptional responses to single and combined abiotic stress in stress-tolerant and stress-sensitive potato genotypes. Front Plant Sci 11:169. https://doi.org/10.3389/fpls.2020.00169
28. Diaz RJ, Breitburg DL (2009) Chapter 1 the hypoxic environment. Fish Physiol 27:1–23. https://doi.org/10.1016/S1546-5098(08)00001-0
29. Farhad M, Kumar U, Tomar V, Bhati PK, J NK, Kishowar-E-Mustarin N, Barek V, Brestic M, Hossain A (2023) Heat stress in wheat: a global challenge to feed billions in the current era of the changing climate. Front Sustain Food System 7:1203721. https://doi.org/10.3389/fsufs.2023.1203721
30. Farooq M, Wahid A, Kobayashi N, Fujita D, Basra SMA (2009) Plant drought stress: effects, mechanisms and management. Springer, Dordrecht. https://doi.org/10.1007/978-90-481-2666-8_12
31. Feng Y, Li Z, Kong X, Khan A, Ullah N, Zhang X (2025) Plant coping with cold stress: molecular and physiological adaptive mechanisms with future perspectives. Cells 14(2):110. https://doi.org/10.3390/cells14020110
32. Ganguly S, Nimitha K, Saha S, Sinha Mahapatra N, Bhattacharya K, Kundu R, Bhattacharyya S (2024) Identification and analysis of low light responsive yield enhancing QTLs in rice. Sci Rep 14(1):210110. https://doi.org/10.1038/s41598-024-71593-y
33. Gebrechorkos SH, Sheffield J, Vicente-Serrano SM, Funk C, Miralles DG, Peng J, Dyer E, Talib J, Beck HE, Singer MB, Dadson SJ (2025) Warming accelerates global drought severity. Nature 642:628–635. https://doi.org/10.1038/s41586-025-09047-2
34. Gong F, Zhou X, Cao K, Xu H, Zhou X (2024) Analyses of the metabolism and assessment of antioxidant activity in Rhododendron chrysanthum Pall. after UV-B Irradiation. Plant Growth Regul 103(3):525–537. https://doi.org/10.1007/s10725-024-01119-z
35. Gowda AK, Mishra S, Barman M, Saikumar M (2020) CRISPR/Cas9: a revolutionary tool for recent advances in crop improvement: a review. Int J Curr Microbiol Appl Sci 9:200–214
36. Gull A, Lone AA, Wani NUI (2019) Biotic and abiotic stresses in plants. In: Abiotic and biotic stress in plants. IntechOpen, London
37. Habib-ur-Rahman M, Ahmad A, Raza A, Hasnain MU, Alharby HF, Alzahrani YM, El Sabagh A (2022) Impact of climate change on agricultural production; Issues, challenges, and opportunities in Asia. Front Plant Sci 13:925548. https://doi.org/10.3389/fpls.2022.925548
38. Hasanuzzaman M, Nahar K, Alam MM, Roychowdhury R, Fujita M (2013) Physiological, biochemical, and molecular mechanisms of heat stress tolerance in plants. Int J Mol Sci 14(5):9643–9684. https://doi.org/10.3390/ijms14059643
39. Hassan MA, Xiang C, Farooq M, Muhammad N, Yan Z, Hui X, Yuanyuan K, Bruno AK, Lele Z, Jincai L (2021) Cold stress in wheat: plant acclimation responses and management strategies. Front Plant Sci 12:676884
40. Huang L, Wang Y, Wang W, Zhao X, Qin Q, Sun F, Li Z (2018) Characterization of transcription factor gene OsDRAP1 conferring drought tolerance in rice. Front Plant Sci 9:94. https://doi.org/10.3389/fpls.2018.00094
41. Hussain HA, Hussain S, Khaliq A, Ashraf U, Anjum SA, Men S, Wang L (2018) Chilling and drought stresses in crop plants: implications, cross talk, and potential management opportunities. Front Plant Sci 9:393. https://doi.org/10.3389/fpls.2018.00393

42. Hwang GH, Song B, Bae S (2021) Current widely-used web-based tools for CRISPR nucleases, base editors, and prime editors. Gene Genome Edit 1:100004. https://doi.org/10.1016/j.ggedit.2021.100004
43. Ilyas M, Nisar M, Khan N, Hazrat A, Khan AH, Hayat K, Fahad S, Khan A, Ullah A (2020) Drought tolerance strategies in plants: a mechanistic approach. J Plant Growth Regul 40(3):926–944. https://doi.org/10.1007/s00344-020-10174-5
44. Jan R, Khan M, Asaf S, Lubna N, Waqas M, Park J, Asif S, Kim N, Lee I, Kim K (2022) Drought and UV radiation stress tolerance in rice is improved by overaccumulation of non-enzymatic antioxidant flavonoids. Antioxidants 11(5):917. https://doi.org/10.3390/antiox11050917
45. Jiang Z, van Zanten M, Sasidharan R (2025) Mechanisms of plant acclimation to multiple abiotic stresses. Commun Biol 8(1):655. https://doi.org/10.1038/s42003-025-08077-w
46. Kai H, Iba K (2014) Temperature stress in plants. In: Encyclopedia of life sciences. Willey
47. Kaur G, Singh G, Motavalli PP, Nelson KA, Orlowski JM, Golden BR (2020) Impacts and management strategies for crop production in waterlogged or flooded soils: a review. Agron J 112(3):1475–1501. https://doi.org/10.1002/agj2.20093
48. Kopecká R, Kameniarová M, Černý M, Brzobohatý B, Novák J (2023) Abiotic stress in crop production. Int J of Mol Sci 24(7):6603. https://doi.org/10.3390/ijms24076603
49. Lal MK, Tiwari RK, Kumar A, Dey A, Kumar R, Kumar D, Singh B (2022) Mechanistic concept of physiological, biochemical, and molecular responses of the potato crop to heat and drought stress. Plants 11(21):2857. https://doi.org/10.3390/plants11212857
50. Latha GM, Raman KV, Lima JM, Pattanayak D, Singh AK, Chinnusamy V, Mohapatra T (2019) Genetic engineering of indica rice with AtDREB1A gene for enhanced abiotic stress tolerance. Plant Cell, Tissue Organ Cul 136(1):173–188. https://doi.org/10.1007/s11240-018-1505-7
51. Lekhana MV, Meena RK, Anirudh TV, Barman M, Kumar A, Reddy MS, Singh SK (2025) Effect of terminal heat stress on stay-green and senescence process could explain genetic variation in grain yield and nutritional profile in wheat (*Triticum aestivum* L.). Plant Growth Regul 1–13. https://doi.org/10.1007/s10725-025-01345-z
52. Leng G, Hall J (2019) Crop yield sensitivity of global major agricultural countries to droughts and the projected changes in the future. Sci Tot Env 654:811–821. https://doi.org/10.1016/j.scitotenv.2018.10.434
53. Li X, Yang R, Li L, Liu K, Harrison MT, Fahad S, Wei M, Yin L, Zhou M, Wang X (2023) Physiological and molecular responses of wheat to low light intensity. Agronomy 13(1):272. https://doi.org/10.3390/agronomy13010272
54. Ma Y, Dias MC, Freitas H (2020) Drought and salinity stress responses and Microbe-Induced tolerance in plants. Front Plant Sci 11:591911. https://doi.org/10.3389/fpls.2020.591911
55. Malik JA, Goyal MR, Kumari A (2023) Food process engineering and technology. In: Food process engineering and technology. Springer, Singapore
56. Mallick A, Dey S, Datta S, Barman M, Samui S, Dutta G (2022) Auxin and cytokinin signaling in plant stress response. In: Signaling and communication in plants. Springer, Cham. https://doi.org/10.1007/978-3-031-05427-3_9
57. Mishra AK, Singh VP (2010) A review of drought concepts. J Hydrol 391(1–2):202–216. https://doi.org/10.1016/j.jhydrol.2010.07.012
58. Muthurajan R, Ramanathan V, Bansilal Shillak A, Madhuri Pralhad S, Shankarrao CN, Rahman H, Madasamy P (2021) Controlled over-expression of AtDREB1A enhances tolerance against drought and salinity in rice. Agronomy 11(1):159. https://doi.org/10.3390/agronomy11010159
59. Oguz MC, Aycan M, Oguz E, Poyraz I, Yildiz M (2022) Drought stress tolerance in plants: interplay of molecular, biochemical and physiological responses in important development stages. Physiologia 2(4):180–197. https://doi.org/10.3390/physiologia2040015
60. Osman KT (2018) Saline and sodic soils. In: Management of soil problems. Springer, Cham. https://doi.org/10.1007/978-3-319-75527-4_10
61. Pati B, Barman M, Kundu R, Datta J, Das A, Saha D, Mukherjee S (2025) Characterization of morpho-physiological and biochemical variations in diverse linseed (Linum usitatissimum L.) genotypes reveals terminal heat stress-tolerant accessions for genetic improvement and insights into tolerance studies. Euphytica 221(6):84. https://doi.org/10.1007/s10681-025-03539-y

62. Patterson BD, Reid MS (1990) Genetic and environmental influences on the expression of chilling injury. In: Chilling injury of horticultural crops. CRC Press, Florida
63. Paul A, Bhatia A (2025) Exploring the effects of contemporary agricultural techniques on soil and water pollution in rice fields. In: Navigating the nexus: hydrology, agriculture, pollution and climate change. Springer Nature, Switzerland
64. Pellegrineschi A, Reynolds M, Pacheco M, Brito RM, Almeraya R, Yamaguchi-Shinozaki K, Hoisington D (2004) Stress-induced expression in wheat of the Arabidopsis thaliana DREB1A gene delays water stress symptoms under greenhouse conditions. Genome 47(3):493–500
65. Pitman MG, Läuchli A (2002) Global impact of salinity and agricultural ecosystems. Salinity: Environ-Plants-Mol 3:20
66. Playne MJ, Smith BR (1983) Toxicity of organic extraction reagents to anaerobic bacteria. Biotechnol Bioeng 25(5):1251–1265. https://doi.org/10.1002/bit.260250508
67. Popa A, Rusu T, ŞIMON A, Russu F, Bărdaş M, Oltean V, Merca NC (2021) Influence of biotic and abiotic factors on maize crop yield in Transylvanian plain conditions. Sci pap Ser A Agron 64(2):103–112
68. Praveen B, Sharma P (2019) A review of literature on climate change and its impacts on agriculture productivity. J Public Affairs 19(4):e1960
69. Radha B, Sunitha NC, Sah RP, TP MA, Krishna GK, Umesh DK, Siddique KH (2023) Physiological and molecular implications of multiple abiotic stresses on yield and quality of rice. Front Plant Sci 13:996514
70. Rady MM, Abdelfattah MA, Desoky EM, Azzam CR, Abdel Latef AA, Mahmoud AEM, Mohamed IAA (2025) A novel group of bio-stimulators: their attenuation of abiotic stress impacts in crop plants. J Plant Nutr 1–16
71. Ramachandran P, Ramirez A, Dinneny JR (2024) Rooting for survival: how plants tackle a challenging environment through a diversity of root forms and functions. Plant Physiol 197(1):kiae586. https://doi.org/10.1093/plphys/kiae586
72. Rathod A, Verma NS (2023) Impact of abiotic stress on agronomical crops. Front Agron. Elite Publishing House, India
73. Rizvi A, Zaidi A, Ameen F, Ahmed B, AlKahtani MDF, Khan MS (2020) Heavy metal induced stress on wheat: phytotoxicity and microbiological management. RSC Adv 10(63):38379–38403
74. Sabagh AE, Islam MS, Skalicky M, Raza MA, Singh K, Hossain M A, Hossain A, Mahboob W, Iqbal MA, Ratnasekera D, Singhal RK, Ahmed S, Kumari A, Wasaya A, Sytar O, Brestic M, Çig F, Erman M, Rahman MHU, Arshad A (2021) Salinity stress in wheat (Triticum aestivum L.) in the changing climate: adaptation and management strategies. Front Agron 3:661932. https://doi.org/10.3389/fagro.2021.661932
75. Saha S, Mahapatra NS, Bhattacharya K, Kundu R, Nimitha K, Ganguly S, Bhattacharyya S (2024) The ratio of A400/A1800 Mapping identifies chromosomal regions containing known photoprotection recovery-related genes in rice. Rice 17(1):62
76. Saharan BS, Brar B, Duhan JS, Kumar R, Marwaha S, Rajput VD, Minkina T (2022) Molecular and physiological mechanisms to mitigate abiotic stress conditions in plants. Life 12(10):1634
77. Salehi-Lisar SY, Bakhshayeshan-Agdam H (2016) Drought stress in plants: causes, consequences, and tolerance. Springer, Cham. https://doi.org/10.1007/978-3-319-28899-4_1
78. Sarma B, Kashtoh H, Lama Tamang T, Bhattacharyya PN, Mohanta YK, Baek KH (2023) Abiotic stress in rice: visiting the physiological response and its tolerance mechanisms. Plants 12(23):3948. https://doi.org/10.3390/plants12233948
79. Shah MH, Saha S, Barman M, Kundu S, Islam S, Pramanik K, Ajaharuddin SM (2023) Ricinus communis and stressful conditions. In: Medicinal plant responses to stressful conditions. CRC Press, USA
80. Shi Y, Ke X, Yang X, Liu Y, Hou X (2022) Plants response to light stress. J Genet Genomics 49(8):735–747. https://doi.org/10.1016/j.jgg.2022.04.017
81. Singh SK, Barman M, Prasad JP, Bahuguna RN (2022) Phenotyping diverse wheat genotypes under terminal heat stress reveal canopy temperature as critical determinant of grain yield. Plant Physiol Rep 27(2):335–344. https://doi.org/10.1007/s40502-022-00647-y

82. Sun L, Lai M, Ghouri F, Nawaz MA, Ali F, Baloch FS, Shahid MQ (2024) Modern plant breeding techniques in crop improvement and genetic diversity: from molecular markers and gene editing to artificial intelligence—a critical review. Plants 13(19):2676
83. Tahjib-Ul-Arif M, Asaduzzaman, M, Shirazy BJ, Khan MSU, Rahman AMS, Murata Y, Abdel Hamed S, Abdel Latef AA (2024) Seed priming improves chilling stress tolerance in rice (*oryza sativa* L.) seedlings. Phyton-Int J Exp Bot 93(11):3013–3027
84. Tahjib-Ul-Arif M, Hasan MT, Rahman MA, Nuruzzaman M, Rahman AMS, Hasanuzzaman M, Haque MR, Hossain MA, Abdel Latef AA, Murata Y, Brestic M (2023) Plant response to combined salinity and waterlogging stress: current research progress and future prospects. Plant Stress 7:100137
85. Wang X, Hu Y, Dong Y, Zhang L, Wang B (2025) Abiotic stress-regulated LEA gene mediates the response to drought, salinity, and cold stress in Medicago sativa L. Plant Cell Physiol 66(5):781–796
86. Wilhite DA, Glantz MH (1985) Understanding: the drought phenomenon: the role of definitions. Water Internat 10(3):111–120
87. Wojtyla Ł, Paluch-Lubawa E, Sobieszczuk-Nowicka E, Garnczarska M (2020) Drought stress memory and subsequent drought stress tolerance in plants. Academic Press, Elsevier. https://doi.org/10.1016/B978-0-12-817892-8.00007-6
88. Xiong L, Yang Y (2003) Disease resistance and abiotic stress tolerance in rice are inversely modulated by an abscisic Acid-Inducible Mitogen-Activated protein Kinase[W]. Plant Cell 15(3):745–759
89. Xu Y, Bu W, Xu Y, Fei H, Zhu Y, Ahmad I, Nimir NEA, Zhou G, Zhu G (2024) Effects of salt stress on physiological and agronomic traits of rice genotypes with contrasting salt tolerance. Plants 13(8):1157
90. Yadav SK (2009) Cold stress tolerance mechanisms in plants. Agron Sustain Dev 30(3):515–527. https://doi.org/10.1051/agro/2009050
91. Yadav S, Modi P, Dave A, Vijapura A, Patel D, Patel M (2020) Effect of abiotic stress on crops. Sustain Crop Prod 3(17):5–16
92. Yang K, Ma Y, Tang Q (2025) Visualisation of key thresholds of crop production potential and their future distribution patterns in China under climate change scenarios. J Geovisualization Spat Anal 9(2):33. https://doi.org/10.1007/s41651-025-00234-3
93. Yue Y, Zhou Y, Wang J, Ye X (2016) Assessing wheat frost risk with the support of GIS: an approach coupling a growing season meteorological index and a hybrid fuzzy neural network model. Sustainability 8(12):1308. https://doi.org/10.3390/su8121308
94. Zahedi SM, Karimi M, Venditti A, Zahra N, Siddique KHM, Farooq M (2024) Plant adaptation to drought stress: the role of anatomical and morphological characteristics in maintaining the water status. J Soil Sci Plant Nut 25(1):409–427. https://doi.org/10.1007/s42729-024-02141-w
95. Zhang Y, Xu J, Li R, Ge Y, Li Y, Li R (2023) Plants' response to abiotic stress: Mechanisms and strategies. Int J Mol Sci 24(13):10915. https://doi.org/10.3390/ijms241310915
96. Zhao C, Liu B, Piao S, Wang X, Lobell DB, Huang Y, Huang M, Yao Y, Bassu S, Ciais P, Durand J, Elliott J, Ewert F, Janssens IA, Li T, Lin E, Liu Q, Martre P, Müller C, Asseng S (2017) Temperature increase reduces global yields of major crops in four independent estimates. Proceed Nat Acad Sci 114(35):9326–9331. https://doi.org/10.1073/pnas.1701762114
97. Zhu Y, Wang H, Xiang X, Hayat K, Wu R, Tian J, Du S (2024) A dose-dependent effect of UV-328 on photosynthesis: exploring light harvesting and UV-B sensing mechanisms. J Hazard Mat 473:134670. https://doi.org/10.1016/j.jhazmat.2024.134670

Overview of Biotic Stress and Its Agricultural Impact

Affan Afzal, Baber Ali, K. G. B. A. Samarasinghe, Muhammad Nauman Khan, Muhammad Ammar Javed, and Aroona Saleem

Abstract Crops are silently robbed of their yields by one of the most relentless and destructive stresses, known as biotic stress. It occurs between plants and other living organisms, such as bacteria, fungi, nematodes, arachnids, and insects, which constantly reduce plant growth, disrupt its metabolism, and lower overall yield. In a time when global populations depend on major crops like wheat, rice, maize, and soybean, these are severely impacted by biotic stresses, threatening both their quantity and quality. This chapter summarizes the management strategies that have been used for years and shows how these traditional techniques have influenced modern technologies to address previous limitations. The prolonged use of conventional methods has led to widespread resistance; however, advanced techniques, including molecular breeding, marker-assisted selection, transgenic technology, genome editing (CRISPR/Cas), and plant–microbe engineering, have broken this cycle of resistance and restored agricultural resilience. With further advances

A. Afzal
Department of Microbiology, Faculty of Veterinary and Animal Sciences, The Islamia University of Bahawalpur, Bahawalpur, Pakistan
e-mail: a.durrani95@gmail.com

B. Ali · K. G. B. A. Samarasinghe
School of Science, Western Sydney University, Penrith, NSW, Australia
e-mail: b.ali@westernsydney.edu.au

K. G. B. A. Samarasinghe
e-mail: bhagya.kg@westernsydney.edu.au

M. N. Khan
Department of Botany, Islamia College Peshawar, Peshawar, Pakistan
e-mail: nomiflora@uop.edu.pk

M. A. Javed
School of Agriculture and Environment, UWA Institute of Agriculture, The University of Western Australia, Perth, WA, Australia
e-mail: ammarjaved94@gmail.com; muhammadammar.javed@research.uwa.edu.au

A. Saleem (✉)
School of Science, Edith Cowan University, Joondalup, WA, Australia
e-mail: aroona.saleem@ecu.edu.au

A. A. H. Abdel Latef (ed.), *Chitosan Nanoparticles for Plant Resilience Under Abiotic and Biotic Stressors*, Smart Nanomaterials Technology,
https://doi.org/10.1007/978-981-95-6422-4_3

and innovations like nanotechnology, agriculture is entering a new era of global food security and sustainable farming. Additionally, combining traditional methods with modern technologies, such as crop rotation with genetic resistance, has the potential to shift the balance back toward biological diversity and natural biological control in agriculture.

Keywords Biotic stress · Plant–pathogen interactions · Yield losses · Sustainable agriculture

1 Introduction

The concept of biotic stress can be explained as a situation where the metabolic process of plants is interfered with by the advent of a living organism [119]. Being sessile organisms, plants must continually confront transforming conditions during their life. Other changes may be a threat to the normal growth and development of plants and, therefore, classified as stresses, i.e., biotic stresses [125]. Biotic stress has been attributed to harmful microbes that not only prevent plants from growing normally but also have many adverse effects on crops throughout the world [24]. Biotic stress belongs to the number of real scourges among the environmental pressures, as a result of which colossal losses of crops are realized. Plants are not able to achieve their full genetic potential in vegetative and reproductive growth because of biotic factors (which include microbial pathogens, weeds, and herbivores) [105]. Organisms such as fungi, bacteria, viruses, nematodes, insects, and weeds impose biotic stress by inducing a loss of nutrients, infection, and significant crop damage. Whereas fungi, nematodes, and viruses affect the growth of the plant by using toxins, feeding, or causing systemic harm, insects and mites also affect the direct plant planting situation and serve as vectors of disease [5].

There are two categories of fungal parasites: necrotrophs, which destroy host cells with toxins, and biotrophs. Coupled with bacteria, they bring about vascular wilts, leaf spots, and cankers, among others, and they may infect various parts of the plant. The nematodes suck the insides of the plant cells and can feed on all parts of the plant, but the plant parasitic nematodes are soil-borne diseases and attack the root system of the plants. They cause symptoms of nutrient deficiency, which include wilting or stunting. Viruses specifically cause both local lesions and long-range systemic effects, which make a plant stunted/yellower and develop malformations in various parts of the plant, although it is quite seldom that they can kill their hosts. On the contrary, there are pests, especially insects and mites. They either feed on or lay eggs on plants, thereby causing damage to plants. Piercing-sucking insects also transmit them to the plants by injecting their stylets [43].

These biotic stress mechanisms cause several disorders, diseases, and crop damage to agriculture, thus reducing its production [120]. These are the most prevalent pathogens that cause the plants to get diseased and would have diverse symptoms like the development of local lesions, stunting, rotting, as well as chlorosis. These microorganisms merge, and this leads to the devastating effects of the disease instead of them being present as single pathogens, causing the infection. Proximity of known stress factors is determined by various aspects like the environment, types of the causal agent, host specificity, and the causal agent [120].

Biotic stresses minimize the plant output, which calls for the need to know the plant pathogen interaction. Plants are exposed to many biotic stresses along with numerous unfavorable environmental factors. They overcome such stresses by a number of morphological, biochemical, and molecular responses, and it has been indicated that these distinct signaling pathways interact with one another [43]. Defense responses are characterized by plants with combined, multi-layered defense mechanisms that co-evolve with pathogens, such as ROS scavenging, regulation of transcription factors, activation of pathogenesis-related (PR) proteins, and accumulation of lignin. Immunity is strengthened by key resistance pathways, such as systemic acquired resistance (SAR) and induced systemic resistance (ISR), which are modulated by genes such as ethylene response factor (ERF) and nonexpressor of pathogenesis-related genes 1 (NPR1). All these defense mechanisms prevent the invasion of the pathogen and the spread of the disease [120]. Physical such as strong cell walls, trichomes, and cuticle thickness are physical barriers or resistance mechanisms, and chemical resistance mechanisms deal with the formation of phytochemicals, antimicrobial proteins, phytoalexins, hormonal signaling, cell wall reinforcement, etc. [3].

Ecologists, physiologists, and agronomists have traditionally divided plant stresses into two broad groups: biotic and abiotic. They have had negative impacts on plant health and crop productivity, yet there is a considerable difference in the origin, nature, and action of the same.

Biotic stress occurs as a consequence of the interaction with living organisms, which can include pathogens, insects, or weeds, and which challenge the health of the plant by either infecting the plant, feeding on it, or competing with it. These stressors interfere with cellular and physiological changes, usually resulting in symptoms of diseases and slowed growth. A study showed that the effect of pre-existing or simultaneous abiotic stress on plant biotic stress responses may either inhibit or boost immunity, depending on the time when exposure to abiotic stress commences and its severity [88]. Saijo and Loo [106] emphasize that a complex network of immune signaling generates the patterns of plant response under biotic stress, such as pattern-triggered immunity and effector-triggered immunity, which are regulated by hormones such as salicylic and jasmonic acids, enabling combined responses to multiple, co-occurring stresses.

Abiotic stress can be caused by abiotic environmental conditions like drought, salinity, extremely high or low temperatures, imbalance of nutrients in a plant, and oxidative environments, since they all negatively impact plant growth and development. These stresses damage critical physiological processes through interruption of cellular homeostasis, water uptake, the production of reactive oxygen species (ROS), and the photosynthesis inhibition [100]. As demonstrated by scientists, abiotic stress is worse when it coincides with biotic stress, and this tends to damage plants more than each individual stress [11]. Ramegowda and Senthil-Kumar [96] stress that the presence of abiotic challenges triggers molecular and physiological changes in plants, such as the signaling of stress hormones, activity of antioxidant defenses, and activation of the stress-responsive genes.

Biotic and abiotic stresses are typically examined in isolation; however, in a natural setting, they often co-exist, and can interact synergistically or antagonistically, compounding or reducing the impact that the other causes. As an illustration, abiotic stresses, such as drought and heat, can undermine plant immune systems and render plants susceptible to disease. On the other hand, the infections by pathogens could affect the functionality of roots, decrease the outflow of water and nutrients, and sensitize abiotic stresses. Atkinson and Urwin [11] demonstrate that these interactions use common signaling pathways, e.g., hormonal crosstalk involving abscisic acid (ABA), salicylic acid (SA), and jasmonic acid (JA), and have the potential to alter plant transcriptional responses.

It has recently been found that the plant reaction to combinations of two or more stress factors is peculiar to them and it cannot be directly inferred by analogy to the plant reaction to each of the different stresses applied separately. Additionally, a combination of various stresses leads to a high level of complexity in plant response, whereby the response to the combination of the stresses is heavily regulated by various and, in some cases, contradicting signaling pathways that could interact with each other [11]. These interactions need to be clearly understood to breed and manage crops with greater resistance to diverse, concurrent stress factors.

2 Historical Background and Relevance to Agriculture

Hundreds of years ago, plants were evolving in response to the growing biotic stress that was stimulated by the growing number of human activities over the past two centuries that reshaped habitats with multiple new harsh stress factors, such as biodiversity reduction and ecosystem pollution. The scientific community has gone to great lengths in the sphere of biotic stresses with the perspective of comprehending, managing, and controlling the plant harm brought by the biotic stress agents, the effects of which will assist in forming sustainable agriculture [43]. In India, losses to crops through a variety of biotic stresses are approximated to be 225,000 crores/year. It is believed that plant diseases and invasive insect pests cost about 290 billion dollars throughout the world. Some crop pests and diseases and their associated causal agents were reported during the pre-independence period. Diseases reached some of them to a level of severe yield loss of epidemic proportions. The alternative methods used were fumigation and chemical/pesticide spray to control the emergence of new pests and diseases, and were widespread during independence and the green revolution periods, with the introduction of high-yield varieties [104].

In ancient Yemen, insects, rodents, and birds acted as a major threat to agricultural produce as they destroyed seeds, foliage, and grains kept in storage houses. Farmers employed convert methods of repelling pests through crop rotation, intercropping, sprinkling of ash or sand without using synthetic pesticides. Smoking of burning herbs, hand picking of pests, and homemade natural repellents containing the use of garlic, neem, and chili were typical. Animal dung was another used as an improvised barrier/spray in wells. Most of the yields are lost when the crop is attacked by pests, including weeds, pathogens, and animal pests. The different control techniques have evolved over the past 100 years, but currently crop losses across the world lie within a 25–40% range. The greatest losses are in the form of weeds, then animal pests and diseases. Losses have not reduced significantly despite the use of more pesticides. Integrated pest management (IPM) is an alternative solution that is sustainable since it aims at minimizing the use of chemicals while keeping the yields at the same time [87].

Due to the efforts of Pasteur and Koch through the discovery of bacteriology late in the nineteenth century, the concept of biotic stress changed in agriculture. Their discoveries that were initially aimed at the affections of the human being defined the bacteria as the makers of plant diseases. Plant pathology developed as American researchers borrowed Koch's laboratory techniques and were able to isolate and study crop pathogens. This transition led pest and disease management to no longer be a matter of conventional practice, but science-based management, a critical point toward integrative management of biotic stress due to a focus on protecting crops through specific management techniques [44, 45].

In the Green Revolution, it was important to combat biotic stress to enhance crop yield. Norman E. Borlaug finished at what was the current significance of breeding varieties with resistance to diseases, especially on most common severe threats such as rusts of wheat. He cautioned that pests and pathogens are tending to constantly changing and so the resistance to them has to be a constant practice, where gains made in agriculture can be sustained with the application of integrated crop management [20, 90]. The Green Revolution was only part of the transformation of world agriculture in the twentieth century, which was induced by the blistering pace of growth of the sciences and huge investments and political backing for agriculture by people. The English wheat phenomenon is common. It required almost 1,000 years to double the wheat production to 2.0 tons per hectare, whereas it maxed out with 7.0 tons per hectare throughout the twentieth century. All of this was spurred on by modern plant breeding, new agronomy, and the creation of inorganic fertilizers and modern pesticides. By the second half of the twentieth century, the majority of industrial nations had been able to attain chronic food surpluses and had eradicated the fear of food scarcity [48].

Pre-existing and contributing to biotic stress to the present extent is the presence of insect pests in the crop system, which currently consumes 5–20% of global grain yields. According to recent research, like that of [35] the increasing level of temperature caused by climate change can also increase this stress. Insect-related yield reductions in major staple grains such as wheat, rice, and maize crops are expected to rise by 10–25% with every degree Celsius, at least in temperate-climate regions. This shows the increased risks of insect pests in the climate change scenario and the importance of considering the dimension of pest climate interactions in the management of biotic stress to world food security. With trade and traveling that comes with globalization, the spread of invasive species and exotic diseases is fast [8]. Plants have evolved complex and delicate interactions to survive the stresses in the long process of evolution. It is worth investigating these mechanisms. Crops and vegetable plants are the two most notable classes of plants that are most directly related to the day-to-day activities of the human being. Understanding the underlying mechanisms will not only shed new light on the stress mechanisms in plants, but also be able to breed new crops or horticultural plants with strong resistance to stress [126].

3 Importance in the Context of Global Food Security

Centrality in global food security. Security of food is based on four legs that include availability, access, utilization, and stability. The development of these pillars will ensure that the world collectively addresses the challenges it focuses on globally to sustainably address food demand, which is increasing [126]. This makes finding sustainable and efficient solutions to the pest management problem of high priority because by the year 2050 it is estimated that there will be a global population of 9.2 billion with an associated increase in food demand by 70% due to dietary changes and the use of grain as livestock food sources, and because new agricultural land new agricultural land will be scarce will be scarce. The minimization of yield losses associated with crop pests continues to be a key challenge in the field of agriculture. These are analyzing the existing losses, productivity, and costs associated with the use of pesticides, and the use of alternative, chemical, biotechnological, and biological methods of combating pests. Although pesticides are vital in the protection of crops, further considerations of environmental and health effects have to be mentioned [91]. Pathogens and pests on crops are a great hindrance to crop production and production quality, a potential risk to food security, both locally and globally. Yield losses associated with 137 pathogens and pests were attributed to a global evaluation of five crops: wheat, rice, maize, potato, and soybean. The global losses are estimated at 17–30% with the worst loss occurring to rice (30%). Aggravating this, losses tend to be worse in food-deficient areas where populations are increasing, and these are typically triggered by new or reemerging pests. The results confirm that there is a necessity to focus more on health management of the crops to promote the sustainability of agriculture and food system resilience [91]. This chapter attempts to summarize sources, mechanisms, and effects of biotic stress on agricultural systems, as well as emergent tools and mitigation options. It tries to marshal existing evidence, while highlighting the gaps both in evidence and the research, to inform future work that could influence one of the oldest and most entrenched problems in world agriculture.

4 Types of Biotic Stresses

Biotic stresses are adverse influences of the living organisms on the health and productivity of plants. These are insects, mites, nematodes, fungi, bacteria, viruses, and parasitic plants as shown in Fig. 1 and Table 1. The kind and the intensity of such interactions vary depending on the kind of organism, the disposition of the plant in terms of its defense ability, and the climate conditions. This section highlights the three most important categories of biotic stresses: insect pests, fungal pathogens, and bacterial pathogens.

4.1 Insect and Pest Attacks

Insects destroy forests, crops in the field, and even commodities stored in their warehouses on an astronomical level. Insects attack the crops in common due to their

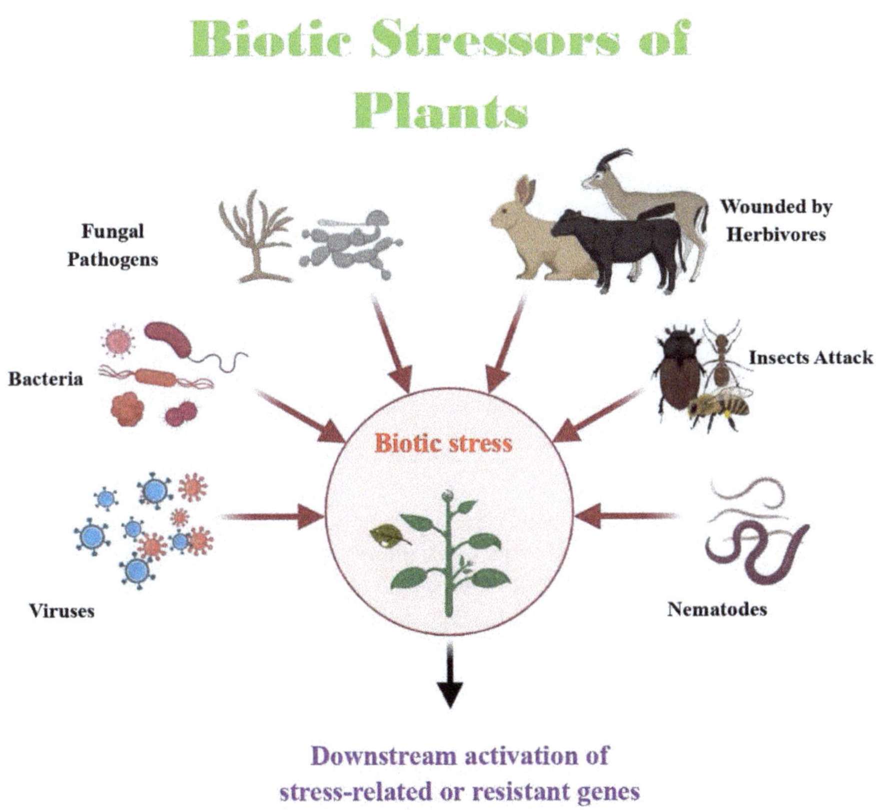

Fig. 1 Types of biotic stresses that affect plants

Table 1 Major biotic stresses in crops

Stress type	Examples	Damage mechanism	Crops affected	Impact	References
Insects	Caterpillars, Beetles, Aphids, Whiteflies, Thrips	Chewing, sap-sucking, virus transmission	Cotton, Cabbage, Brinjal, Okra, Tomato, Onion	18–20% global crop loss (~US$470B)	[66]
Fungi	*Magnaporthe oryzae*, *Fusarium spp.*, *Puccinia spp.*, *Aspergillus flavus*	Biotrophs (haustoria), Necrotrophs (toxins, ROS), Mycotoxins	Rice, Wheat, Maize, Soybean, Potato, Oilseeds, Nuts	14% yield loss; $60B/yr damage; aflatoxins	[61]
Bacteria	*Xanthomonas oryzae*, *Ralstonia solanacearum*, *Erwinia amylovora*	Type III secretion, effector proteins, immunity suppression	Rice, Citrus, Tomato, Potato, Apple	Rapid outbreaks; severe tropical losses	[19]
Viruses	Potyviridae, Geminiviridae, Closteroviridae	Vector-borne (aphids, whiteflies, leafhoppers, seeds)	Vegetables, Grapevine, Fruits, Rice	Emerging viral diseases; climate-linked epidemics	[36]
Parasitic weeds	*Striga spp.*, *Orobanche spp.*, *Phelipanche spp.*	Haustoria attachment, nutrient drain	Cereals, Legumes, Solanaceae, Brassicaceae, Sunflower	Up to 100% yield loss	[122]
Nematodes	*Meloidogyne spp.* (root-knot)	Root galls, vascular disruption, nutrient loss	Vegetables, Cereals, Legumes	Stunting, wilting, chlorosis, reduced yields	[118]

attraction and as part of their life cycle. The plants and their parts are chewed on regardless of leaves, stems, and roots, suck juices, lay eggs, and transfer bacterial, fungal, and viral diseases by insects. More than several hundred diseases of plants have been known to be caused by insects. Therefore, insects trigger the vigor of plants, leading to severe crop losses [66].

Biotic stress is an important factor causing crop production losses all over the world. One of the major reasons behind the low crop yield is the damage that is inflicted by the insect pest [74]. Limited evidence suggests that arthropods are destroying an estimated 18–20% of what is produced by plant crops globally, valued at more than US$470 billion annually [110]. The scarce information available about molecular mechanisms behind plant perception of insects and the signaling cascades specifically involved is emerging, and our understanding of this highly specialized process [18]. The sucking and chewing are very destructive pests of various crops such as cotton, cabbage, brinjal, okra, tomato, and onion all over the world [79].

4.1.1 Chewing Pests

The caterpillars and the beetles are major families of insects that damage plants through their chewing mouthparts. Grasshoppers, katydids, termites, cockroaches, and relatives belong to other chewing insects that are harmful to plants. Insects with chewing mouthparts include dragonflies and damselflies (order Odonata), termites (order Isoptera), adult lacewings (order Neuroptera), beetles (order Coleoptera), ants (order Hymenoptera), cockroaches (order Blattaria), grasshoppers, crickets and katydids (order Orthoptera), and caterpillars (order Lepidoptera).

Caterpillars, beetles, and grasshoppers have strong mouthparts/mandibles that enable them to bite, tear, and consume different parts of plants. They usually consume and devour the leaves, stems, flowers, fruits, and even roots of plants, forming a variety of observable multi-modal damage. The most usual signs of their feeding are holes and notches in the leaves, flowers, or fruits, and as a result, many of them may be defoliated. In most instances, these insects feed on soft tissue between leaf veins, making them appear at the end of laticifers with a skeletonized look against them. The occurrence of such damage not only decreases the photosynthetic capability of the plant but also makes it more vulnerable to secondary infections and environmental stress [23].

4.1.2 Sucking Pests

Sucking pests are aphids, leafhoppers, thrips, whiteflies, flies, bugs, and mites. Sucking pests pierce the plants with delicate and sharp-tipped mouthparts and suck the plant juices. The sap removal leaves tiny white, brown or red spotting in the leaves, fruits or stems of the plant. It can also result in curled leaves, distorted fruit, general wilting, browning, and drying of the whole plant. Most of such pests also show pesticide resistance. Certain of these pests are important vectors of plant viruses, of which many are transmitted [108].

Sucking insect pests are said to be one of the greatest biotic constraints in India on vegetable production. These pests also produce direct damage by aspirating the sap through highly modified mouth structures and exude the sugary honeydew film on top of the plants and form the black Hume-like layer known as butter, thus impeding further normal photosynthesis of the plants. Besides directly destroying them, they are also vectors of various viral diseases [98]. The insects of such mouth parts pierce the tissues and suck out the contents (sap/blood/nectar) with the help of the mandibular stylets, with the action of the pharyngeal and cibarial muscles.

4.2 Fungal Pathogens

Fungal pathogens fungi constitute the highest plant pathogen, which destroys crop yields up to 14% every year; a factor that makes the fungi one of the key issues in attaining global food security and agricultural sustainability. Fungal diseases decrease the overall crop output and productivity, thus depleting the economy of the globe in general [98]. Fungal diseases destroyed at least 125 million tons of crops such as wheat, rice, soybeans, maize, and potatoes in 2012. The damage to rice, wheat, and maize caused by fungi worldwide reaches 60 billion dollars a year [113].

4.2.1 Major Fungal Diseases

A survey of molecular plant pathology identified the following as the most important pathogens: *Magnaporthe oryzae* (rice blast), *Botrytis cinerea* (grey mold), *Puccinia* spp. (rusts), *Fusarium* spp. (wilt, head blight), *Blumeria graminis* (powdery mildew), *Mycosphaerella graminicola* (leaf spot), *Colletotrichum* spp. (anthracnose), *Ustilago maydis* (corn smut), and *Melampsora lini*. Many attack cereals such as rice and wheat and endangering staple food security. Antifungal chemicals, transgenic resistance, and biological control are some of the control strategies used in the control of it [61]. All the potential plant pathogens that can interfere with crop yields can be ranked with the leading risk that mycotoxigenic fungi pose to global food security, before the generalist pathogens and pests become dominant, which contaminates the primary sources of nutrition for all world citizens. One of the biggest challenges is to reduce mycotoxin contamination of produce during the postharvest period [107]. Initially, e.g., *Aspergillus flavus* has an extended host range as an opportunistic pathogen/saprobe. It is the most common fungus of the soil. Much apprehension exists over the production of toxins by this fungus, which are known to be highly carcinogenic aflatoxins in agriculture, as these toxins are a health hazard to animals. *A. flavus* is mainly an issue in the oilseed crops of maize, peanuts, cottonseed, and tree nuts in the field. *A. flavus* has the potential to grow in practically any crop seed and produce aflatoxin under improper storage conditions. It, too, is an animal and insect pathogen. It is mostly an opportunistic pathogen in immunosuppressed people in humans [65].

4.2.2 Lifecycle and Infection Modes

Fungal pathogens acquire nutrients as either biotrophs that use living plant tissue with little or no host cell death or as necrotrophs that kill host cells and subsist on dead tissue. Others are hemi-biotrophs; they initially have a biotrophic stage, but later on change to necrotrophy [40].

Other fungi, e.g., *Blumeria gramiens, Uromyces fabae, Ustilago maydis, Cladosporium fulvum, Puccinia graminis,* and *Phytophthora infestans* are biotrophic and live by extracting nutrients and energy out of living cells of the plants and exist in the interstitial spaces. When the plant is attacked by the pathogen, and it tries to absorb the available nutrients, it forms its external structures, such as appressoria or haustoria, which attach to the plant cell wall, collapse plant cell layers, and release effector molecules to avoid the plant's immune defense system [37].

Necrotrophic fungi cause host cells to be killed and overcome plant defenses through the release of toxins, enzymes, and small RNA, and trigger plant defenses that are overcome by receptors such as BOTRYTIS-INDUCED KINASE 1 (BIK1) and SUPPRESSOR OF BIR1-1 (SOBIR1) to finally complete infection and reproduction [4]. Necrotrophic pathogens are infamous in their friendly strategies of virulence that enable them to dine to the death or dead cells. These fungi are known to secrete diverse phytotoxins, cell wall-degrading enzymes and reactive oxygen species induce cell necrosis which facilitating penetration and prefer cell leakage to nutrient leakage [75].

4.3 Bacterial Pathogens

Although there are fewer bacterial pathogens in number as compared to fungi, they can bring about destruction in crops. They can mostly be seed-borne, or they can enter through wounds or stomata, or hydathodes, and spread rapidly in moist conditions [115].

4.3.1 Host-Pathogen Interactions

Host-pathogen interactions although fewer in number as compared to the other biologic agents, bacterial pathogens contribute to some of the worst crop diseases anywhere in the world, especially in the warmer, moist environment of tropical and subtropical areas. Prominent in these are bacterial blight of rice (*Xanthomonas oryzae* pv. oryzae), bacterial wilt (*Ralstonia solanacearum*), citrus canker (*Xanthomonas axonopodis* pv. citri), and Huanglongbing (*Candidatus Liberibacter* asiaticus). The important diseases of temperate regions are fireblight (*Erwinia amylovora*) and soft rot (*Pectobacterium carotovorum*). The pathogens are of great concern to yield reduction and economic losses via different disease pathways, and their biology is critical in the development of more effective control measures [19].

A considerable number of phytopathogenic bacteria transfer virulence effector proteins into plant cells through an Hrp type III secretion mechanism (TTSS). In the absence of the TTSS, these pathogens are unable to overcome the basal mechanisms, multiply in plants, cause disease symptoms in the host, or induce the hypersensitive response in nonhosts. Bacteria secrete effector proteins to evade detection and to hijack host metabolism. To respond to these invaders, plants in turn utilize pattern-triggered immunity (PTI) and effector-triggered immunity (ETI) [6].

Beneficial microbes suppress the growth of bacterial plant diseases under the sustainable control of biological control. Effectiveness requires agents that persist on crops, outcompete the native microflora, and work under field conditions. Proper formulation, favorable environment, and molecular tools may enhance their efficacy, and thus, they would constitute part of integrated pest management [83].

4.4 *Viral Infections*

In the world, plants normally experience parasitic and non-parasitic diseases. Viruses constitute one of the significant categories of pathogens that infect plants, and all of them are obligate parasites that cannot grow outside of a living cell [36]. Plant viruses are very small in comparison to other plant pathogens like fungi and bacteria, which can be visualized with microscopes, but plant viruses can only be seen under transmission electron microscopes and are comprised of a coat protein and either a DNA or RNA-based system of nucleic acid as the source of genetic information [58].

4.4.1 Emerging Plant Viruses and Transmission Vectors

New plant viruses and transmission agents. They are usually insect or seed-borne, and global warming-induced changes are likely to have an impact on both the level of incidence and severity. Emerging disease is a term used to describe new virus diseases threatening agricultural production in the world to occur on different vegetable crops, grapevine, fruit crops, and rice. Some recently emerging viruses include RNA or DNA genomes and are included in Potyviridae, Bunyaviridae, Geminiviridae, Closteroviridae, Betaflexiviridae, and Secoviridae families [36]. Viruses that have emerged in the last two decades of the world limit the production of important vegetable crops in tropical, subtropical, and temperate regions, and many of the causal agents are whitefly-transmitted (order Hemiptera, family Aleyrodidae [82]).

4.5 *Parasitic Weeds and Nematodes*

The parasitic weeds or heterotrophic plants are easily categorized into several distinct sets: hemi/partial- or holo/total-parasites, facultative or obligate parasites, root or stem parasites, and annual or perennial parasites. Of the 4000 or so species of parasitic plants, some species belonging to a small number of plant families (Orobancheaceae, Convolvulaceae, Loranthaceae, and Viscaceae) are thought to be the most economically disastrous species that can cause agricultural menace. The levels of infestation, the susceptibility of crops, and the environments determine up to a 100% reduction in the yield of host crops incurred. They are very thickly distributed, and this adversely affects host-crop farming [84]. Striga spp., mainly *S. hermonthica*, are obligatory root-hemiparasites that highly affect cereal crops in more than 25 countries in Africa. Their complicated life cycle and robust host interactions create a problem in controlling them [84]. Associates Parasitic plants parasitize hosts with haustoria and are either root- or stem-haustorial, facultatively or obligately. The number of genera that damage crops is about 25, including Orobanche, Phelipanche, and Striga. The weeds have a severe impact on legumes, Solanaceae, Brassicaceae, and sunflowers, whereby the impacts are mostly prevalent in the Mediterranean and the Middle East. It is hard to control, and biological techniques are rather ineffective [122].

Root-knot nematodes (*Meloidogyne* spp.) (RKN) are ubiquitous, pan-phytic endoparasites creating specialized feeding structures within host roots using effectors, and degrade plant growth and defence. Overuse of chemical nematicides is risky to health and the environment, and in recent times, governments have mandated the utilization of environmentally friendly alternatives [122]. Some of the sustainable management practices that have the potential to restrict the population and reproduction of nematodes include the implementation of biological control (bacteria, fungi, predators, and parasites), organic amendments (botanicals, oil cakes), cultural procedures, soil solarization, and resistant cultivars [63]. Since RKN attacks the vascular system of their hosts, they cause plant nutrient starvation and failure in water transport to the plant, resulting in top growth symptoms of stunting, wilting, chlorosis, and lighter crop yields. Scholars frequently refer to psychotherapy as a school of thought because it is a theory of how to conduct psychotherapy [118].

5 Physiological and Molecular Responses to Biotic Stress

Biotic (pests, pathogens, weeds) stresses affect plants and reduce crop yield. Since they are immobile, they have developed ways of sensing stress and stimulating defense mechanisms. Insight into the processes contributes to the creation of stress-resistant crops [120]. This section offers details of four crucial features, namely: hormonal signaling routes, reactive oxygen species (ROS) and programmed cell death, pattern-related and effector-related immunity (PTI and ETI), and epigenetic and transcriptional management.

5.1 Hormonal Signaling Pathways

In medicinal plants, secondary metabolism is controlled by plant hormones through biosynthetic pathways, signal transduction, and interaction networks that modulate secondary metabolite synthesis and accumulation of secondary metabolite [70]. Jasmonic acid (JA) and salicylic acid (SA), which regulate growth, defense, and secondary metabolism, are plant hormones. JA regulates terpenoid indole alkaloid, glucosinolate, anthocyanin, and phenylpropanoid biosynthesis, whereas SA activates anthocyanins, alkaloids, and phenolics. GA and ABA can regulate the production of monoterpenes and sesquiterpenoids, and increase stress resistance through them [72].

5.2 Jasmonic Acid (JA)

Jasmonic acid (JA) and its analogs, jasmonates, are important plant phytohormones controlling plant growth and development as well as defense against pathogenic and environmental stress. Due to climate change, the pressure of disease caused by fungi, bacteria, and viruses has accelerated the importance of the whole process of immunity in general and JA in particular. A biosynthesis and signaling Pathogenesis-Related (PR) genes, phytoalexin production, and defense proteins are activated, with modulation of pathogen effectors through the activity of JA, as well as crosstalk with salicylic acid (SA), ethylene (ET), and abscisic acid (ABA). Exogenous JA can boost the immunity via Systemic Acquired Resistance (SAR) against Induced Systemic Resistance (ISR), which can help safeguard harvest and quality [102].

5.3 Salicylic Acid (SA)

NPR1 oligomers in the cytosol are transformed to pathogen-driven active SA-improved redox-strained monomers. These monomers migrate to the nucleus and bind to TGACG-Binding Factor (TGA) proteins and initiate a SA-dependent Pathogenesis-Related (PR) gene expression, one of the most important indicators of Systemic Acquired Resistance (SAR). Nonexpressor of Pathogenesis-Related Protein 3 (NPR3) and Nonexpressor of Pathogenesis-Related Protein 4 (NPR4) serve as SA receptors and control the NPR1 degradation and functionality [16].

5.4 Ethylene

Ethylene (ET) is a gaseous secondary hormone elicited by abiotic and biotic stresses that acts in a central role in plant defense by the production of defensive volatiles and phenolics. It controls the immunity-associated genes, helping in resistance against pathogens and herbivores. Its interaction with other phytohormones is not a secret; however, its autonomous effect on insect herbivory defense is understudied. In this review, the biosynthesis, signaling, and mechanisms of ET contribution to plant response to feeding were mentioned [21].

5.5 Abscisic Acid

One of the stress hormones activating this process is the abscisic acid (ABA), which makes stomata close to prevent water loss and the admission of pathogens. It provokes defense metabolites, interacts with other hormones, and activates the signaling through ROS, RCS, NO, and Ca^{2+}, resulting in prompt defense and long-term adjustment. Although pathogens have tried to eliminate these impacts, ABA is paramount to the stress tolerance and immunity of plants [21].

Plant responses to biotic and abiotic stresses are mediated by a complex network of hormone synergy, antagonism, and addition termed hormone crosstalk. This crosstalk constrains immune responses to be complex and broad-based in coverage, yet also produces specificities against varied microbes and insects. Recent studies indicate its regulation at several layers, such as the stability of proteins and the transcription of genes, as well as hormonal levels, although the primary mediation of the jasmonic acid pathway in Arabidopsis thaliana occurs through transcription factors MYC2 and ORA59 [2].

5.6 *Reactive Oxygen Species (ROS) and Cell Death*

Cell Death Stressors activate ROS, which are the hallmark of oxidative stress in the plant pathogen relationship. Adequate ROS production and scavenging are necessary because ROS are also considered secondary messengers, triggering defence-related genes, hypersensitive response (HR), and systemic signaling. Attack by pathogens causes a temporary and permanent increase in ROS primarily through the activity of superoxide dismutase (SOD) and the NADPH oxidase. ROS are critical in plant defense mechanisms, as they mediate HR-based programmed cell death (PCD) and systemic acquired resistance (SAR) [80].

They include direct antimicrobial effect against bacteria and parasites, and redox regulation of innate immune signaling and the induction of inflammasomal activation. Nevertheless, not the immunity that results in the pattern triggered immunity (PTI) reaction is initiated in the plant cell surface when detecting microbe-derived signals referred to as microbe- or pathogen- associated molecular patterns or pathogen (damage)-associated molecular patterns of cells damaged in the plant. This recognition is by membrane-bound receptor proteins called pattern recognition receptors. Though a large part of the PTI machinery is preserved, there is natural variability in the response to PTI both between and within species in relation to the factors involved in pattern detection, response activation, and response magnitude. Many studies have been conducted to determine detailed ROS analysis, and even less has been made to determine the actual redox control target molecules [49].

ROS are important signaling molecules involved in plant responses to pathogens, and responses vary depending on pathogen type: necrotroph or biotroph. ROS are signaling molecules at moderate levels, triggering defence responses, but at extreme levels, ROS destroy cellular constituents, denature biomolecules, and hinder yield. Antioxidant systems help plants to maintain homeostasis of ROS, whereas endophytes promote antioxidant synthesis and provoke host defences. Endophyte-facilitated regulation of ROS under biotic pressure has received little review, thus necessitates further research [105]. HAMPs activate Ca^{2+} and ROS signaling through wounding and salivary elicitors, which strengthen plant defence-for instance, planthopper saliva activates both pathways to increase resistance [89].

5.7 *Pattern-Triggered and Effector-Triggered Immunity (PTI & ETI)*

To survive in the harsh ecosystem, to protect themselves against invasive, harmful pathogens, plants have developed a two-tiered innate immune system to defend themselves, and this immune system is engaged through two types of receptors, those located in the plasma membrane (PM), e.g., pattern-recognition receptors (PRRs), and an intracellular one, e.g., the nucleotide-binding leucine repeat (NLR) resistome [73].

The Pattern-Triggered Immunity

The immunity that results in the pattern-triggered immunity (PTI) reaction is initiated in the plant cell surface when detecting microbe-derived signals referred to as microbe- or pathogen-associated molecular patterns or pathogen (damage)-associated molecular patterns of cells damaged in the plant. This recognition is by membrane-bound receptor proteins called pattern recognition receptors. Though a large part of the PTI machinery is preserved, there is natural variability in the response to PTI both between and within species in relation to the factors involved in pattern detection, response activation, and response magnitude triggered [52, 112].

Moreover, in the past years, it has been elucidated that signaling peptides, like those produced by PepT7, are detected by plasma membrane (PM) receptor kinases (RKs), which structurally resemble pattern recognition receptors (PRRs), activating pattern-mediated immunity (PTI) by recognizing the microbe-associated molecular patterns (MAMPs) [103]. This group of proteins includes receptor-like kinases and receptor-like proteins whose pro-activation induces a wave of signaling responses, including production of ROS, mitogen-activated protein kinase (MAPK) cascades, and induction of pathogenesis response (PR) genes [86]. The PTI is part of the two perception systems in the plant immune system, the other of which is monitored by the intracellular immune system receptors in the nucleotide-binding leucine-rich repeat receptor (NLR). PTI signaling is under active investigation, and significant advances have been made regarding the pathways of ligand detection, as well as the regulation of particular downstream outcomes [32].

TI is the recognition of virulence factors that are encoded by the pathogen, also known as effectors. Pathogens synthesize effectors that interact with the host to establish a replicative environment and/or counter host immunity. In contrast to PAMPs, effectors tend to be heterogeneous and fast-evolving and are therefore not suitable objects of direct detection by germline-encoded receptors. Effector detection is instead commonly achieved indirectly when they are recognized due to their pathogenic processes. In theory, the use of ETI remains a viable method of pathogen sensing and is employed in a wide range of phyla, including plants, but the biochemistry behind ETI is complicated in comparison to the simple receptor/ligand approach of PAMPs [32]. In ETI, pathogen effectors are recognized directly or indirectly by host plants through resistance proteins, and a stronger and quicker defense reaction ensues. NLRs are the typical schematized forms to be found in resistance proteins, usually released in the form of nucleotide-binding and leucine-rich-repeat-containing receptors (NLRs). Recognition by effectors induces a structural alteration in an NLR and triggers the formation of NLR associations. This engagement triggers a cycle of events, ultimately leading to the phenomenon of the epicentric response (ETI), a large group of immune responses that include cell death (hypersensitive response, or HR). TI causes reactive oxygen species, antimicrobial compounds, and other signaling molecules (including salicylic acid) that serve elsewhere in the amplification module to regulate immune responses [56].

There has been increased evidence of interaction and collaboration between ETI and PTI and how they can work synergistically and trigger a cascade of defense responses [86]. Alignment between the processes of PTI and ETI is also close, but the molecular complexity is less clear [123].

5.8 Epigenetic and Transcriptional Regulation

Phenotypic plasticity is, to varying degrees, evolved by plants to ensure that they can face various pathogens as well as pests in their immobile life. Microbial infection triggers the various layers of host immune response, and plant gene expressions are reprogrammed at both the transcriptional and post-transcriptional levels, exemplifying swift and accurate changes. In a recent direction, epigenetic regulation in response to biotic stresses is important [51]. The study of changes in gene functions that can be invariantly passed on through cell division or inherited in the germline is known as epigenetics. Epigenetic controls in plants involve the manipulation of chromatin structure via DNA methylation and histone modification, as well as small RNA, to regulate gene expression and ensure the stability of genomes, which helps in the adaptation of the plants to external stresses [94]. There are pieces of evidence of stress inducing fast and transient rearrangements of epigenetics to affect the expression of genes associated with secondary metabolic routes. These alterations not only enable short-term metabolic adaptations, but they may also form part of the stress memory. This memory can also sometimes be carried over and passed directly to the subsequent generation, affecting the stress responses of the progeny [54]. Epigenetic modifications provide plants with the benefit of prolonged adaptation to stress or generation inheritance of genes. DNA methylation is one of the major epigenetic marks that occur on a long-term basis and are passed on through many generations. The transgenerational epigenetic dimension of the biotic stresses in plants is less understood than the abiotic ones [59].

Different transcribing factors (TFs), as well as their cofactors and cis-regulatory factors, are significant in the plant defense mechanism. Transcriptional regulation by

TFs plays a critical role in the development of plant defenses and related functions against viral and bacterial attacks. Transcriptional factors key to stress responses, including WRKY, bHLH, bZIP, AP2/ERF, and MYB gene families, are implicated in regulating defense-related genes and coordinating throughout the network of responses to stress in adverse stress conditions. It is the only one that can decipher those other answers [93].

TFs Key TFs that modulate SM biosynthesis, such as AP2/ERF, WRKY, bHLH, bZIP, MYB, and NAC, are all involved in a complex combination of effects on SMs biosynthetic pathway genes. The stressful conditions can change SM levels substantially in the adaptive response of the plant. In this process, TFs play a crucial role, regulating SM biosynthesis-related gene expression. Regulation of several signaling signatures has allowed TFs to coordinate numerous defense processes in a plant, and allow these processes to accumulate their desired SMs at the right time and in the right amounts to counter the effects of the stress response [92], e.g., in mulberry, MiWRKY53 (Group III WRKY transcription factor), though being SA-inducible, relocates to the nucleus in yeast but does not activate transcription. It increases PR-1 synthesis and tolerance to *Pseudomonas syringae* in overexpressing plants, and promoter analysis shows important SA-responsive sequences [85].

6 Biotic Stress Impact on Crop Yield and Productivity: Case Studies from Major Crops

6.1 Wheat

Wheat (*Triticum aestivum* L.) is one of the major food commodities that provides nutrition to people in various locations around the world. Bread wheat is a major crop on a global scale, constituting 20% of calories [15]. Potent biotic stress refers to significant fungal infections, including rusts, smut, bunt, tan spot, fusarium head blight, foot rot, and false eyespot, among others [62].

Wheat Rust: The fungi causing rust diseases include numerous important crop pathogens, such as the Puccinia fungi that belong to the order Pucciniales (phylum Basidiomycota) [38]. Wheat rusts were the cause of major disease epidemics in the past and continue to pose threats to wheat production, as well as the advancement of fungicidal chemicals and resistant cultivars [62]. Rust diseases—Leaf rust, stem rust, and yellow rust—negatively affect productivity and cost hundreds and millions of US dollars per year. During pathogen epidemics, yellow rust causes 10–70%, leaf rust 50%, and stem rust 100% yield losses, the severity of which is dependent on the cultivar, timing of the infections, and the virulence of the pathogen [15].

Fusarium head blight (FHB) is one of the most common diseases of wheat. They can trigger heavy epidemics, resulting in excessive losses of the yield per field, or they can annihilate entire areas where not only the yield is compromised, but the remaining yield is also in bad condition due to toxins [76]. FHB incurs substantial losses in wheat yields because the affected grains are small, shriveled, low in mass and quality, and contaminated with mycotoxins that are poisonous to human beings and in animal feeds. The most prominent mycotoxins of the genus Fusarium are trichothecenes, zanabre felt hats zearalenones, fumonisins, and novel toxins, i.e., beauvericin, enniatins, fusaproliferin, and moniliformin [7].

In this regard, insect pests are the major biotic stakeholders. Insects are experienced in various phenological phases of wheat plants all over the world and sometimes lead to the economic destruction of the wheat plant by consuming the root, stem, leaf, spike, flower, and grain of the plants. It is known that a potential global crop loss as a result of harmful insects in wheat is on average 7.9%. The subject of discrimination is people with breast cancer, as the issue of inequality in the provision of medical care to patients with breast cancer has become a focus of attention of society and the healthcare sector. As an example, Russian wheat aphids (*Diuraphis noxia*) are one of the most significant insect pests of wheat crops in Ethiopia [42].

6.2 Rice

With a total production of 770 million tons, rice (*Oryza sativa*) is a commonly consumed staple cereal in Indonesia, China, and other nations in Asia [95]. Stem-borers, hoppers, and defoliators attack rice production. The effect on yields of a complex of stem borers in Asia was also calculated as a meta-analysis, in which each 1% increase in whiteheads was associated with a yield loss of 4% [29]. Farmers also incur an average of 37% losses of rice crop annually because of the insect pests and disease incidences [78]. The major rice diseases that are affecting the productivity of the plant include Bacterial Leaf Blight (*Xanthomonas oryzae*), Rice Blast (*Pyricularia oryzae*), etc. [55].

Rice Blast: Rice blast is one of the fungal diseases of the rice plant; globally, it is estimated that fungal disease alone costs 14% of the world's annual rice production [114].

Bacterial blight is among the riskiest killers of rice. When this disease penetrates the field, there is an even increased loss of yield. Bacterial blight of rice causes a yield loss of 70% when the susceptible varieties are used and the environmental conditions have created favorable conditions for this pathogen of the disease. The disease-causing pathogen is *Xanthomonas oryzae* Pv. oryzae [46].

BPHs are very destructive pests of rice, particularly in Southeast Asia, that tend to cause immense damage to rice [71]. A severe outbreak was witnessed in Southern Vietnam between 2005 and 2006, where over 485,000 ha of paddies were hit by viral infections, apparently brought in by BPH and other vector insects, that cost US$120 million [46].

6.3 Maize

Maize (*Zea mays*) is one of the most important cereal crops. It comes in the 3rd position after wheat and rice. In the yield losses case, various factors are at work [121]. The losses in maize are 9.4% annually worldwide due to the large amounts of pathogenic microorganisms (fungi, bacteria, viruses) and insects destroying grains and plants. Insects are reported to be the leading damaging and low-yielding producers of maize, followed by fungi [109].

The fall armyworm *Spodoptera frugiperda* is a pest in the western hemisphere that has caused severe damage in South Africa since 2016, when it was established and damaged maize crops. In a report in Zimbabwe, around 25–50% of maize fields are estimated to have a high infestation of the FAW, and the result was a 11.57% drop in yield [25].

Maize lethal necrosis (MLN) is a synergistic mixed infection between any one of several potyviruses infecting maize and the unrelated machlomovirus, Maize chlorotic mottle virus (MCMV). The loss on crop yields of LN-affected plots amounts to US$ 53.2, 180, and 198 million in 2012, 2013, and 2014, respectively [17].

Fusarium ear rot (FER) caused by the fungal pathogen *Fusarium verticillioides* [53]. In Ethiopia, FER causes lower yield, quality, and safety of maize, and affects the economic situation of the maize producers negatively [34].

6.4 Soybean

Soybeans are one of the most prominent crops around the world. Both vegetable oil and protein meal can find a useful application in soybean seeds [47]. It is less productive due to fungal diseases, nematodes, and insect pests [77].

Soybean is a legume crop that has its origin in Southeast Asia, but its cultivation is hampered by soybean cyst nematodes (*Heterodera glycines*). It has led to over 30% loss in the soybean production when it acts alone or together with others, causing loss of soybean yields under favorable environmental conditions [124].

The major focus is to study Asian Soybean Rust (ASR), which is caused by *Phakopsora pachyrhizi*—a biotrophic fungus with an extensive host range—that causes significant yield losses (10–100%) across Asia [50].

Soybean looper SBL, *Chrysodeixis includens*, ranks as one of the most destructive insect pests of soybean in the United States. Soybean production in the South region represented 15.5% of the total soybean production in the United States in 2022, and yield losses as a result of invertebrate pests were 5.8%, or 1.09 million metric tons, in 2022 [31]. The effect of biotic stresses on major crops has been shown in Fig. 2 and compiled in Table 2.

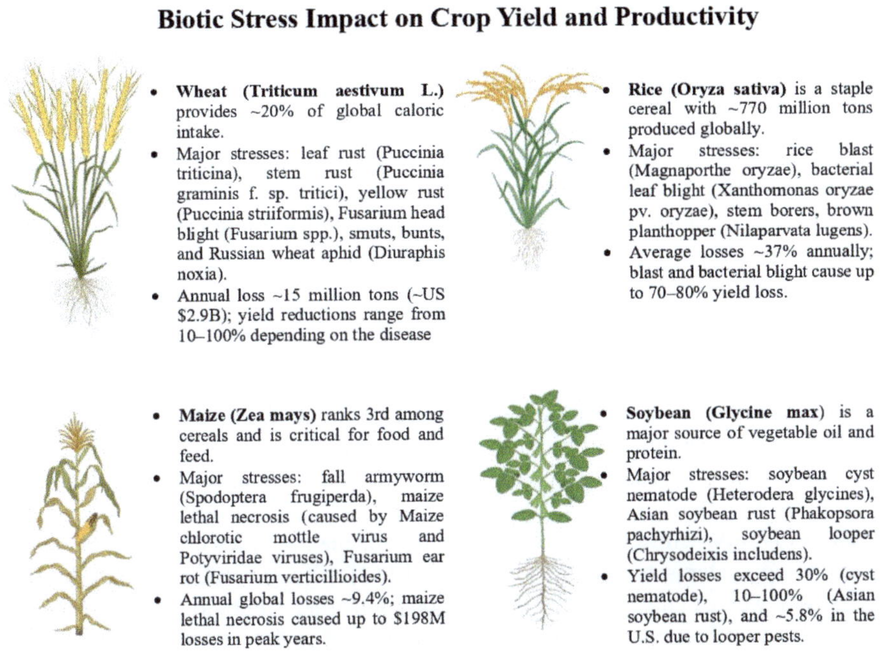

Fig. 2 Illustrate the impact and diseases caused by different biotic stress

Table 2 Biotic stress impact on major crops

Crop	Major biotic stress	Key pathogens/pests	Impact on yield	Economic/global significance	References
Wheat	Fungal diseases (Rusts, Fusarium head blight), Insect pests	*Puccinia spp.*, *Fusarium spp.*, Russian wheat aphid	Rusts: 10–100% loss; ~15 Mt loss/year (~US$2.9B). FHB causes yield + quality loss via mycotoxins	Global staple; 20% caloric intake; severe epidemics threaten food security	[62]
Rice	Fungal and bacterial diseases, Insect pests	*Magnaporthe oryzae* (Blast), *Xanthomonas oryzae* (Blight), Brown planthopper	Blast: up to 70–80% loss; Blight: up to 70% loss; Pests: 37% avg. losses/year	770 Mt global production; heavy losses in Asia; SE Asia BPH outbreak caused US$120 M loss	[55]
Maize	Insects, Viruses, Fungal pathogens	Fall armyworm, MLN viruses (*MCMV* + Potyviridae), *Fusarium verticillioides*	Avg. 9.4% annual losses; FAW infestation causes ~11.5% loss; MLN caused US$53–198 M loss; Fusarium ear rot reduces yield and quality	3rd most important cereal; major food + feed crop; wide pathogen diversity	[110]
Soybean	Nematodes, Fungal pathogens, Insect pests	Soybean cyst nematode (*Heterodera glycines*), *Phakopsora pachyrhizi* (ASR), Soybean looper (*Chrysodeixis includens*)	SCN: >30% loss; ASR: 10–100% loss; SBL: 5.8% loss in US (1.09 Mt, 2022)	Key source of protein and oil; major global economic crop, critical for food/feed industries	[77]

7 Conventional Approaches to Managing Biotic Stress

The loss of about USD 290 billion per annum to plant diseases and pests is experienced globally. In India, the transition of management strategies was from relying on pesticides under the Green Revolution to integrated and environmentally friendly management like bio-pesticides, resistant species, and IPM. The next challenges are pest resistance, transboundary threats, and sustainability solutions [104].

7.1 Cultural Practices

One of the surest and earliest known approaches to controlling pests that farmers use in their farm gardens is the cultural strategy of controlling pests. The advantages of these practices are that they propagate natural pest control, owing to unfavorable conditions, they will be subjected to growth, as well as improving the competitive capacity of crops being cultivated. Loosely, it is the manipulation of cultural practices or cropping systems to achieve pest control [14]. The generally advised cultural methods of weed, plant disease, and arthropod control are: sanitation, soil, cultivation, manipulation of alternative hosts, optimization and maintenance of plant health, planting and crop harvest time (disruption of crop-pest phenology), crop rotation (enhancing and maintaining temporal diversity), interplanting or multiple cropping systems, (maintenance and enhancement of spatial diversity), trap cropping, non As an example: The cultural practices can be readily incorporated with other cultural practices and therefore holds enormous potential when it comes to other cotton pests management in most countries. They can help to minimize the use of insecticides to maintain the production of cotton in Africa on small-scale farms. Nonetheless, there are still some cultural methods that are underexploited. There is a greater need to do more research and adapt more culturally-based insect pests of cotton in tropical Africa [57]. A survey of cultural practices by farmers in South Nyanza indicates that immediately, crop protectors should focus their research and extension programs on crop sanitation and avoidance of adjacent planting of successive crops. Farmers need to understand the mechanism of movement of the sweet potato weevils and butterflies; thus, the life cycle and behavior of these insects should be explained to them. They need to appreciate how the insects move, and as such, the need for sanitary practices and adjacent planting should become more obvious to them [116]. Cultural practices make up a significant component of IPM and involve ecological methods to control pests, such as crop rotation, intercropping, resistant varieties, and the altered time of planting and sanitation. Model cases, like rice-fish culture, through intercropping maize and legumes, and trap cropping, demonstrate alleviated pest resolution, reduced pesticide applications, higher yields, and increased soil health, diversity, and resource economy [125].

7.2 Chemical Pesticides

Pesticides play their role in present-day agriculture to meet the population's needs. Application of pesticides does not only end in the field, but also in domestic settings against pests, the distribution of disease-causing insectivores, and in domestic gardening [99]. Intensive farming techniques have become heavily reliant on using synthetic chemical pesticides to control pests and diseases, and the methods have become unsustainable over time. Continued use of increasing amounts of pesticides results in the emergence of resistance in the beneficial pests and contributes to adverse effects on biodiversity [97]. The pesticides are mostly categorized based on the target pests, mode of action, and chemical nature. Moreover, the pesticides were subdivided into two large categories, i.e., synthetic pesticides and biopesticides. To abate and eradicate the pests that attack the crops, pesticides could be used to protect the crops in pre-harvest, harvest, and even after harvesting/storage stages. It is of utmost importance to prevent crops during their pre-harvest stage, and it can lead to more crop production. Therefore, the pesticide has a major role in safeguarding the crops and agricultural produce against the infestation of the pests either by killing them or by rendering the plant to become resistant to the pests, thereby boosting the crop yield and enhancing high-quality and pest-free crops [26].

7.3 Biological Control

Biological control refers to the ecologically based pest control technique that involves one group in the use of another organism, referred to as the natural enemies, to control the pest species. Natural enemies are parasitoids, predators, entomopathogenic nematodes, pathogens, competing microorganisms with or hyperparasites of plant pathogens, herbivores feeding on weeds and weed seeds, competitors over resources, and organisms producing toxins, termed antibiosis or allelopathy. Biological control may be biotic, as naturally occurring invading organisms, repopulating by invading again after establishment, augmented in population by importing agents, or maintained in numbers by conservation and/or enhancement of an agent [12].

7.4 Crop Rotation and Intercropping

Sequential cropping/Crop rotation is among the environmentally friendly agricultural management practices used in controlling several biotic stresses that reside in the soil ecosystem. The primary theory of the sequential cropping cycle is the rotation with nonhost alternative crop plants against biotic stress. Host and non-annually set procedures require a 3–7 year periodicity of rotation under the management of biotic stresses [27]. A pesky habit that is as old as it is effective, crop rotation translates to the repetitive planting of various crops in the same direction to increase soil fertility, reduce pests and diseases, and maximize the use of available resources. Wonderful examples of crop rotation, like three-year and 4-year rotations, and examples like legumes, grains, and cover crops, show how and why crop rotation is so flexible. In intercropping, the farmer plants two or more crops in the same piece of land, including complementary relationships to achieve higher yields, resource-use efficiency, and resilience. Examples of the various techniques that have been introduced to tap the potential synergies of crops include row intercropping, strip intercropping, and agroforestry [111].

7.5 *Integrated Pest Management*

Since the introduction of the concept of integrated control, that is, applied pest control that combines and integrates biological and chemical control, Project Management Professional (PMP) and its integration into Internet Control Message Protocol (ICMP) technology have progressed quite a long way. The idea first arose when there was indiscriminate use of broad-spectrum insecticides and insect outbreaks due to the destruction of natural enemies and the development of resistance to pesticides among entomologists [13]. To reduce the adverse impact of chemical pesticides, integrated pest management through different environment-friendly strategies and the reasonable application of pesticides is required to ensure that the level of losses due to pests is maintained at a realistic limit [60]. IPM approaches apply a combination of pest management methods that are aimed at reducing or eliminating the use of synthetic pesticides. It is a very old method that has been applicable, and it is a sustainable option for pest management. When IPM is used, there is control of low pests that is at a level that cannot destroy the economy. It entails the identification of measures that are realistic, cost-effective, and environmentally friendly. The IPM entails the utilization of numerous crop management methods to ensure that the level of pest is not at an economically harmful level. The approach is a systematic process of integrating numerous pest-curtailing procedures in a single conglomerate [9].

7.6 *Physical or Mechanical Control*

This methodology involves the application of mixed physical or mechanical methods of pest exclusion, trapping (in some instances comparable to behavior control), elimination, or destruction. Examples of physical or mechanical control methods include pest exclusion by using netting or row covers, handpicking or vacuums to remove the pest, mechanical tools to control weeds, rodent traps, changing environmental conditions such as greenhouse heat or dryness, steam sterilization or solarization, and visual or tactical bird deterrents, which include using reflective materials or sonic or other devices [30].

8 Modern Approaches in Biotic Stress Management

Although traditional tools such as crop rotations, pesticides continue to be significant, resistance pleasing, environmental risks, and climate-related pest changes have made advanced molecular and biotechnological techniques attractive. In a world where crop resilience is ever more critically needed, recent discoveries in genomics, genome editing, and plant-microbiome interactions are offering new opportunities to build crop resilience to levels never seen. This section outlines four key/recent practices in modern agriculture: molecular breeding, genetic engineering, genome editing, and plant-microbiome engineering, which altogether constitute the future of crop protection.

8.1 Molecular Breeding and Marker-Assisted Selection

Markers molecular breeding comprises marker-assisted selection, marker-assisted backcross breeding, and other more recent breeding methods, including marker-assisted recurrent selection and genomic selection. The application of marker-assisted selection enables the determination of the existence and absence of genes in lines, cultivars, and breeding populations, therefore, speeding up the selection process compared to of other traditional measures. Some genes have been identified and mapped accurately by association with DNA markers. Pyramiding of two or more resistance genes with marker assistance offers advantages to create resilience to critical diseases and insects. The molecular-breeding methods provide sufficient opportunities to plant breeders in the development of stress-tolerant, high-yielding plants [41]. Genetic linkage maps based on molecular markers and identifying genomic regions essential to yield under biotic stresses, such as Fusarium wilt (FW), have opened the door to marker-assisted breeding and selection of chickpea cultivars.

8.2 Genetic Engineering and Transgenics

Genetic engineering is a type of recombinant DNA technology and entails the artificial insertion, removal, or restructuring of stretches of base of a DNA molecule to modify the shape and performance of an organism. It supplements plant breeding in vastly expanding the number of genes and germplasm that may be used by breeders and by reducing the number of years it takes to develop new varieties and hybrids. Since conventional breeding can only be conducted between sexually compatible crops, genetic engineering can be used as an alternative way of developing varieties that would have better resistance to biotic stresses; this process is time-intensive and random [1].

Even the common measures involved in breeding resistant varieties and the use of agrochemicals and biocontrol agents in curbing diseases and pests can be said to have had lean success in the past. Over the past decade, genetic engineering developments have produced transgenic crop types with resistance to biotic stress. There are many genetically modified crops in the pipeline, but their realization in the fields is still low. Transgenic crops produced in insect resistance/or tolerance to herbicides proved to be a triumph in the marketplace place such as the Bt cotton genre.

8.3 Genome Editing (e.g., CRISPR/Cas9)

In genome editing, a strong genetic modification tool, DNA sequences can be altered precisely with molecular scissors and artificial nucleases. History goes back to 1985–1986 when successful gene insertions in mammalian homogames were established. The three prominent gene editing technologies are: Zinc finger nucleases (ZFNs), transcription activator-like effector nucleases (TALENs), and CRISPR/Cas. Genome editing in biology alters the genome of entities to improve crop characteristics, such as high yield, disease and pest resistance, shortening of breeding periods, ensuring food security, and sustaining agriculture practices due to the changing environment [67]. Genome editing involves the insertion, deletion, or replacement of a foreign gene into the DNA organism. Efficient conversion will incorporate this new string into the host genome. This fixation of specific DNA sequences is provided by several processes that are engaged with the assistance of nucleases. Plant breeding cannot accomplish the necessary traits by itself, but, with the help of the CRISPR-associated enzymes (CRISPR/Cas and CRISPR/Cpf1), the need to fulfill the necessary traits in crops can be attained efficiently [1].

8.4 Plant Microbe Engineering and Biocontrol Agents

Increasing understanding and appreciation of microbial diversity and biology have also spurred the rapid growth in microbiome research. The presence of the microbiome plays a decisive role in the food production system because it can positively contribute to the health of the environment and crop resistance to pathogens [33]. The use of beneficial microorganisms may provide a good alternative to fulfill this goal. Several soil microorganisms have been reported to promote plant growth. These microorganisms would make crops better through increasing nutrient uptake, nutrient cycling, nitrogen fixation, phytohormone production, and resistance to abiotic stress and biotic stress. Thus, it is conceivable to find and characterize these useful microorganisms to come up with a microbial formulation or consortia to enhance crop productivity using sustainable biotechnology at low cost [69].

The effect on the growth and development of a plant by plant growth-promoting microbes (PGPM) is positive. However, in addition to their growth promotion by various mechanisms, PGPM are also known to afford protection to plants against diseases and control phytopathogens and hence can remedy for biotic stresses in plants due to sustainable agriculture. The number of mechanisms that GPM must defend and protect the plants against pathogens is actually varied. Other PGPM, including Bacillus, Pseudomonas, and Trichoderma, defend plants by stimulating them to develop resistance, or priming, without actually attacking the host pathogen [39]. Recalling the need to manage yield losses caused by plant stresses, which increasingly grow intense because of climate change, the use of biological control agents (BCAs) and biostimulants can therefore be considered as a practical solution. CAs and biostimulants have been advised because it has a dual effect of lessening both stress factors at the same time. Despite the fact that protection against biotic stresses is not included in the generally accepted definition of biostimulant, the biocontrol feature, providing a reduction in the biotic pressure on crops, is a characteristic feature of some microbial and non-microbial biostimulants. There is little application of synergisms between the use of biostimulants and BCAs to manage crop stresses. At present, a potential application through a combined application of the two agents presents a good alternative in enhancing the yield and growth of crops and to cope with stresses [10]. The key strategies used against biotic stress and their pros and cons have been mentioned in Table 3.

9 Role of Nanotechnology in Biotic Stress Alleviation

The use of crop biotic stressors poses serious risks to the attainment of sustainable agricultural productivity and food security across the entire globe. The use of nano-agricultural technology as a sustainable way of controlling the biotic stress of crops has more economic and environmental advantages than the conventional analogs [22]. In an agricultural context, nanotechnology has transformed agriculture and has the potential to manage diseases in plants, even though the area of nanotechnology remains quite nascent and requires further analysis [64]. Metallic degenerates can be found as promising doings when used against plant pests and diseases, and to enhance plant growth and health under differing conditions. Thus, these nanoparticles (NPs) promote the plant biomass and the crop production yield of crops with a variety of mechanisms, such as crop protection (nano pesticides), stress tolerance, and soil enhancement, as well as crop production. These NPs may both act on the pathogens themselves but also increase the resistance of the plants to diseases through improvement of nutrition (plant nutrition and quality), inhibitions of pathogen attachments (bacteria, fungi, viral), and directly enhancing nutritional quality (plant nutrition and quality) as well as crops (crop productivity).

Chitosan and chitosan nanoparticles (CNPs) may enhance PR-proteins and antioxidant genes up-regulation and are likely to significantly contribute to successful biocontrol [28]. CNPs are multi-effective in nature, i.e., they possess direct antimicrobial effect, the ability to induce plant defense, and the capacity to modulate microbe-gene expression. It possesses extensive use in suppressing the disease and

Table 3 Key strategies to mitigate the biotic stress from plants and their significance, and challenges

Approach	Key strategies	Examples/notes	Significance	Limitations/challenges	References
Conventional approaches					
Cultural practices	Crop rotation, intercropping, sanitation, planting time, trap crops	Cotton pests in Africa, sweet potato weevil (sanitation)	Eco-friendly, reduces pesticide use, enhances biodiversity	Underutilized requires farmer awareness	[14]
Chemical pesticides	Synthetic & biopesticides, pre-/post-harvest use	Classified by target pest and mode of action	Quick pest suppression	Resistance, biodiversity loss, health risks	[97]
Biological control	Natural enemies (parasitoids, predators, microbes)	*Bacillus*, *Trichoderma*, nematodes	Sustainable pest suppression	Field survival and consistency issues	[12]
Crop rotation and intercropping	Sequential cropping, host–nonhost rotation, multiple cropping	3–7 year cycles; agroforestry	Improves soil fertility, lowers pest pressure	Long rotation period, land limitations	[27]
Integrated pest management (IPM)	Combines bio, cultural & chemical methods	Threshold-based pesticide use	Sustainable, minimizes pesticide harm	Knowledge-intensive, adoption gap	[60]
Physical/ mechanical control	Netting, row covers, handpicking, traps, solarization	Bird deterrents, steaming	Safe, non-chemical	Labor-intensive, limited scale	[30]
Modern approaches					
Molecular breeding & MAS	Marker-assisted selection, gene pyramiding	Chickpea FW resistance	Fast breeding of resilient cultivars	Costly, requires lab facilities	[41]

(continued)

Table 3 (continued)

Approach	Key strategies	Examples/notes	Significance	Limitations/challenges	References
Genetic engineering/transgenics	DNA modification, foreign gene insertion	Bt cotton, insect-resistant crops	Broad-spectrum resistance	Regulatory, ethical and market concerns	[1]
Genome Editing (CRISPR, TALENs, ZFNs)	Precise DNA edits, insertions/deletions	CRISPR/Cas9 in crops	Accurate trait improvement	Off-target effects, biosafety issues	[1]
Plant–microbe engineering and biocontrol agents	PGPM, BCAs, biostimulants	*Bacillus*, *Pseudomonas*, *Trichoderma*	Boosts nutrients and induces resistance	Variable field success, formulation limits	[69]

making the plant healthier. The methods, viz. ionic gelation, emulsion cross-linking, and solvent evaporation, are widely employed to fabricate CNPs and exhibit ease of generating smaller-sized particles, maintaining stability, and biocompatibility. The current review emphasizes the synthesis of CNPs, their possible use as crop protectants, the mechanism of action against pathogens in plants, and their toxicity on plants [81].

10 Climate Change and Biotic Stress Interactions

Agriculture in the twenty-first century is challenged with a need to fulfill food demands, as well as satisfy the sustainability objectives. This is further complicated by climate change, which influences the pattern of crop pests (intended here as insects, plants, and pathogenic agents) and the magnitude of their outbreaks. The growing alarm about health and environmental issues and new laws limiting pesticide application, especially in the European Union, gives stimulus to seek environmentally friendly alternatives to pest management based on pesticides [68]. Climate acts as a major determinant of pest populations; thus, climate change will necessitate adaptive management efforts to handle the new situation of pests and pathogens. A stringent chain of analytical tools is essential in undertaking risk analysis, policy-making, and landscape design for pest management at the regional and landscape field levels [117]. Although technological and crop yield potential improvements are enormous, crop production is still largely affected by climate conditions, because the primary factors that affect crop growth are solar radiation, temperature, and precipitation. Climate affects plant diseases and pest infestations, the supply of irrigation water, and the demand for irrigation water. It is anticipated that individual farmers and consumers will be impacted by the changes in the global and regional climates [101]. Climate change-affected crops and agricultural pests are a serious problem. Insect pests are generally adjustable living organisms; thus, they respond to different causes of climate change in different ways. As temperature is the most influential environmental factor that affects insect population dynamics, it becomes important to suggest that climate warming on a global scale may lead to the enlargement of their geographic distribution, overwintering survival, an increment in the number of generations, augmented risk of insect species invasive species and invasive insect species or insect transmits the plant diseases, and altered interactions with the host plants and natural enemies [101].

11 Future Perspectives and Research Needs

Devising strategies on how best to deal with biotic stress in agriculture in the wake of the fast-changing climate needs approaches that are multidimensional, beyond the traditional pest control practices. Future management frameworks should be plausible not only to genetic, technological, ecological, and policy-depth considerations, but also should be sustainable in food production. The nature of the biotic stress itself is dynamic in nature due to the interplay of crop-pest-pathogen relationships, climatic changes, global commerce, and transformations occurring in the landscape. Therefore, resiliency should be both crop-based and at a system level.

Among the most promising directions is the development of the molecular breeding approach and the evolution of genome editing. MAS and genomic selection are already increasing the speed of quantitative trait loci (QTL) identification relevant to fungal, bacterial, and viral resistance. This has been transformed with the use of genome editing tools, including CRISPR/Cas9, which will enable the introduction of high precision and efficiency into the development of resistance. As an example, in both rice and wheat, there is evidence of increased resistance to pathogens such as blast and rust through target editing of the corresponding susceptibility genes. Such strategies offer the prospect of coming up with resistant varieties faster than conventional breeding, which is vital given the rate at which pests come up with new virulent strains. Attention should now be directed toward pyramid resistance genes, or gene pyramid, to ensure longer-term protection, and at the same time, to considerations of off-target effects and regulatory acceptability.

The exciting prospect is the plant-microbiome engineering field, where beneficial plant–microbe interactions could be used to develop resilience to biotic stresses. Endophytic fungi, rhizobacteria, and arbuscular mycorrhizal fungi have demonstrated great promise in elevated levels of plant immune response, nutrient uptake, and multiple stress tolerance. It may be feasible to supply plants with an initial defense against pathogens and herbivores by manipulating root exudates or inoculating seeds with well-chosen microbe consortia. Besides, the development of metagenomics and synthetic biology allows arranging microbial communities with specific functionalities. With regard to this research frontier, this promises not only to increase the protection of crops, but also to decrease the footprint of chemical pesticides on the environment.

New digital technologies are revolutionizing the monitoring and management of biotic stress at the same time. The emergence of artificial intelligence (AI), machine learning (ML), drones, and remote sensing platforms permits stress to be detected, in real-time, at field and regional scales. Climatic, genomic, and field-level data can be combined in AI-based models to predict pest and disease outbreaks with improved levels of accuracy. Indicatively, fall armyworm outbreak predictive analytics in Africa have already proved how early warning systems can assist farmers and policymakers to be prepared to pre-empt interventions. Multispectral and hyperspectral imaging employing remote sensing technology can identify stress symptoms before they can be detected by the human eye and provide an opportunity to react swiftly and effectively. These technological advancements make precision farming possible, where certain kinds of interventions like pesticide spraying, irrigating, and/or fertilizing only occur when it is needed in specific areas, meaning that resources are used efficiently and harm to the environment is minimized.

Policy frameworks and extension services are an important part of the process of putting innovation into practice despite these technological advances. The safe use of nanotechnology-based pesticides, biopesticides, and GM crops needs to be more strongly supported with regulations. Governments and global organizations need to implement risk evaluation measures that are risk–benefit expressive to facilitate innovation and biosafety, and the avoidance of the insecurity of farmers and consumers against new technologies. At the same time, integrated stress management approaches to farm management need to be given a center stage under farmer-centered extension services. The focus in such training should be ecological sustainability, less use of broad-spectrum chemicals, and more adoption of eco-friendly equivalents. Traditional knowledge combined with modern practices is most likely to be adopted maximally by the smallholder farmers.

In addition, players need to be interconnected globally in addressing trans-boundary pest and pathogen risks, locust plagues, wheat rust races, or viral outbreaks transmitted by migratory insects. These threats have no respect for political borders and require integrated surveillance and data sharing, fast-response mechanisms. Models that can be expanded to manage other biotic stresses lie on international platforms such as the Desert Locust Information Service provided by the FAO or the Borlaug Global Rust Initiative.

Moving forward, there is the challenge of bringing theoretical processes into laboratories into the practical world. Research should not merely be practical about the success of novel solutions but also their long-term ecological and socio-economic viability. Other questions, like what the consequences of the constant use of nanoparticles on soil would be for microbiota, or are CRISPR-edited seeds are even affordable to smallholder farmers, have to be answered before their general use. Further research should also pay more attention to multi-stress interactions as natural agricultural conditions may be simultaneously challenging crops with drought, salinity, and disease.

In summary, the way forward in biotic stress management is an integrative system that identifies and combines the best science with solutions that help to meet farmer needs. Genetic innovation, microbiome development, digital solutions, and favorable policies can combine the forces of sustainable agriculture. Yet, it is sustainable only when it is interdisciplinary work, international collaboration, and a desire to be ecologically and socially responsible.

12 Conclusion

Biotic stress is a substantial threat to world food security and has been linked with considerable loss of yields and economies in most cropping systems. The traditional methods-including cultural practices, crop rotations, resistant varieties, and chemical pesticides, have been important in the management of biotic stresses. Nonetheless, these solutions are proving to be of short-lived reprieve because of the emergence of pesticide resistance, ecological degradation, and overdependence on synthetic inputs. There has been an increasing demand for sustainable and long-term solutions, and therefore, there has been a rising interest in innovative new mechanisms that would integrate modern solutions with ancient methods.

The recent achievements in molecular breeding, marker-assisted selection, and genetic engineering have created an opportunity to create resistant crop varieties faster and more specifically than in the past. The development of genome editing technologies, especially with CRISPR/Cas9, has radically changed how crops might be improved since it is now possible to make precise edits that increase resistance to pathogens or insect pests. Paired with these genetic advances, there are technological advances with nanotechnology, including the use of chitosan, zinc oxide, and silver nanoparticles to control pests and deliver agrochemical targeting. These methods contribute to the decrease of chemicals on a large scale and increase efficiency, but their environmental safety is still a valuable topic of future research. On the same note, plant-microbiome engineering has come to the forefront with the ability to target the beneficial microbes to fortify plant defense, enhance nutrient capture, and resist a broad range of stresses at a time.

Climate change is an added complication to the challenges of biotic stress, changing pest-pathogen relationships, and extending the distribution of invasive species. Increasing temperatures, changing rainfall patterns, and more extreme weather would all cause a corresponding rise in pest pressure in the tropics and temperate regions. This growing unpredictability demands such notions as integrated solutions and adaptive solutions that reference innovative biotechnologies and ecological stability. Techniques like intercropping, agroforestry, and biological control continue to be part and parcel of a sustainable agricultural system, particularly when complemented with the recent advances in digital agriculture, remote sensing, and AI. The combination of these tools will facilitate real-time analysis of stress, highly targeted actions, and better policy-level decisions.

Moving forward, future biotic stress management will need cross-functional input between plant scientists, microbiologists, geneticists, nanotechnologists, data scientists, and policy makers. This is just as vital as the role of farmers, whose move to uptake innovations will be the ultimate characteristic of its impact. Ensuring that technologies are affordable, accessible, and customizable to local use is an important aspect toward ensuring that there is a gap between scientific progress and its application in the field. Interventions that support input sustainability, stimulate ecological farming, and investment in extension services will also help to enhance resilience.

All in all, a multifaceted solution to this problem that is based on sustainability and innovation will provide the best way forward in reducing biotic stress in agriculture. Through the integration of modern science into traditional knowledge and the development of collaboration at both international and local levels, agricultural systems can be turned into being more resilient. Not only do such practices protect the productivity of crops and stabilize food supplies, but they also help the ecosystem's health and prepare the climate. Moving forward on escalating global challenges, there is certainly a scientific need, but also a food security foundation to integrate biotic stress management.

References

1. Abebe A, Tafa Z (2022) A review on: Impact of genetic engineering in biotic stresses resistance crop breeding. Int J Res Appl Sci Biotechnol 9:243–250
2. Aerts N, Pereira Mendes M, Van Wees SCM (2021) Multiple levels of crosstalk in hormone networks regulating plant defense. Plant J 105:489–504
3. Afzal A, Zubair H, Ali U, Nawaz M, Mansoor A, Zaib M, Marium A (2023) Mechanisms of disease resistance in plants for sustainable agriculture. Curr Agric Res 4:1–13
4. Akhtar N, Perween S, Ansari AM, Ahmad MR (2020) Life style of fungi from biotrophy to necrotrophy and saprotrophy. Int J Agric Appl Sci 1:93–103
5. Al-Khayri JM, Rashmi R, Toppo V, Chole PB, Banadka A, Sudheer WN, Nagella P, Shehata WF, Al-Mssallem MQ, Alessa FM (2023) Plant secondary metabolites: the weapons for biotic stress management. Metabolites 13:716
6. Alfano JR, Collmer A (2004) Type III secretion system effector proteins: double agents in bacterial disease and plant defense. Annu Rev Phytopathol 42:385–414
7. Alisaac E, Mahlein A-K (2023) Fusarium head blight on wheat: Biology, modern detection and diagnosis and integrated disease management. Toxins (Basel) 15:192
8. Anderson PK, Cunningham AA, Patel NG, Morales FJ, Epstein PR, Daszak P (2004) Emerging infectious diseases of plants: pathogen pollution, climate change and agrotechnology drivers. Trends Ecol Evol 19:535–544
9. Angon PB, Mondal S, Jahan I, Datto M, Antu UB, Ayshi FJ, Islam MS (2023) Integrated pest management (IPM) in agriculture and its role in maintaining ecological balance and biodiversity. Adv Agric 2023:5546373
10. Anuar MSK, Hashim AM, Ho CL, Wong M-Y, Sundram S, Saidi NB, Yusof MT (2023) Synergism: biocontrol agents and biostimulants in reducing abiotic and biotic stresses in crop. World J Microbiol Biotechnol 39:123
11. Atkinson NJ, Urwin PE (2012) The interaction of plant biotic and abiotic stresses: from genes to the field. J Exp Bot 63:3523–3543
12. Baker BP, Green TA, Loker AJ (2020) Biological control and integrated pest management in organic and conventional systems. Biol Control 140:104095
13. Barzman M, Bàrberi P, Birch ANE, Boonekamp P, Dachbrodt-Saaydeh S, Graf B, Hommel B, Jensen JE, Kiss J, Kudsk P (2015) Eight principles of integrated pest management. Agron Sustain Dev 35:1199–1215
14. Bashyala S, Poudela D, Gautamb B (2022) A review on cultural practice as an effective pest management approach under integrated pest management. Trop Agroecosyst (TAEC) 3:34–40
15. Basnet B, Juliana P, Bhattarai K, Upreti U (2022) A review on major rust resistance gene and amino acid changes on wheat (Triticum aestivum L). Adv Agric 2022:7419326
16. Benjamin G, Pandharikar G, Frendo P (2022) Salicylic acid in plant symbioses: beyond plant pathogen interactions. Biology 11:861

17. Biswal AK, Alakonya AE, Mottaleb KA, Hearne SJ, Sonder K, Molnar TL, Jones AM, Pixley KV, Prasanna BM (2022) Maize Lethal Necrosis disease: review of molecular and genetic resistance mechanisms, socio-economic impacts, and mitigation strategies in sub-Saharan Africa. BMC Plant Biol 22:542
18. Bonaventure G (2012) Perception of insect feeding by plants. Plant Biol 14:872–880
19. Borkar SG, Yumlembam RA (2016) Bacterial diseases of crop plants. CRC Press
20. Borlaug NE (2002) The green revolution revisited and the road ahead. Nobelprize. org Stockholm
21. Bungala LT, da C, Park C, Dique JEL, Sathasivam R, Shin SY, Park SU (2024) Ethylene: a modulator of the phytohormone-mediated insect herbivory network in plants. Insects 15:404
22. Cao X, Wang Z (2022) Application of nano-agricultural technology for biotic stress management: mechanisms, optimization, and future perspectives. Environ Sci Nano 9:4336–4353
23. Chapman RF (2009) Mouthparts. In: Encyclopedia of insects. Elsevier, pp. 663–668
24. Chaudhary P, Agri U, Chaudhary A, Kumar A, Kumar G (2022) Endophytes and their potential in biotic stress management and crop production. Front Microbiol 13:933017
25. Chen C (2022) The evaluation of the toxicity of Bacillus thuringiensis Cry proteins against reference and field populations of fall armyworm, Spodoptera frugiperda (JE Smith), in South Africa
26. Chengala L, Singh N (2017) Botanical pesticides—a major alternative to chemical pesticides: a review. Int J Life Sci 5:722–729
27. Choudhary V, Kumar KK, Sahu B, Muthappa S, Das ADA (2021) Agronomic innovations in biotic stress management and its combined effect with abiotic stresses in crop production. Ind J Agron 66:237–257
28. Chun S-C, Chandrasekaran M (2019) Chitosan and chitosan nanoparticles induced expression of pathogenesis-related proteins genes enhances biotic stress tolerance in tomato. Int J Biol Macromol 125:948–954
29. Conde S, Catarino S, Ferreira S, Temudo MP, Monteiro F (2025) Rice pests and diseases around the world: literature-based assessment with emphasis on Africa and Asia. Agriculture 15:667
30. Dara SK (2019) The new integrated pest management paradigm for the modern age. J Integr Pest Manag 10:12
31. Debnath R, George J, Gautam M, Shafi I, Kariyat R, Reddy GVP (2024) Ecological interactions, host plant defenses, and control strategies in managing soybean looper, Chrysodeixis includens (Lepidoptera: Noctuidae). Front Insect Sci 4:1480940
32. DeFalco TA, Zipfel C (2021) Molecular mechanisms of early plant pattern-triggered immune signaling. Mol Cell 81:3449–3467
33. del Carmen Orozco-Mosqueda M, del Carmen Rocha-Granados M, Glick BR, Santoyo G (2018) Microbiome engineering to improve biocontrol and plant growth-promoting mechanisms. Microbiol Res 208:25–31
34. Deressa T, Adugna G, Suresh LM, Bekeko Z, Opoku J, Vaughan M, Proctor RH, Busman M, Burgueño J, Prasanna BM (2024) Biophysical factors and agronomic practices associated with Fusarium ear rot and fumonisin contamination of maize in multiple agroecosystems in Ethiopia. Crop Sci 64:827–845
35. Deutsch CA, Tewksbury JJ, Tigchelaar M, Battisti DS, Merrill SC, Huey RB, Naylor RL (2018) Increase in crop losses to insect pests in a warming climate. Science 361(6405):916–919
36. Ertunc F (2020) Emerging plant viruses. In: Emerging and reemerging viral pathogens. Elsevier, pp 1041–1062
37. Fei W, Liu Y (2023) Biotrophic fungal pathogens: a critical overview. Appl Biochem Biotechnol 195:1–16
38. Figueroa M, Dodds PN, Henningsen EC, Sperschneider J (2022) Global landscape of rust epidemics by Puccinia species: current and future perspectives. Plant Relat Fungal-Plant Interact 391–423

39. García-Montelongo AM, Montoya-Martínez AC, Morales-Sandoval PH, Parra-Cota FI, de los Santos-Villalobos S (2023) Beneficial microorganisms as a sustainable alternative for mitigating biotic stresses in crops. Stresses 3:210–228
40. García-Guzmán G, Heil M (2014) Life histories of hosts and pathogens predict patterns in tropical fungal plant diseases. New Phytol 201:1106–1120
41. Gazal A, Dar ZA, Wani SH, Lone AA, Shikari AB, Ali G, Abidi I (2016) Molecular breeding for enhancing resilience against biotic and abiotic stress in major cereals. SABRAO J Breed Genet 48
42. Gemeda SR (2023) Evaluation of wheat genotypes against Russian wheat aphid, Diuraphis Noxia (RWA) under irrigated condition in Ethiopia. Appl Sci Res Period 1:8–12
43. Gimenez E, Salinas M, Manzano-Agugliaro F (2018) Worldwide research on plant defense against biotic stresses as improvement for sustainable agriculture. Sustainability 10:391
44. Gossel PP (2000) Pasteur, Koch and american bacteriology. Hist Philos Life Sci 81–100
45. Guardino RF (2005) Early history of microbiology and microbiological methods. Encycl Rapid Microbiol Methods 1:1–27
46. Hadi M, Abbas A, Ahsan M (2025) Bacterial blight of rice (xanthomonas oryzea pv oryzea) disease and its management options: an overview. J Life Soc Sci 2025:37
47. Hamza M, Basit AW, Shehzadi I, Tufail U, Hassan A, Hussain T, Siddique MU, Hayat HM (2024) Global impact of soybean production: a review. Asian J Biochem Genet Mol Biol 16:12–20
48. Hazell PBR (2009) The Asian green revolution. Int Food Policy Res Inst 911
49. Herb M, Schramm M (2021) Functions of ROS in macrophages and antimicrobial immunity. Antioxidants 10:313
50. Hossain MM, Sultana F, Yesmin L, Rubayet MT, Abdullah HM, Siddique SS, Bhuiyan MAB, Yamanaka N (2024) Understanding Phakopsora pachyrhizi in soybean: comprehensive insights, threats, and interventions from the Asian perspective. Front Microbiol 14:1304205
51. Huang C-Y, Jin H (2022) Coordinated epigenetic regulation in plants: a potent managerial tool to conquer biotic stress. Front Plant Sci 12:795274
52. Hudson A, Mullens A, Hind S, Jamann T, Balint-Kurti P (2024) Natural variation in the pattern-triggered immunity response in plants: investigations, implications and applications. Mol Plant Pathol 25:e13445
53. Hudson O, Meinecke CD, Brawner JT (2024) Comparative genomics of Fusarium species causing Fusarium ear rot of maize. PLoS ONE 19:e0306144
54. Ibragić S, Dahija S, Karalija E (2025) The good, the bad, and the epigenetic: stress-induced metabolite regulation and transgenerational effects. Epigenomes 9:10
55. Iqbal J, Yousaf U, Asgher A, Dilshad R, Qamar FM, Bibi S, Rehman SU, Haroon M (2023) Sustainable rice production under biotic and abiotic stress challenges. In: Sustainable agriculture in the era of the OMICs revolution. Springer, pp 241–268
56. Islam F, Khan MSS, Chen H, Chen J (2024) Utilizing effector-triggered immunity (ETI) as a robust priming agent to protect plants from pathogens. Stress Biol 4:51
57. Javaid I (1995) Cultural control practices in cotton pest management in tropical Africa. J Sustain Agric 5:171–185
58. Jeong J, Ju H, Noh J (2014) A review of detection methods for the plant viruses. Res. Plant Dis. 20:173–181
59. Joseph JT, Shah JM (2022) Biotic stress-induced epigenetic changes and transgenerational memory in plants. Biologia (Bratisl). 77:2007–2021
60. Karuppuchamy P, Venugopal S (2016) Integrated pest management. In: Ecofriendly pest management for food security. Elsevier, pp 651–684
61. Kaur L (2019) A review: top ten fungal pathogens. Int J Res Anal Rev 6
62. Kayim M, Nawaz H, Alsalmo A (2022) Fungal diseases of wheat. In: Wheat-recent advances. IntechOpen
63. Khan A, Khan A, Ali A, Fatima S, Siddiqui MA (2023) Root-knot nematodes (Meloidogyne spp.): biology, plant-nematode interactions and their environmentally benign management strategies. Gesunde Pflanz 75:2187–2205

64. Khan M, Khan AU, Hasan MA, Yadav KK, Pinto MMC, Malik N, Yadav VK, Khan AH, Islam S, Sharma GK (2021) Agro-nanotechnology as an emerging field: a novel sustainable approach for improving plant growth by reducing biotic stress. Appl Sci 11:2282
65. Klich MA (2007) Aspergillus flavus: the major producer of aflatoxin. Mol Plant Pathol 8:713–722
66. Kumar R, Rathor VS (2020) Nature and types of damage by insect pests. J Entomol Res 44:639–646
67. Kumar T, Tiwari N, Saxena DR, Istanbuli T, Hamwieh A (2024) Genetic and genomic resources for improving biotic stress tolerance in chickpea. In: Genomics-aided breeding strategies for biotic stress in grain legumes. Springer, pp 13–48
68. Lamichhane JR, Barzman M, Booij K, Boonekamp P, Desneux N, Huber L, Kudsk P, Langrell SRH, Ratnadass A, Ricci P (2015) Robust cropping systems to tackle pests under climate change. A review. Agron Sustain Dev 35:443–459
69. Lau S-E, Teo WFA, Teoh EY, Tan BC (2022) Microbiome engineering and plant biostimulants for sustainable crop improvement and mitigation of biotic and abiotic stresses. Discov Food 2:9
70. Li W, Lin S, Wang R, Chen C, Ni L, Wang W, Liang Z (2025) Regulation of plant hormones on the secondary metabolism of medicinal plants. Med Plant Biol 4
71. Liu X-D, Zhang B, Liu Y (2025) A new method to monitor the brown planthopper population size and its damage using canopy temperature of rice plants. Precis Agric 26:74
72. Lv Z-Y, Sun W-J, Jiang R, Chen J-F, Ying X, Zhang L, Chen W-S (2021) Phytohormones jasmonic acid, salicylic acid, gibberellins, and abscisic acid are key mediators of plant secondary metabolites. World J Tradit Chin Med 7:307–325
73. Ma Y, Zhao Y, Huang H, Zhao Y, Cao R, He K, Zhou L, Ye Y (2025) Decursin, identified via high-throughput chemical screening, enhances plant disease resistance via two independent mechanisms. Mol Plant Pathol 26:e70101
74. Manosathiyadevan M, Bhuvaneshwari V, Latha R (2017) Impact of insects and pests in loss of crop production: a review. Sustain Agric Towar Food Secur 57–67
75. Mengiste T (2012) Plant immunity to necrotrophs. Annu Rev Phytopathol 50:267–294
76. Mesterhazy A (2024) What Is Fusarium head blight (FHB) resistance and what are its food safety risks in wheat? problems and solutions—a review. Toxins (Basel). 16:31
77. Mishra R, Tripathi MK, Sikarwar RS, Singh Y, Tripathi N (2024) Soybean (Glycine max L. Merrill): a multipurpose legume shaping our world. Plant Cell Biotechnol Mol Biol 25:17–37
78. Mohapatra SD, Banerjee A, Senapati R, Prasanthi G, Mohapatra M, Nayak PK, Nayak AK, Maiti D (2021) Current status and future prospects in biotic stress management in rice
79. Murtaza G, Ramzan M, Ghani MU, Munawar N, Majeed M, Perveen A, Umar K (2019) Effectiveness of different traps for monitoring sucking and chewing insect pests of crops. Egypt Acad J Biol. Sci A Entomol 12:15–21
80. Nadarajah KK (2024) Defensive strategies of ROS in plant–pathogen interactions. In: Plant pathogen interaction. Springer, pp 163–183
81. Nandhini R, Rajeswari E, Harish S, Sivakumar V, Gangai Selvi R (2025) Role of chitosan nanoparticles in sustainable plant disease management. J Nanopart Res 27:13
82. Navas-Castillo J, Fiallo-Olivé E, Sánchez-Campos S (2011) Emerging virus diseases transmitted by whiteflies. Annu Rev Phytopathol 49:219–248
83. Nega A (2014) Review on concepts in biological control of plant pathogens. J Biol Agric Healthc 4:33–54
84. Negewo T (2024) Challenges and management of parasitic weeds in Ethiopia. CABI Agric Biosci 5:114
85. Negi N, Khurana P (2021) A salicylic acid inducible mulberry WRKY transcription factor, Mi WRKY53 is involved in plant defence response. Plant Cell Rep 40:2151–2171
86. Nguyen Q-M, Iswanto ABB, Son GH, Kim SH (2021) Recent advances in effector-triggered immunity in plants: new pieces in the puzzle create a different paradigm. Int J Mol Sci 22:4709
87. Oerke E-C (2006) Crop losses to pests. J Agric Sci 144:31–43

88. Pandey D, Rajendran SRCK, Gaur M, Sajeesh PK, Kumar A (2016) Plant defense signaling and responses against necrotrophic fungal pathogens. J Plant Growth Regul 35:1159–1174
89. Parmagnani AS, Maffei ME (2022) Calcium signaling in plant-insect interactions. Plants 11:2689
90. Patel R (2013) The long green revolution. J Peasant Stud 40:1–63
91. Popp J, Pető K, Nagy J (2013) Pesticide productivity and food security. A review. Agron Sustain Dev 33:243–255
92. Rabeh K, Hnini M, Oubohssaine M (2025) A comprehensive review of transcription factor-mediated regulation of secondary metabolites in plants under environmental stress. Stress Biol 5:15
93. Rai N, Sarma BK, Rai SP (2024) Transcriptional regulation of biotic and abiotic stress responses: challenges and potential mechanism for stress tolerance and chickpea improvement. Trop Plant Biol 17:83–107
94. Rajpal VR, Rathore P, Mehta S, Wadhwa N, Yadav P, Berry E, Goel S, Bhat V, Raina SN (2022) Epigenetic variation: a major player in facilitating plant fitness under changing environmental conditions. Front Cell Dev Biol 10:1020958
95. Ramadass M, Thiagarajan P (2021) Current trends and emerging technologies for pest control management of rice (Oryza sativa) plants. Environ Biotechnol 4:125–179
96. Ramegowda V, Senthil-Kumar M (2015) The interactive effects of simultaneous biotic and abiotic stresses on plants: mechanistic understanding from drought and pathogen combination. J Plant Physiol 176:47–54
97. Rana DS, Dass A, Rajanna GA, Kaur R (2016) Biotic and abiotic stress management in pulses. Ind J Agron 61:238–248
98. Rani AT, Vasudev K, Pandey KK, Singh B (2020) Sucking pests of vegetable crops. In: Sucking pests of crops. Springer, pp 307–340
99. Rani L, Thapa K, Kanojia N, Sharma N, Singh S, Grewal AS, Srivastav AL, Kaushal J (2021) An extensive review on the consequences of chemical pesticides on human health and environment. J Clean Prod 283:124657
100. Rathod A, Verma NS (2023) Impact of abiotic stress on agronomical crops. Front Agron 27
101. Rosenzweig C, Iglesius A, Yang X-B, Epstein PR, Chivian E (2001) Climate change and extreme weather events-Implications for food production, plant diseases, and pests
102. Roychowdhury R, Hada A, Biswas S, Mishra S, Prusty MR, Das SP, Ray S, Kumar A, Sarker U (2025) Jasmonic acid (JA) in plant immune response: unravelling complex molecular mechanisms and networking of defence signalling against pathogens. J Plant Growth Regul 44:89–114
103. Rzemieniewski J, Stegmann M (2022) Regulation of pattern-triggered immunity and growth by phytocytokines. Curr Opin Plant Biol 68:102230
104. Sahu B, Choudhary VK, Sahu MP, Kumar KK, Sujayanand GK, Gopi R, Prakasam V, Sridhar J, Mallikarjuna J, Singh HK (2023) Biotic stress management. In: Trajectory of 75 years of indian agriculture after independence. Springer, pp 619–653
105. Sahu PK, Jayalakshmi K, Tilgam J, Gupta A, Nagaraju Y, Kumar A, Hamid S, Singh HV, Minkina T, Rajput VD (2022) ROS generated from biotic stress: Effects on plants and alleviation by endophytic microbes. Front Plant Sci 13:1042936
106. Saijo Y, Loo EP (2020) Plant immunity in signal integration between biotic and abiotic stress responses. New Phytol 225:87–104
107. Sarrocco S, Vannacci G (2018) Preharvest application of beneficial fungi as a strategy to prevent postharvest mycotoxin contamination: a review. Crop Prot 110:160–170
108. Sarwar M (2020) Insects as transport devices of plant viruses. In: Applied plant virology. Elsevier, pp 381–402
109. Seyi-Amole DO, Onilude AA (2021) Microbiological control: a new age of maize production. In: Cereal grains, vol 2. IntechOpen
110. Sharma S, Kooner R, Arora R (2017) Insect pests and crop losses. In: Breeding insect resistant crops for sustainable agriculture. Springer, pp 45–66

111. Shekhar M, Shivashankar E, Pandey SK (2024) Crop rotation and intercropping techniques. In: Modern concept agron (Edition I). ND Global Publication House. https//doi. org/10 5281
112. Shu L, Kahlon PS, Ranf S (2023) The power of patterns: new insights into pattern-triggered immunity. New Phytol 240:960–967
113. Shukla S, Upadhyay D, Mishra A, Jindal T, Shukla K (2022) Challenges faced by farmers in crops production due to fungal pathogens and their effect on Indian economy. In: Fungal diversity, ecology and control management. Springer, pp 495–505
114. Simkhada K, Thapa R (2022) Rice blast, a major threat to the rice production and its various management techniques. Turk J Agric Sci Technol 10:147–157
115. Singh PK, Vishwakarma RC, Kaur J, Singh R (2023) Growth and survival of plant pathogens. In: Mishra RC, Singh R (eds) Principles of plant pathology, p 72
116. Smit N, Matengo LO (1995) Farmers' cultural practices and their effects on pest control in sweetpotato in South Nyanza. Kenya Int J Pest Manag 41:2–7
117. Sutherst RW, Constable F, Finlay KJ, Harrington R, Luck J, Zalucki MP (2011) Adapting to crop pest and pathogen risks under a changing climate. Wiley Interdiscip Rev Clim Chang 2:220–237
118. Tapia-Vázquez I, Montoya-Martínez AC, De los Santos-Villalobos S, Ek-Ramos MJ, Montesinos-Matías R, Martínez-Anaya C (2022) Root-knot nematodes (Meloidogyne spp.) a threat to agriculture in Mexico: biology, current control strategies, and perspectives. World J Microbiol Biotechnol 38:26
119. Tyagi J, Chaudhary P, Jyotsana UB, Bhandari G, Chaudhary A (2022) Impact of endophytic fungi in biotic stress management. Plant Prot Chem Biol 447–462
120. Umar OB, Ranti LA, Abdulbaki AS, Bola AL, Abdulhamid AK, Biola MR, Victor KO (2021) Stresses in plants: biotic and abiotic. In: Current trends in wheat research. IntechOpen
121. ur Rehman F, Adnan M, Kalsoom M, Naz N, Husnain MG, Ilahi H, Ilyas MA, Yousaf G, Tahir R, Ahmad U (2021) Seed-borne fungal diseases of maize (Zea mays L.): a review. Agrinula J Agroteknologi dan Perkeb 4:43–60
122. Vurro M (2024) Are root parasitic broomrapes still a good target for bioherbicide control? Pest Manag Sci 80:10–18
123. Wang D, Wei L, Liu T, Ma J, Huang K, Guo H, Huang Y, Zhang L, Zhao J, Tsuda K (2023) Suppression of ETI by PTI priming to balance plant growth and defense through an MPK3/MPK6-WRKYs-PP2Cs module. Mol Plant 16:903–918
124. Wendimu GY (2022) Cyst nematode (Heterodera glycines) problems in soybean (Glycine max L.) crops and its management. Adv Agric 7816951
125. Yadav RS, Yadav S, Sharma P, Seemar MK (2025) Role of cultural practices in pest management. In: Integrated pest management strategies
126. Yang P (2025) Abiotic and biotic stress of the crops and horticultural plants. Plants

Salinity Stress Management Using Chitosan Nanoparticles

Khizar Hayat, Nasir Assad, Marzia Batool Laila, Muhammad Nauman Khan, Barkat Ullah, Tabassum Yaseen, Sana Wahab, Alevcan Kaplan, and Amjad Ali

Abstract Salinity stress is a major abiotic factor that significantly reduces crop yields by affecting plant growth and physiological processes. Current common approaches to manage salinity stress are inadequate and largely ineffective for ecological reasons. In this context, chitosan nanoparticles (CNPs) have gained popularity as an environmentally friendly and effective means of combating salt-related disorders. Chitosan, a naturally occurring chitin- derived biopolymer, is biocompatible, biodegradable, and inherently bioactive, and exists in an optimized nanoparticulate form that offers greater surface area and targeting capabilities. This chapter explores the potential of CNPs in combating salinity stress by modulating physiological, biochemical, and molecular processes in plants. The application of CNPs can enhance seed germination, nutrient uptake, ion homeostasis, and antioxidant defense system under salinity conditions. They also promote osmolyte accumulation and regulate the expression of stress-responsive genes, thereby boosting plant tolerance. CNPs have been shown to improve the growth and yield of crops such

K. Hayat (✉) · M. N. Khan · B. Ullah
Department of Botany, Islamia College Peshawar, Peshawar, Pakistan
e-mail: khizarjann@hotmail.com

N. Assad · M. B. Laila
Institute of Chemistry, University of Sargodha, Sargodha, Pakistan

M. N. Khan
Biology Laboratory, University Public School, University of Peshawar, Peshawar, Pakistan

T. Yaseen
Department of Botany, Bacha Khan University, Charsadda, Khyber Pakhtunkhwa, Pakistan

S. Wahab
Department of Plant Sciences, Quaid-I-Azam University, Islamabad, Pakistan

A. Kaplan
Department of Crop and Animal Production, Sason Vocational School, Batman University, Batman, Turkey

A. Ali
Department of Sustainable Crop Production, Universita Cattolica del Sacro Cuore, Milan, Italy

A. A. H. Abdel Latef (ed.), *Chitosan Nanoparticles for Plant Resilience Under Abiotic and Biotic Stressors*, Smart Nanomaterials Technology,
https://doi.org/10.1007/978-981-95-6422-4_4

as rice, wheat, tomatoes, and maize when applied as a foliar spray or soil additive in response to salt stress. In addition, CNPs can serve as carriers for growth regulators, nutrients, and other bioactive compounds, offering further avenues for tailored stress regulation. Although promising, issues such as dosage optimization, plant-nanoparticle interactions, and long-term ecological effects need to be further investigated. Overall, CNPs have the potential to be a sustainable and innovative approach to alleviating salinity stress, helping to secure global food supplies and promote sustainable agriculture.

Keywords Antioxidant defense · Chitosan nanoparticles · Salinity stress · Sustainable agriculture · Plant resilience

1 Introduction

The problem of salinization in agriculture is one of the biggest issues for agriculture worldwide. This problem emerges when the concentration of dissolved salts in irrigation water or soil exceeds a level that is detrimental to plant growth and development and even yield. The problem of stress affects food production and the sustainability of agriculture around the world [49]. This is especially common in arid and low rainfall areas where the soil is naturally more saline and water resources are limited. This condition diminishes the growth of plants when water is scarce due to the osmotic pressure of salts, leading to water loss and nutrient deficiency in plant tissue. The accumulation of harmful ions such as sodium (Na^+) and chloride (Cl^-) is also toxic to plants as they disrupt cellular functions and normal metabolism. The effects of salinity stress are devastating as they, greatly reduce plant productivity, degrade cropland and threaten food supply systems worldwide [21].

The salinity stress has two primary causes, which can be both natural and anthropogenic. Natural salinity is known to occur in saline coastal and desert regions with insufficient drainage systems where salts accumulate over time. Meanwhile, the increase in salinity has become a problem mainly due to human behavior, especially with regard to irrigation [32]. In some regions of the world, salt accumulation occurs in the root zone because too much saline water or water of poor quality is used to irrigate fields. Secondary salinization is also the process by which the water applied during irrigation evaporates and the resulting salt is deposited in the soil. Further, the problem is exacerbated by water management issues such as the absence of proper drainage. This means that if a water source is used repeatedly without the salts being sufficiently flushed out, the salinity of the soil will continue to rise, making the land unusable and inhospitable for most crops. Soil salinization is exacerbated by improper irrigation systems and fertilizer runoff, creating a self-destructive cycle that reduces agricultural activity [36].

It is estimated that more than 20% of the total irrigated area in the world is salinized to some extent, which in turn corresponds to millions of hectares of land with impaired agricultural productivity [49]. The impact on main crops like rice, wheat, maize and vegetables is particularly severe, as these categories of crops are sensitive to the toxicity of salt in the soil. Salinity stress can alter crop composition nutrient content, flavor and perishability and reduce crop yields. This has a direct impact on food security aspects in all countries where agriculture is the main source of income in countries and arable land is scarce. As the world's population grows and food demand increases, managing salinity is becoming an increasingly pressing issue. As climate change accelerates, the effects of salinization are predicted to become even more severe. Therefore it is necessary to find viable and sustainable ways to control the increasing food hazards in several parts of the world [51].

The older methods of combating salt stress, like leaching the soil, using chemical recipes and developing salt-resistant crop strains, were not very fruitful. Soil leaching requires massive use of water to wash various salts from the soil, but is mainly water-intensive and it may not be effective in areas with limited water resources. Moreover, leaching could also contaminate, neighboring water sources, leading to further environmental concerns. To decrease the harmful impact of salinity, chemical additives such as gypsum or sulfur can be used, but this is costly and can be detrimental to soil health in the long term [12]. Breeding salt-tolerant crops using conventional methods has also not yielded many results, as the process of genetic modification to make the plants salt-tolerant is hectic, time-consuming and costly. Such varieties are in most cases not adapted to the various salinity conditions in the different environments, so their functionality is limited.

The need to address salt stress in the agriculture increases the need for new solutions that can contribute to a sustainable solution to the problem. Environmentally friendly, effective and durable technologies are of interest as approaches to combat salinity are required. The use of chitosan nanoparticles (CNPs) is one of the solutions that has received much attention in recent years [19]. A biopolymer called chitosan is naturally synthesized from chitin, the structural component of the exoskeleton of crustaceans (shrimp and crabs), and much research has been conducted on its ability to improve health or limit the effects of stress, such as salinity. Chitosan has a number of attractive properties, such as being biodegradable, biocompatible, and non-toxic, making it a perfect product to use sustainable agriculture.

The activity of chitosan in promoting plant growth and stress reduction dramatically is increased when chitosan is formulated in the form of nanoparticles. Nanoparticles (NPs) are so small that they increase the surface area and enable the NP*s* to interact more effectively with the plant cells, basically delivering nutrients, growth regulators, and other bioactive compounds to the target areas. CNPs used in salinity stress improve seed germination and nutrient uptake, maintenance of ion homeostasis, and activation of the plant's antioxidant defense mechanisms. In addition, CNPs could be used as carriers for growth regulators and other molecules and would have a higher effect on plant stress and resistance. The new strategy is a significant advance on the old systems, which offer a sustainable and environmentally friendly alternative to chemicals and salt-tolerant plant varieties [22].

There is huge potential for the use of CNPs to combat salt stress in agriculture and to further enhancing agricultural production in saline. The resistance of plants to the harmful effects of salt can potentially be improved by the properties of CNPs, which would lead to stronger plant development, higher production, and better quality. The use of CNPs is also in agreement with the growing need for environmentally friendly agricultural practices as a working approach concentrates on the methods that will be more environmentally friendly in terms of minimal environmental impact and optimal yield. The more the application of NPs in agriculture is explored, the more evident the potential of CNPs to play a pivotal role in combating salinity stress and overall solving the world food security dilemma becomes by the day. Therefore, the aim of this chapter is to explore the potential of CNPs in controlling salinity stress in the area of their properties, the action of CNPs, and the possible applications of CNPs in agriculture.

2 Chitosan Nanoparticles: Properties and Mechanisms

Chitosan is a biopolymer of natural origin that is produced by the degradation of chitin and that is predominantly found in the shells of crustaceans such as shrimps and crabs. It consists of the monomers glucosamine and N-acetylglucosamine and is considered biocompatible, biodegradable, and non-toxic polysaccharide. These properties make chitosan a desirable candidate in many respects, particularly in the agricultural sector. Chitosan has attracted much interest attracted much interest over the years attracted massive interest, due to its ability to stimulate plant growth, increase resistance to disease and alleviate diverse abiotic stress factors such as salt stress [3]. Its chemical structure can also be modified through structural changes to improve its utility characteristics, including its solubility, ability to form films, gels and nanoparticles. It has paved the way for its nanotechnological application, where CNPs serve as an example for improving the resilience and production of plants.

CNPs increase their beneficial impacts, creating new opportunities for improving agricultural activities. Nanoparticles are small; they have a large surface to volume ratio compared to their bulk counterparts, which gives them different physical, chemical and biological properties. This transformation allows chitosan more soluble, stable, and bioavailable and these properties are crucial for its ability to interact with plant cells. Different methods can be used to produce CNPs, including solvent evaporation and coacervation as well as ionic gelation [2]. These processes result in a variety of NPs with different, shapes and surface properties and can be adapted to specific agricultural applications. The size and surface charge of CNPs could be manipulated so that they can be used very flexibly, e.g., for drug delivery, nutrient release and reducing plant stress levels. In the agricultural system, CNPs can penetrate plant tissue much more easily than large CNPs, hence fertilizers, growth-inducing or -regulating compounds, or other bioactive molecules can generally be distributed systematically and directly into the plant cell where they can be used effectively.

The relationship between the CNPs and the plants takes place at the cellular level. The ability of CNPs to passively enter plant cells is due to their small size, which allows them to be taken up by plant tissues through passive diffusion or endocytosis. When CNPs are found inside the plant, they can control different physiological and biochemical processes. They can also influence nutrient uptake by altering the permeability of cell membranes, which enhances the movement of important minerals and nutrients from the soil into the plant. Additionally, ion homeostasis can be ensured with the help of CNPs by ensuring that ions are moved across cell membranes and that sodium (Na^+), and chloride (Cl^-) ions are not stored in a detrimental pool in a saline environment. This is the ionic balance control that is important for normal cell functioning and healthy plant growth in stressful situations. Besides, CNPs can stimulate the plant defense system, e.g., the activity of antioxidant enzymes that help to relieve animals in soils with high-soils salt content [22].

The most remarkable attributes of CNPs are that they have the ability to increase the plant's molecular and biochemical responses to stress. When plants are subjected to environmental stressors such as salinity, they are likely to upregulate a complex network of signaling pathways to restrict the effects of stress. These signaling pathways also detect the responses to CNPs that lead to elevated levels of stress-responsive genes and proteins that mediate the negative effects of salinity on plants. An example of this is the induction of the synthesis of osmolytes by CNPs, small molecules that help maintain cells turgor and protect enzymes and proteins from the denaturation they can cause under high salt conditions [7]. Also, CNPs can increase the production of reactive oxygen species (ROS) scavengers such as superoxide dismutase and catalase, which inactivate the toxic effects of the oxidative stress induced by salinity. In addition to ascertaining salt resistance in the plant, CNPs strengthen the plant's salt tolerance by modulating these biochemical pathways, leading to improved growth and higher yields in salt-prone regions.

CNPs can, therefore, be regarded as an outstanding invention in the field of sustainable agriculture. They have unique properties of the natural biopolymer chitosan and can act on plant cells to stimulate growth, increase stress-tolerance and improve overall plant health. The coating properties of CNPs enhance the effects of the agent to deliver their bioactive products and extend their functions in physiological processes; therefore, CNPs are a potential means to control salt stress under stressful conditions and address other environmental problems [27]. Research is still ongoing to fully unlock the potential of CNPs in agriculture, but given the promise CNPs have in modulating plant responses at the molecular and biochemical level, CNPs will make a great leap towards enhancing crop production and food security as environmental demands increase.

3 Chitosan Nanoparticles in Mitigating Salinity Stress

CNPs have created immense opportunities to reduce salt stress, which is an environmental hazard to agriculture as a whole. They can be used to promote the growth of seeds, increase the growth of plants and generally enable them to withstand a highly saline environment. The unique properties such as size, immense surface area, and the ability to interact with plant cells at the molecular level make CNPs useful tools to improve plant the performance under stress. Plants have numerous physiological, biochemical and molecular signaling pathways that help them to survive in saline environments and are influenced by CNPs [46].

The other significant impacts of CNPs include the fact that they stimulate seed germination and other early growth under saline conditions. Salt is a general inhibitor of seed germination because the salinity decreases the water potential of the soil and the seed cannot readily take up water. Under saline conditions, the osmotic pressure exerted by the salts prevents the seeds from absorbing sufficient water adequately and their germination capacity is normally low. However, this water uptake can be improved by the use of CNPs, which increase the water uptake of the seeds and activate the metabolism necessary for germination [19]. The CNPs could also be applied to increase the permeability of the seed layers, increase hydration and to promote water uptake. This leads to an increase in germination and early seedling development. In addition, CNPs can trigger the production of crucial enzymes for germination including amylases, which can break down starch into sugars and thus support seedling growth [47]. These benefits allow seeds to overcome the early challenges caused by salinity, resulting in improved seed germination and healthy seedlings.

Improving nutrient uptake and maintaining ion homeostasis in saline-stressed plants also plays a significant role in CNPs. Salty water can reduce the ionic balance of plant cells, primarily by increasing the concentration of toxic ions such as Na^+ and Cl^-, and inhibiting the uptake of essential nutrients such as potassium (K^+) and calcium (Ca^{2+}) [33]. Under these circumstances, CNPs can modulate the uptake of nutrients by increasing the permeability of root cell membranes and facilitating transport through active nutrient transport. The equilibrium between the required cations and anions is also controlled by CNPs without the accumulation of CNPs without accumulating harmful ions in the plant tissue. In addition, CNPs have been shown to support the selective uptake of potassium, which competes with sodium for uptake and helps to mitigate the toxic consequences of sodium in the plant [26]. CNPs contribute to plant growth and health by improving ion movement and nutrient uptake even in high salinity conditions.

In addition to the benefit of increased nutrient uptake, CNPs also stimulate the plant's antioxidant defense, which proves to be important in combating oxidative stress in a salt-laden environment. Salt stress leads to an overproduction of ROS, highly reactive molecules that can break down plant cells such as lipids, proteins, and DNA. Normally, plants have antioxidant systems that counteract ROS to limit cell damage. However, they adapt to saline conditions, where these systems are commonly overwhelmed, leading to oxidative stress and cell damage. CNPs have been demonstrated to enhance the activity of important antioxidant enzymes such as superoxide dismutase (SOD), catalase (CAT) and peroxidase (POD) to scavenge ROS and protect the plant from oxidative stress [41]. By enhancing these protective systems, CNPs can help plants to better cope with the harmful impact of salinity on plants and improve their overall resilience to environmental stress.

The second important way in which CNPs assist plants cope with salt stress is by increasing the synthesis of osmolytes, which are essential for maintaining cell activity under unfavorable conditions, including proline, glycine betaine, and sugars. Osmolytes assist plants promote osmotic balance by preventing dehydration, stabilizing proteins, and protecting cell structures [6]. These compounds are normally accumulated in plants under saline environments as one of the adaptive measures to osmoregulatory stress. These osmolytes can be synthesized and accumulated with the assistance of CNPs so that plants can maintain cell turgor and avoid cell damage due to water loss [38]. CNPs help with osmotic regulation and can therefore be used to induce plants to survive in saline conditions an achieve better growth and higher yields. Furthermore, osmolytes stabilize cell membranes and enzymes and improve the plant's ability to withstand salinity induced stress.

In this way, CNPs are a multitasking solution to mitigate the adverse effects of salt stress in plants. They enhance seed germination and nutrient uptake, ion homeostasis, the antioxidant defense system, and the build-up of osmolytes that help plants develop and endure in salty environments. Therefore, CNPs represent a viable and sustainable measure to cope with salt stress in regions where saline soils are present. Their ability to enhance numerous physiological and biochemical activities makes them a modern weapon against one of the most vexing environmental factors of modern agriculture. As research on the use of CNPs is still at an early stage of development, there is great hope that this could lead to a change in agricultural practice, especially because the future of agriculture will be promising and more fruitful with the use of CNPs.

4 Molecular and Biochemical Responses Modulated by CNPs

CNPs are of interest because of their ability to regulate molecular and biochemical plant responses to increased stress, particularly salinity stress. The ability of CNPs to regulate gene expression, protein activity, and metabolism is necessary to enhance plant resistance and growth under unfavorable conditions. The interaction mechanisms of CNPs at the cellular and molecular levels have enabled plants to respond more efficiently to stress; CNPs have enabled plants to withstand salinity and other stress factors in the environment. These biochemical and molecular adaptations help to improve survival and growth compared to a scenario that could hinder plant.

One of the most significant salinity-induced changes in plant response by CNPs is gene regulation. The most common adaptation of salt-stressed plants is that they activate a series of stress-response genes to adapt to the adverse effects of excessive salinity. These genes are associated with the following mechanisms: Ion transport, osmolyte accretion, and antioxidant defense. Salt-tolerant genes utilized by CNPs have been shown to be present at higher levels. In one example, CNPs can lead to the detection of osmoregulatory effects, such as the substance proline, which controls cell turgor and prevents desiccation of plant cells [5]. Similarly, CNPs can upregulate genes related to antioxidant cell enzymes associated with scavenging ROS, such as the enzymes SOD and CAT, which are also involved in mitigating oxidative damage [18]. CNPs also increase the ability of plants to balance the homeostasis of their cells and salinity stress toxicity by controlling the expression of these stress-responsive genes. In this way, CNPs provide plants with a genetic advantage and improve their overall survival under unfavorable environmental factors.

CNPs regulate not only the expression of genes, but also the activity of proteins that play an important role in the response of plants to salinity. Salinity stress conditions are always characterized by a change in the activity of major mechanical proteins and enzyme networks that control various physiological processes. The presence of toxic ions on the plant cells could lead to a disruption of the normal functions of proteins, resulting in a slowdown of metabolism and growth. This is compensated by CNPs, which control the action of proteins involved in stress tolerance. In one instance, CNPs were found to enhance the activity of antioxidant enzymes that protect the plant from oxidative stress caused by ROS [8]. Besides, CNPs may stimulate the functions of transporters that regulate ion content, including the Na^+/K^+ pump, to ensure an appropriate ion balance in plant cells [37]. Such regulation of protein activity is important to reduce the deleterious effects of salinity and contribute to the overall physiology of the plant.

CNPs also regulate the different metabolic pathways in plants which play an important role in the response to environmental stress. To respond to high osmotic pressure and ion toxicity, plants undergo metabolic reprogramming in the presence of salinity stress. This includes changes in primary and secondary metabolism and the activation of metabolic pathways involved in the stress response. One important metabolic pathway that is disturbed by salinity stress is the synthesis of the osmolytes, small molecules that maintain the osmotic balance of the plant cells. The accumulation of osmolytes proline, glycine betaine, and sugar which protect cells from dehydration and stabilize proteins and cell structures in saline environments, is increased by CNPs [5]. CNPs support the cells in synthesizing these substances and thus enable plants to maintain turgor pressure in their cells, protect themselves from osmotic stress and generally maintain cell processes.

CNPs have been found to influence the formation of phenolic compounds, flavonoids, and other antioxidants that have important functions in protecting against excessive salinity-induced oxidative stress [16]. These secondary metabolites serve not only as antioxidants, but also as signaling cascades in which other stress responses are then engaged. CNPs may also control the action of certain enzymes that mediate the biosynthesis of these compounds and lead to increased production of protective metabolites.

Besides their impact on osmolytes and secondary metabolites, CNPs have the potential to affect important metabolic pathways related to nutrient uptake and ion transport. Plants in salty environments need to maintain a balance between the uptake of essential nutrients and the excretion of harmful ions. CNPs can enhance the activity of ion transporters and channels that facilitate the uptake of K^+ and Ca^{2+}, which are dangerous for the maintenance of cellular functions, and counteract the toxic impact of Na^+ [37]. CNPs improve the efficiency of nutrient uptake and ion homeostasis, thereby supporting cellular processes like photosynthesis, respiration, and protein synthesis, which are essential for plant growth and survival.

The contact of CNPs with metabolic processes is also based on their impact on plant growth regulators. This is the case as salt stress tends to disrupt the production and mechanism of action of growth hormones such as auxins, cytokinins, and gibberellins, the latter playing a crucial role in maintaining cell division, elongation and growth [14]. Even in saline environments the CNPs can restore the balance of these hormones so that growth and development can take place. The CNPs control the rate of plant growth regulators, allowing plants to show their normal developmental patterns despite the negative effects of salinity.

Overall, CNPs play a central role in regulating the molecular and biochemical reactions of crop plants to salt stress. They can impact the expression of genes, the activity of proteins and metabolic processes, which is also the key to improving the stress tolerance of plants. CNPs facilitate cellular homeostasis, oxidative and osmotic stress by regulating stress-responsive genes, stimulating the activity of antioxidant enzymes, and increasing the amount of osmolytes. Otherwise, CNPs stimulate important metabolic pathways, such as the uptake of nutrients and the transport of ions and the production of secondary metabolites, which increases growth and survival under saline conditions. CNPs are increasingly promising new and effective interventions to optimize crop growth and stress tolerance as research gains new insights into the entire spectrum of molecular and biochemical pathways involved in CNPs to produce a plant effect on salinity, as summarized in Table 1.

Table 1 Molecular and biochemical responses of plants modulated by CNPs under salinity stress

Aspect	Description	References
Gene regulation	CNPs up-regulate salt-tolerant genes, such as those involved in osmolyte synthesis (e.g., proline) and antioxidant enzymes (SOD, CAT), enhancing plant tolerance to salinity stress	(Mohammadi and Keivanian)
Protein activity modulation	CNPs enhance antioxidant enzymes (SOD, CAT, peroxidases) and ion transporters (e.g., Na^+/K^+ pump), improving cell protection and ion balance under stress	Hernández-Hernández et al. [23]
Metabolic pathways	CNPs stimulate the synthesis of osmolytes (proline, glycine betaine, sugars) to maintain turgor pressure and stabilize cell structures under saline conditions	Rasheed et al. [43]
Secondary metabolites	CNPs influence the accumulation of phenolic compounds, flavonoids, and other antioxidants, providing defense against oxidative stress caused by salinity	Quitadamo et al. [42]
Nutrient uptake & ion transport	CNPs improve ion transporter efficiency, facilitating the uptake of K^+ and Ca^{2+}, which help counteract the toxic effects of Na^+ and maintain cellular functions	Saeedi et al. [45]
Growth regulators	CNPs regulate plant growth hormones (auxins, cytokinins, gibberellins) to maintain normal growth and development even under salinity stress	Khan et al. [31]

5 Application of Chitosan Nanoparticles in Crop Management

The use of CNPs in crop management has become a sign of major changes in the agricultural industry, especially in dealing with salt stress. The effectiveness of CNPs lies in the fact that they can increase plant growth and resilience to several environmental stressors such as salinity [19]. They may be executed in various ways, each of them offering specific benefits related to the results and the nature of stress to which the plant is exposed. Foliar spraying, soil application of the method and also targeted application to key crops are the most widely used methods of testing and implementation. All methods enable the CNPs to have direct contact with the plant tissue, increasing their capacities to control stress and their overall efficiency.

Foliar sprays are one of the most effective methods of treating plants with CNPs. CNPs solution of are sprayed CNPs directly onto the leaves are be internalized via the stomata and the epidermal cells. By spraying the leaves, the CNPs can directly reach the surface and interior of the plant and rapidly release the nutrients and bioactive compounds, as shown in Fig. 1. One of the most important benefits of applying CNPs application in foliar sprays is the strengthening of a plant [25]. Under salt stress, plants natural defenses are prone to oxidative damage, which is related to disturbance of the ROS balance, and can affect cell structures and growth. CNPs, when used as a foliar spray, are known to increase antioxidant defense systems in the plant, and enhance the activity of enzymes such as SOD, CAT, and POD. This strengthens the plant's ability to counter ROS and contain oxidative stress caused by high salt levels [55, 56].

Additionally, the use of CNPs through foliar application controls the uptake of nutrients, especially in stressful environments where salt hinders the plants' process

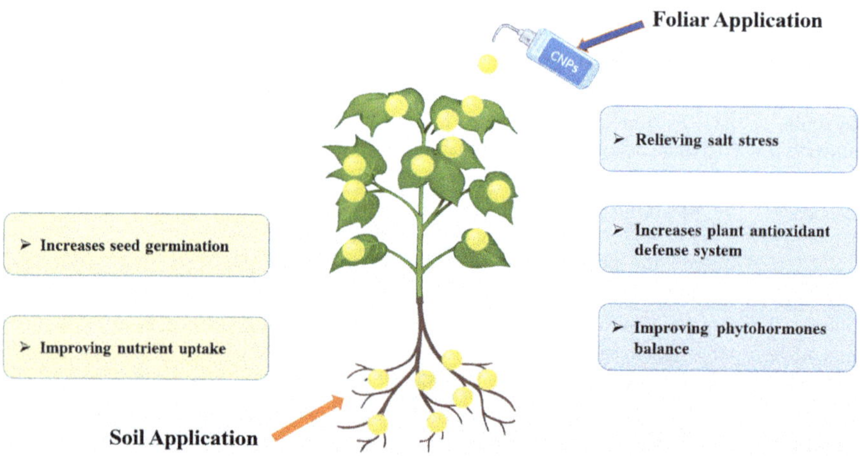

Fig. 1 Effects of foliar and soil application of CNPs on plants

of utilizing other nutrients in the soil. CNPs also increase the permeability of plant membranes, resulting in improved uptake of essential nutrients [39], including K^+ and Ca^{2+} which are important for healthy cell function and prevent the incidence of harmful effects due to Na^+ accumulation. CNPs in the form of foliar sprays improve ion balance and nutrient availability, promoting better growth and higher photosynthetic efficiency as well as the ability to resist salt-induced damage. Also, foliar sprays can offer an immediate response to stress, with the readily absorbed NP*s* having an immediate impact on plant health. For this reason, foliar fertilizer applications are of great benefit when short-term stress situations need to be managed or acute salt stress has occurred [9].

CNPs have shown improved performance in the main salt-sensitive crops, namely rice, wheat, maize and tomatoes. For example, it was recently found that the use of CNPs in rice increased the percentage of seed germination, root biomass, and salinity tolerance [50]. The NPs maintain the relative water status of rice plants by increasing ion transport and antioxidant protective systems, leading to improved growth and yield in salty soils. In wheat, CNPs applied by spraying were found to promote seedling vigor and salt tolerance, which is attributed to the improvement of nutrient uptake, and metabolism in the plants. The accumulation of osmolytes like proline triggers the osmotic stress tolerance of the wheat plant to high salinity with the help of CNPs [20].

CNP treatment by soil and foliar application in maize had led to a significant increase in growth enhancement under salt stress [48]. CNPs improve nutrient uptake, antioxidant capacity, and synthesis of stress-related proteins; all of which contribute to the enhanced resistance of maize to salinity. In tomato, CNPs have been effectively used to increase plant growth and fruit yield under salt stress conditions by increasing nutrient utilization and reducing oxidative damage [55, 56]. The CNPs-based fertilizers can increase flexibility and effectiveness in combating salt stress in different plant systems [9].

The application of NPs [10], specially CNPs in crop production, is a sustainable, eco-friendly, and highly effective approach to reducing salt stress and improving agricultural productivity. Whether through foliar sprays, soil applications, or targeted use in major crops, CNPs often provide benefits for augmenting plant health, nutrient uptake, and strengthening stress resistance. As the global agricultural community strives to develop more resilient farming practices to cope with salinity and other environmental challenges, CNPs are a valuable tool that can help secure food production and ensure long-term agricultural sustainability.

6 CNPs as Delivery Systems for Growth Regulators and Nutrients

Cross-linked CNPs provides a distinct delivery system for growth regulators, nutrients and other bioactive molecules, which is a key advantage of CNPs in modern agronomy. Their effect is greatly enhanced by the ability to precisely and selectively package and deliver these important compounds to plant tissue. CNPs are vectors for most materials used for the delivery of growth regulators and nutrients, making them highly effective in improving plant health and vigor [31]. This delivery system not only promotes plant growth and development, but also counteracts the negative impact of stress factors such as salinity.

The ability of CNPs to serve as carriers of growth regulators is one of the most important advantages of CNPs as they regulate plant growth, development, and stress responses. Phytohormones like growth regulators (auxins, cytokinins, gibberellin, and abscisic acid) are important players in the processes of cell division, cell elongation, root development, flowering, and stress response. The application of growth regulators in agriculture has long been limited due to their stability, solubility, and bioavailability [58].

However, CNPs however, because they are small, have a large surface area, and are biocompatible, so they offer a solution to these problems. CNPs preserve these compounds in their structure by ensuring that they are not destroyed and deliver them to the plant tissue in the most specific way possible. This slow-release ensures that the growth regulators are delivered at the right time and in the right concentration, which promotes plant growth and stress resistance.

As an example, the use of CNPs to provide auxin, an important plant regulator of root growth and elongation can be used to promote root growth especially under the unfavorable conditions of salt stress [4]. Likewise, cytokinins, which promote cell division and prevent senescence, can be effectively used to promote shoot growth and increase leaf longevity through CNPs. CNPs can also be used to control the levels of gibberellins involved in seed germination and flowering [31]. CNPs lead to better plant growth, development, and yield by increasing the effectiveness of these growth regulators, which is particularly effective in environments where the use of conventional growth hormones is less effective or more expensive.

Besides the ability to provide growth regulators, CNPs also serve as effective nutrient delivery systems that are useful in regulating the nutrient needs of stressed plants. Salinity stress can typically hinder a plant's uptake and movement of essential soil nutrients, leading to nutrient deficiencies that may disrupt growth and development. CNPs ensure that this problem is mitigated by including the necessary nutrients, such as phosphorus, nitrogen, calcium, potassium, and magnesium required for plant development [40]. Just like the carriers, CNPs can also transport these nutrients to the plant cells, allowing them to be efficiently absorbed and utilized.

One example is the ion (K^+), which competes with Na^+ ions under salty conditions and is needed in plant cells to maintain ionic balance. CNPs can supplement the uptake of K^+ so that plants can maintain cell activity and ion balance in saline habitats. Speaking of CNPs, they can also promote the uptake of calcium, which is needed to stabilize cell membranes and regulate enzyme activity. This specific nutrient transport helps plants to cope with nutrient shortages that might occur due to high salt concentrations in the soil. In addition, CNPs also lead to better growth and yield of plants in poor or saline soils by increasing the efficiency of nutrients [53].

Another notable advantage of using CNPs as nutrient delivery systems is that they enhance the efficiency of fertilizers [17]. Fertilizer application is inefficient, and a large proportion of nutrients are lost either through leaching, volatilization or soil fixation. By packaging nutrients in CNPs, the nutrients are released in a regulated manner and nutrient losses are reduced so that plants can use them when they are needed most. This not only promotes plant growth, but also leads to a more sustainable form of agriculture, as the production of excessive amounts of chemical fertilizers, which can have a negative impact on the environment, is no longer necessary.

Additionally, other bioactive molecules such as antimicrobial agents, stress-protection molecules, and growth-promoting molecules may also be delivered and released by CNPs. The multifunctionality of CNPs as vectors enables them to take up a wide range of bioactive molecules by increasing their stability and ensuring successful delivery to the plant tissue. Taking the example of antimicrobial compounds, such as fungicides and pesticides, CNPs have the potential to better protect plants from pathogens with a minimum of harmful chemicals that have a negative impact on both nature and humans [52]. The CNPs have the potential to reduce the risk of pesticide resistance and minimize environmental pollution by enabling more localized and controlled delivery of these bioactive molecules.

In addition, CNPs may deliver osmoprotectants or antioxidants or other adjuvants and help plants resist environmental stressors such as drought, salt, and extreme temperatures [19]. This implies that once these compounds are embedded in CNPs, they are no longer degraded and can enter the plant with great ease, making it more resistant to stress and allowing it to continue its normal physiological activity. It is a special feed that enhances the plant's flexibility to survive and thrive in unfriendly conditions, leading to better growth and yield.

The opportunities of integrated pest and stress systems also arise from the possibility of CNPs to utilize different bioactive molecules [1]. One of these is the use of growth regulators, nutrients and stress-protection molecules in a single nanoparticle formulation,this can provide a holistic approach to plant health and productivity. Such an approach decreases the necessity to employ multiple chemical inputs and ensures that plants receive a comprehensive solution to the issue of environmental stress.

CNPs have therefore proven to be very efficient transport systems for growth regulators, nutrients and other bioactive molecules. CNPs are able to stabilize these compounds by encapsulation, improve bioavailability and deliver the compounds specifically to the plant tissue. Growth regulators applied as CNPs increase the growth and drought resistance of plants and their use as nutrient delivery systems increases nutrient uptake and efficiency, especially under salt stress. Moreover, the ability of

CNPs to deliver and transport additional bioactive molecules opens up new perspectives for combined pest and stress management, leading to more sustainable and efficient food production. With ongoing research and technological development, the use of CNPs in crop protection could continue to grow and increase the potential to improve plant health, yield and resistance to environmental stresses.

7 Challenges and Future Directions

Despite the enormous potential of CNPs as a solution to salt stress and to increase agricultural productivity, many challenges and aspects of their application must first be clarified before they can be used on a large scale in agricultural systems. These concerns overlap with optimizing the dosage of CNPs to ensure the ecological and environmental sustainability of their use in the long term [28]. Further experiments on the relationships between plants and NPs are also needed, and the relationships can depend on the plant species, environmental conditions, and formulation of the nanoparticles. These are the challenges that will be the focus of future study to make the application of CNPs in agriculture safe, efficient and sustainable.

The first big challenge in the application is the optimization of the CNPs dosage. One of the most important factors for the effectiveness of this type of practitioner is the concentration of CNPs [31, 32]. Too high a concentration can lead to toxicity, too low a concentration can lead to a lack of the desired effect. Depending on the plant species, stress (salinity, etc.), application type (foliar, soil, etc.), and environmental conditions, soil type, and temperature, the most effective CNPs dosage may vary. Although CNPs are effective in increasing plant growth and stress-resilience, no standard dosage has been proposed for all treatments in agriculture. This lack of standardization poses a challenge to their extensive use. Researchers need to determine the optimal concentrations that maximize benefits without harming plants, soil health, or the environment [24].

Understanding the interactions between plants and nanoparticles is one of the most important areas to consider. Even though the beneficial effects of CNPs on plants, especially their growth and stress tolerance, are well studied, it is not entirely comprehended how the interactions actually happen. Only the penetration of CNPs into the plant cells the contact of the plant cell structures with the CNPs and their transport in the plant tissue need to be investigated in more detail. To illustrate, what happens to the NPs on the plant cell wall, and where do they end up in the plant after application? Do they remain in certain tissues or organs or are they distributed evenly? Moreover, it is important to know how CNPs can affect plant expression, protein activity, and metabolism in order to optimize their use in the agricultural industry [34].

In addition, it is unclear whether nanoparticles can be toxic and whether long-term impacts on the well-being of plants are conceivable, which could also be negative. However, most research studies found that CNPs improve stress tolerance and development. Further research is needed to understand their long-term impacts on plant physiology productivity, and overall health. This knowledge will be useful in developing more precise applications of CNPs to minimize the risks and maximize the benefits.

The second, but not least important, consideration that should be given to the far-reaching implementation of CNPs is the long-term ecological and environmental impacts. NPs are small, there is the possibility that they will encounter soil microorganisms, water systems, and other environmental components that are not yet well-understood. The unintended effects of the continued accumulation of NPs in the environment can include soil fertility, microbial diversity, and ecosystem well being [29]. For an example, although CNPs are biodegradable and based on natural sources (e.g., chitosan), their degradation products and interactions with soil organisms should be studied in more detail. An important measure is whether the use of CNPs leads to bioaccumulation in plants, soil organisms, or aquatic systems, and whether such bioaccumulation can affect the wider food chain. CNPs can also leach into groundwater or surface water and impact water quality, which is a cause for concern. Extensive environmental impact scrutinies are mandatory to investigate the fate of CNPs in the environment, especially for large-scale agricultural applications [54]. Only by understanding these long-term ecological risks can we ensure that CNPs are a safe and sustainable alternative to traditional chemical inputs.

Finally, the study of CNPs study within the agricultural industry must focus on a few aspects in the future to improve their application and effectiveness. The development of more technologically advanced, multifunctional NPs is one of the most promising directions. Currently, CNPs are mainly used for nutrient delivery and stress relief, but a wide range of active ingredients could be incorporated into the nanoparticle scaffold, such as pesticides, growth regulators, or disease-resistant agents. This would lead to smart NPs that can successfully provide a holistic approach to plant stress by responding to diverse challenges. CNPs could be combined with biological control agents to combat abiotic and biotic stress and produce more robust crops that do not rely on the use of chemical pesticides [13]. Research on adapting CNPs formulations to plant species and environmental factors is also important for improving their efficacy. The second area where CNPs have immense potential is the development of location-specific delivery systems that release NPs in response to specific environmental stimuli (e.g., high salinity).

Research on long-term CNPs and soil health and fertility also needs to be explored. Despite the positive impacts on plant development, and stress tolerance, CNPs and their interaction with the soil microbiota and effects on soil parameters, e.g., pH, nutrients, and microbial diversity, have not yet been fully studied. It is recommended that studies be conducted to determine whether CNPs have detrimental effects on beneficial soil organisms such as nitrogen-fixing bacteria, mycorrhizal fungi and other beneficial microorganisms that contribute to soil thriving and health [44]. Further, the interaction of CNPs with soil organic matter will be explored to establish their role in soil carbon sequestration and overall soil structure. Such information will prove helpful in ensuring that CNPs do not have a detrimental effect on soil sustainability in the long term and that their application is in line with the postulates of sustainable agriculture [57]. Although the future of CNPs as growth regulators and to alleviate environmental stress seems promising, there are still challenges in optimizing their application. Studies on the optimization of dosing, the interactions of CNPs with plants, and the environmental sustainability of CNPs in the long term are essential to ensure that their benefits outweigh the risks they may pose [15].

Also, more sophisticated and multi-purpose NPs tailored to meet the requirements of particular agricultural tasks will be created, and the impact of their environmental effects will be more thoroughly considered, paving the way for a comprehensive and sustainable introduction of CNPs in agriculture. As these areas of research continue to evolve in the present, the future of CNPs in crop production holds enormous potential to improve food security, agricultural productivity and sustainable farming practices in agricultural activities around the world.

8 Conclusion

CNPs have gained more focus in agriculture, where they are expected to help relieve salinity stress and make plants more tolerant to adverse environmental influences. CNPs have special properties such as biodegradability, biocompatibility, and the ability to enhance the delivery of bioactive compounds, which makes them a better system compared to conventional solutions. The significant conclusions of this work are that CNPs enhanced seed germination, plant growth and yield in saline environments due to improved nutrient uptake, stability of ion homeostasis, antioxidant systems and eventual accumulation of osmolytes in the plant, helping plants to better adapt to salt stress. CNPs can also serve as vehicles for growth regulators, fertilizers, and other bio-active substances, improving the efficiency and precision of nutrient delivery and reducing the need to apply large quantities of chemicals. CNPs are an ecologically friendly and sustainable substitute for increasing crop production and ensuring food security in the face of increasing environmental effects on agriculture, soil erosion, and dwindling freshwater supplies. Their ability to increase crop tolerance to stresses such as salinity is a real benefit to farmers in very salty regions where farmers cannot cope with traditional soil management and irrigation techniques. The precise delivery of nutrients and growth regulators through CNPs also promises to make more efficient use of resources, reduce the environmental footprint of fertilizers and pesticides, and promote more sustainable agricultural systems. By improving crop yields and their ability to withstand ever-changing conditions, CNPs can make a significant contribution to growing the world's food supply and feeding an ever-increasing population in an increasingly harsh climate. CNPs promise to be useful in agriculture in the future. Ongoing research to optimize CNP formulations, determine optimal dosage and study their interaction with plants, and long-term effects on the environment will help ensure that these NP*s* are used to protect and promote crops. As knowledge of crop-specific needs and the impact of the type of nanoparticle increases, CNPs may be utilized agriculture for in numerous applications ranging from crop stress relief to improved nutrient delivery. Moreover, in combination with other biotechnological developments, including precision agriculture, CNPs offer interesting prospects for improved and sustainable agriculture. Although currently still in the development stage, CNPs could be important for future sustainable agriculture by producing stress-resistant and high-yielding crops, as increasing the robustness of a crop system and the efficiency of nutrient utilization will be possible regardless of environmental limitations.

References

1. Abenaim L, Conti B (2023) Chitosan as a control tool for insect pest management: A review. Insects 14:949
2. Alehosseini E, Shahiri Tabarestani H, Kharazmi MS, Jafari SM (2022) Physicochemical, thermal, and morphological properties of chitosan nanoparticles produced by ionic gelation. Foods 11:3841
3. Alenazi MM, El-Ebidy AM, El-Shehaby OA et al (2024) Chitosan and chitosan nanoparticles differentially alleviate salinity stress in *Phaseolus vulgaris* L. plants. Plants 13:398
4. Alhaithloul HAS, Awad NS, Shoala T et al (2024) Comparison of various hormone combinations and the potential of chitosan nanoparticles for growth stimulation in *Astragalus* spp. Plant Cell Tiss Organ Cult 157:74. https://doi.org/10.1007/s11240-024-02790-w
5. Ali EF, El-Shehawi AM, Ibrahim OHM et al (2021) A vital role of chitosan nanoparticles in improvisation the drought stress tolerance in *Catharanthus roseus* (L.) through biochemical and gene expression modulation. Plant Physiol Biochem 161:166–175
6. Ali S, Ullah S, Khan MN et al (2022) The effects of osmosis and thermo-priming on salinity stress tolerance in *Vigna radiata* L. Sustainability 14:12924
7. Al-Khayri JM, Rashmi R, Surya Ulhas R et al (2023) The role of nanoparticles in response of plants to abiotic stress at physiological, biochemical, and molecular levels. Plants 12:292
8. Arif Y, Siddiqui H, Hayat S (2021) Role of chitosan nanoparticles in regulation of plant physiology under abiotic stress. In: Faizan M, Hayat S, Yu F (eds) Sustainable agriculture reviews 53. Springer International Publishing, Cham, pp 399–413
9. Ashraf U, Zafar S, Ghaffar R, et al (2022) Impact of nano chitosan-NPK fertilizer on field crops. In: Role of chitosan and chitosan-based nanomaterials in plant sciences. Elsevier, pp 165–183
10. Assad N, Ansa, Laila MB et al (2025a) Nanotechnology in oilseed crops: A new frontier for abiotic stress adaptation. In: Abdel Latef AAH (ed) Oilseed crops under abiotic stress. Springer Nature Singapore, Singapore, pp 507–536
11. Assad N, Laila MB, Naeem-Ul-Hassan M et al (2025b) Phytohormones as key players: Boosting oilseed crop tolerance to nutrient deficiency stress. In: Abdel Latef AAH (ed) Oilseed crops under abiotic stress. Springer Nature Singapore, Singapore, pp 415–440
12. Bello SK, Alayafi AH, Al-Solaimani SG, Abo-Elyousr KA (2021) Mitigating soil salinity stress with gypsum and bio-organic amendments: A review. Agronomy 11:1735
13. Bouhadi M, Javed Q, Jakubus M et al (2025) Nanoparticles for sustainable agriculture: assessment of benefits and risks. Agronomy 15:1131
14. Choudhary P, Pramitha L, Rana S et al (2021) Hormonal crosstalk in regulating salinity stress tolerance in graminaceous crops. Physiol Plant 173:1587–1596. https://doi.org/10.1111/ppl.13558
15. De Oliveira ALB, Cavalcante FTT, Da Silva Moreira K et al (2021) Chitosan nanoparticle: Alternative for sustainable agriculture. In: Nascimento RFD, Neto VDOS, Fechine PBA, Freire PDTC (eds) Nanomaterials and nanotechnology. Springer Singapore, Singapore, pp 95–132
16. Elkarmout AF, Yang M, Hassan FA (2022) Chitosan treatment effectively alleviates the adverse effects of salinity in *Moringa oleifera* Lam via enhancing antioxidant system and nutrient homeostasis. Agronomy 12:2513

17. Elshayb OM, Ghazy HA, Wissa MT et al (2024) Chitosan-based NPK nanostructure for reducing synthetic NPK fertilizers and improving rice productivity and nutritional indices. Front Sustain Food Syst 8:1464021
18. Fang J, Peng Y, Zheng L et al (2024) Chitosan–se engineered nanomaterial mitigates salt stress in plants by scavenging reactive oxygen species. J Agric Food Chem 72:176–188. https://doi.org/10.1021/acs.jafc.3c06185
19. Farooq T, Nisa ZU, Hameed A et al (2022) Priming with copper-chitosan nanoparticles elicit tolerance against PEG-induced hyperosmotic stress and salinity in wheat. BMC Chemistry 16:23. https://doi.org/10.1186/s13065-022-00813-1
20. Gholizadeh F, Gohari G, Pál M et al (2025) Enhancing wheat resilience to salt stress through an integrative nanotechnology approach with chitosan proline and chitosan glycine. Sci Rep 15:11126
21. Hailu B, Mehari H (2021) Impacts of soil salinity/sodicity on soil-water relations and plant growth in dry land areas: A review. J Nat Sci Res 12:1–10
22. Hassan FAS, Ali E, Gaber A et al (2021) Chitosan nanoparticles effectively combat salinity stress by enhancing antioxidant activity and alkaloid biosynthesis in *Catharanthus roseus* (L.) G. Don. Plant Physiology and Biochemistry 162:291–300
23. Hernández-Hernández H, Juárez-Maldonado A, Benavides-Mendoza A et al (2018) Chitosan-PVA and copper nanoparticles improve growth and overexpress the SOD and JA genes in tomato plants under salt stress. Agronomy 8:175
24. Hidangmayum A, Dwivedi P (2022) Chitosan based nanoformulation for sustainable agriculture with special reference to abiotic stress: A review. J Polym Environ 30:1264–1283. https://doi.org/10.1007/s10924-021-02296-y
25. Hong J, Wang C, Wagner DC et al (2021) Foliar application of nanoparticles: mechanisms of absorption, transfer, and multiple impacts. Environ Sci Nano 8:1196–1210
26. Huy VN, Methela NJ, Al-Azawi TNI et al (2025) Fulvic acid-releasing chitosan nanoparticles promote the growth and salt stress tolerance of soybean plants. Physiol Plant 177:e70254. https://doi.org/10.1111/ppl.70254
27. Ingle PU, Shende SS, Shingote PR et al (2022) Chitosan nanoparticles (ChNPs): A versatile growth promoter in modern agricultural production. Heliyon 8
28. Isibor PO, Kayode-Edwards II, Onwaeze OO (2024) Prospects and challenges of nanochitosan application in aquaculture. In: Isibor PO, Adeogun AO, Enuneku AA (eds) Nanochitosan-based enhancement of fisheries and aquaculture. Springer Nature Switzerland, Cham, pp 301–320
29. Karamchandani BM, Dalvi SG, Bagayatkar M et al (2024) Prospective applications of chitosan and chitosan-based nanoparticles formulations in sustainable agricultural practices. Biocatal Agric Biotechnol 58:103210
30. Khan BA, Nadeem MA, Javaid MM et al (2022) Chemical synthesis, characterization, and dose optimization of chitosan-based nanoparticles of clodinofop propargyl and fenoxaprop-*p*-ethyl for management of *Phalaris minor* (little seed canary grass): First report. Green Process Synth 11:1118–1127. https://doi.org/10.1515/gps-2022-0096
31. Khan MN, Khan Z, Zhu L et al (2022) Impact of chitosan and chitosan-based metallic nanoparticles on the regulation of plant hormones. In: Role of chitosan and chitosan-based nanomaterials in plant sciences. Elsevier, pp 343–356
32. Khondoker M, Mandal S, Gurav R, Hwang S (2023) Freshwater shortage, salinity increase, and global food production: A need for sustainable irrigation water desalination-A scoping review. Earth 4:223–240
33. Lindberg S, Premkumar A (2023) Ion changes and signaling under salt stress in wheat and other important crops. Plants 13:46
34. Mohamed NG, Abdel-Hakeem MA (2023) Chitosan nanoparticles enhance drought tolerance in tomatoes (*Solanum lycopersicum* L) via gene expression modulation. Plant Gene 34:100406
35. Mohammadi M, Keivanian M Investigating the effect of chitosan on gene Apx, Sod, Pal expression in rapeseed (*Brassica Napus* L.) under salt stress
36. Mohanavelu A, Naganna SR, Al-Ansari N (2021) Irrigation induced salinity and sodicity hazards on soil and groundwater: An overview of its causes, impacts and mitigation strategies. Agriculture 11:983

37. Moosavi SNV, Saadatmand S, Jamei R, Darvishzadeh R (2024) Chitosan nanoparticles (CSNPs) conferred salinity tolerance in maize by upregulating E3 ubiquitin-protein ligase, P5CS1, HKT1, NHX1, and PMP3 genes. Environ Sci Pollut Res 31:61210–61219. https://doi.org/10.1007/s11356-024-34933-z
38. Mulaudzi T, Nkuna M, Sias G et al (2022) Antioxidant capacity of chitosan on sorghum plants under salinity stress. Agriculture 12:1544
39. Nandini T, Sudhalakshmi C, Sivakumar K et al (2025) A review-chitosan nanoparticles towards enhancing nutrient use efficiency in crops. Int J Biol Macromol 141433
40. Pal P, Singh AK, Singh MP et al (2025) Role of chitin-based nanoparticles for crop nutrition and fertilization. In: Bachheti RK, Bachheti A, Husen A (eds) Chitin-based nanoparticles for the agriculture sectors. Springer Nature Singapore, Singapore, pp 103–131
41. Qu DY, Gu WR, Zhang LG et al (2019) Role of chitosan in the regulation of the growth, antioxidant system and photosynthetic characteristics of maize seedlings under cadmium stress. Russ J Plant Physiol 66:140–151. https://doi.org/10.1134/S102144371901014X
42. Quitadamo F, De Simone V, Beleggia R, Trono D (2021) Chitosan-induced activation of the antioxidant defense system counteracts the adverse effects of salinity in durum wheat. Plants 10:1365
43. Rasheed Y, Khalid F, Ashraf H et al (2024) Enhancing plant stress resilience with osmolytes and nanoparticles. J Soil Sci Plant Nutr 24:1871–1906. https://doi.org/10.1007/s42729-024-01821-x
44. Rojas-Pirela M, Carillo P, Lárez-Velásquez C, Romanazzi G (2024) Effects of chitosan on plant growth under stress conditions: similarities with plant growth promoting bacteria. Front Plant Sci 15:1423949
45. Saeedi R, Seyedi A, Esmaeilizadeh M et al (2025) Efficiency of nanostructures containing chitosan-selenium in grafted citrus seedlings under salinity stress: element uptake, biochemical and morphological changes. J Soil Sci Plant Nutr 25:1813–1829. https://doi.org/10.1007/s42729-025-02239-9
46. Sangwan S, Sharma P, Wati L, Mehta S (2023) Effect of chitosan nanoparticles on growth and physiology of crop plants. In: Engineered nanomaterials for sustainable agricultural production, soil improvement and stress management. Elsevier, pp 99–123
47. Sen SK, Das D (2024) A sustainable approach in agricultural chemistry towards alleviation of plant stress through chitosan and nano-chitosan priming. Discov Chem 1:44. https://doi.org/10.1007/s44371-024-00046-2
48. Sharifi Rad J, Karimi J, Mohsenzadeh S et al (2014) Evaluating SiO2 nanoparticles effects on developmental characteristic and photosynthetic pigment contents of *Zea mays* L. Bull Environ Pharm Life Sci 3:194–201
49. Singh A (2022) Soil salinity: A global threat to sustainable development. Soil Use Manag 38:39–67. https://doi.org/10.1111/sum.12772
50. Soni AT, Rookes JE, Arya SS (2023) Chitosan nanoparticles as seed priming agents to alleviate salinity stress in rice (*Oryza sativa* L.) seedlings. Polysaccharides 4:129–141
51. Sultan MT, Mahmud U, Khan MZ (2023) Addressing soil salinity for sustainable agriculture and food security: Innovations and challenges in coastal regions of Bangladesh. Future Foods 8:100260
52. Suwanchaikasem P, Idnurm A, Selby-Pham J et al (2024) The impacts of chitosan on plant root systems and its potential to be used for controlling fungal diseases in agriculture. J Plant Growth Regul 43:3424–3445. https://doi.org/10.1007/s00344-024-11356-1
53. Toksha B, Joshi S, Chatterjee A (2024) The potential of nanocomposite fertilizers for sustainable crop production. In: Abd-Elsalam KA, Alghuthaymi MA (eds) Nanofertilizers for sustainable agroecosystems. Springer Nature Switzerland, Cham, pp 99–124
54. Tortella G, Rubilar O, Pieretti JC et al (2023) Nanoparticles as a promising strategy to mitigate biotic stress in agriculture. Antibiotics 12:338
55. Ullah N, Basit A, Ahmad I et al (2020) Mitigation the adverse effect of salinity stress on the performance of the tomato crop by exogenous application of chitosan. Bull Natl Res Cent 44:181. https://doi.org/10.1186/s42269-020-00435-4

56. Ullah S, Khan M, Idrees M et al (2020) 32. Enhancement effect of Hydroxyl Benzoic Acid (HBA) foliar application on growth performance of *Triticum aestivum* L. (Wheat) under induced chromium toxicity. Pure Appl Biol (PAB) 9:2529–2542
57. Wang X, He M, Wang X et al (2024) Emerging nanochitosan for sustainable agriculture. Int J Mol Sci 25:12261
58. Zhumanova N, Akimbayeva N, Myrzakhmetova N et al (2024) A comprehensive review of new generation plant growth regulators. ES Food & Agroforestry 17:1190

Using Chitosan Nanoparticles in the Alleviation of Drought Stress in Plants

Aryadeep Roychoudhury, Urjashi Chatterjee, Aishani Bhattacharya, and Spandan Dutta

Abstract Increasing food demands and low crop yield, along with environmental issues caused by conventional agrochemicals, underscore the need for advanced solutions such as nanoparticles (NPs) in agriculture. NPs significantly enhance nutrient delivery and function as nanopesticides and nanofertilizers for targeted agrochemical applications. They allow for controlled and sustained release of agrochemicals, reducing overall chemical use and environmental contamination. NPs can also boost plant resistance to stresses such as drought, salinity, and pathogen attacks. Chitosan nanoparticles (CNPs) are emerging as highly effective tools in sustainable agriculture due to their ability to enhance nutrient and agrochemical delivery with high precision and efficiency. Their small size and large surface area facilitate better absorption and mobility within plant tissues, improving plant productivity under stress. CNPs also activate plant defense mechanisms, protecting crops from pathogens and pests while decreasing the need for chemical pesticides. They are biodegradable, biocompatible, environmentally friendly, and represent a safe, essential technology for modern sustainable farming. Among major abiotic stressors, drought significantly limits plant growth and productivity by decreasing water availability, causing stomatal closure and reducing photosynthesis. It delays germination, inhibits cell division, and causes wilting, leaf rolling, and early aging. Drought can also lead to the formation of excess reactive oxygen species (ROS), which damage cellular components. Prolonged drought episode changes root structure and, these effects collectively impair plant metabolism, growth, and yield, making drought a major threat to global food security. This chapter explores how applying CNPs through various methods can help alleviate drought-induced stress by promoting the overall plant growth. They do so through mechanisms such as modulating the antioxidant defense system,

A. Roychoudhury (✉)
Discipline of Life Sciences, School of Sciences, Indira Gandhi National Open University, Maidan Garhi, New Delhi, India
e-mail: aryadeep.rc@gmail.com

U. Chatterjee · A. Bhattacharya · S. Dutta
Department of Biotechnology, St. Xavier's College (Autonomous), Kolkata, West Bengal, India

A. A. H. Abdel Latef (ed.), *Chitosan Nanoparticles for Plant Resilience Under Abiotic and Biotic Stressors*, Smart Nanomaterials Technology,
https://doi.org/10.1007/978-981-95-6422-4_5

scavenging ROS, maintaining osmotic balance, and preserving membrane integrity. Therefore, CNPs have the potential to protect plants and enhance their physiological resilience under drought stress.

Keywords Chitosan nanoparticles · Drought stress · Reactive oxygen species · Antioxidant defense system · Osmotic balance · Plant growth and productivity · Sustainable agriculture

1 Introduction

Drought stress severely impacts agricultural productivity. Ensuring food security and adopting sustainable, less harmful techniques to help crops endure periods of intense abiotic stress has become an urgent need. Furthermore, the worsening climate crisis and other stressors such as salinity and heavy metal toxicity threaten crop production worldwide [28, 40]. As these challenges grow, developing a sustainable and resilient crop production system has become vital to meet global food demands. Traditional farming methods often use excessive amounts of water, energy, and other natural resources and fail to fulfill the increasing global food needs [31]. Relying heavily on agrochemicals and pesticides has also led to soil degradation, reduced soil microbiota, and environmental pollution [26]. It is crucial to recognize that current chemical-based agricultural practices must be immediately replaced by sustainable and more efficient methods to maintain long-term ecological balance. Nanotechnology has emerged as a promising new tool in transforming various agricultural practices, from soil remediation to food packaging [7]. Different nanoparticles (NPs), especially metal NPs like titanium dioxide, iron oxide, and zinc oxide, are gaining attention due to their non-detrimental impact when used in agriculture [67]. Drought reduces photosynthesis, nutrient absorption, and the osmotic and antioxidant activities in plants [68]. Carbon-based nanomaterials such as graphene and fullerene, along with metal-based NPs, have shown promising results in alleviating drought stress by enhancing water and nutrient uptake, promoting stress tolerance, and upregulating genes involved in cell growth [100, 146]. Incorporating nanotechnology into agriculture is now essential. NPs have unique physico-chemical properties that allow them to penetrate cellular membranes and interact efficiently with biological systems [28]. These NPs are effective even at low concentration and can be administered through various methods, including seed treatment, foliar spraying, and hydroponic delivery [122].

NPs in agriculture can be used as fertilizers, pesticides, herbicides, insecticides, growth regulators, nanocarriers, and nano-sensors [31]. Agricultural nanotechnology is aimed at improving crop productivity by studying the effects of regulated inputs of these NPs in real time [77]. Unlike traditional fertilizers that leach into groundwater and can even become surface run off causing pollution, NPs provide a controlled and efficient form of nutrient release to plants [48].

Chitosan is a linear cationic polysaccharide composed of glucosamine and N-acetylglucosamine linked through β-(1,4) glycosidic linkage and can be extracted from marine crustaceans like shrimp, crab, and pinfish or by deacetylating chitin present in insect exoskeleton. It is safe, nontoxic, inexpensive, and is characterized by biocompatibility, biodegradability, and bioactivity [88]. Chitosan was found to induce the expression of a gene cascade involved in plant defense responses, which showed a spike in secondary plant metabolite synthesis [139]. Chitosan treatment also resulted in upregulation of genes involved in photosynthesis and certain storage proteins [96]. Implementing the use of chitosan NPs to provide a sustainable way of improving drought tolerance in crops can thus open new avenues in the field of agricultural nanotechnology. Chitosan nanoparticles (CNPs) are easily absorbed by the epidermis of leaves and translocated to stems, which facilitates the uptake of bioactive molecules enhancing growth and productivity of several crop plants [104].

This chapter explores the benefits and positive effects of NPs in improving the ability of crop plants to combat drought stress. It highlights the potential of CNPs as an eco-friendly and highly effective tool for improving crop resilience to drought and other abiotic stress. It is essential to understand the principles of NPs, their diverse forms and characteristics that confer their unique physico-chemical properties, dictating crop stress management. Extensive discussions regarding the application of NPs in agriculture, with special emphasis on CNPs, have been made in this chapter.

2 Nanoparticles

Nanotechnology deals with matter at the scale of 1 billionth of a meter and NP is the most fundamental component in the fabrication of a nanostructure. In general, the size of NP is in the range between 1 and 100 nm [3, 72]. NPs can be of different shapes, sizes and structures. They can be spherical, cylindrical, conical, tubular, hollow core, spiral or irregular. They can also be crystalline or amorphous, loose or aggregated [84]. There exists a unique correlation between the size of NPs and their unexpected behavior, which translates to their potential for various technological applications [69]. Due to these unique properties, NPs and nanomaterials are used in catalysis, water treatment, energy storage, medicine, and agriculture, [59]. NPs are able to exhibit enhanced properties as compared to their parent material of larger dimensions because of surface effects. Dispersed nanomaterials have a very large surface area and a high particle number per mass unit; the fraction of atoms at the surface is increased and these atoms have fewer direct neighbors. Larger surface areas and larger surface-to-volume ratios generally increase the reactivity of NPs. Secondly, due to the size shrinkage of materials, the quantum effects become more pronounced. As a result of quantum confinement, some non-magnetic materials in bulk such as palladium, platinum, and gold can become magnetic in the nanoscale. Quantum confinement also alters the electron affinity of elements which directly affects their catalytic properties [84].

3 Classification of NPs

Based on their composition, NPs can be classified under three broad categories.

3.1 Carbon-Based NPs

Combining the properties of sp2 hybridized carbon atoms along with the unique physico-chemical properties that arise at the nanoscale, including electrical conductivity, high strength, electron affinity, optical, and thermal properties, carbon-based NPs are viable candidates in drug delivery, energy storage, bioimaging, photovoltaic devices, and environmental biosensors [6, 30, 126]. Prominent examples of carbon-based NPs are fullerenes, carbon black NPs and carbon quantum dots.

3.2 Organic NPs

NPs made of proteins, carbohydrates, lipids, polymers, and other organic compounds like chitin form this subclass. These NPs are non-toxic, biodegradable, and biocompatible. Due to the presence of non-covalent intermolecular interactions, organic nanoparticles are thermally labile and can also be removed easily from the body. Organic nanoparticles are being widely utilized in the biomedical field in targeted drug delivery and cancer therapy [47, 65]. Some examples are dendrimers, liposomes, micelles, and chitosan nanoparticles.

3.3 Inorganic NPs

Inorganic NPs include metal, ceramic and semiconductor NPs. Metal NPs due to localized surface plasmon resonance characteristics exhibit unique optical and electrical properties along with thermal, magnetic, and biological properties [47]. Metal NPs are being widely used in biomedical and pharmaceutical fields. Common examples include gold, silver, and copper NPs. Semiconductor NPs possess properties intermediate between metals and non-metals and their unique bandgaps make them important components in photocatalysis, optics, and electronic devices [66]. ZnO, ZnS and CdS are common examples. Ceramic NPs are metal carbonates, phosphates, oxides, and carbides which exhibit high stability and high load capacity [114]. For these reasons, they find wide application in the biomedical industry. Silicon carbide and corundum are examples of ceramic NPs.

4 Synthesis of NPs

The synthesis of NPs can be broadly classified into the top-down method and the bottom-up method. The approach of NP synthesis, where an external force is applied to a larger parent material, breaking it up into smaller particles is called the top-down approach. It is also called the physical method. The chemical and biological approaches are clubbed under the bottom-up synthesis approach where NPs are produced from atoms of gas and liquid after performing atomic transformation or molecular condensation, [11]. The various approaches of NP synthesis have been given in Fig. 1.

4.1 Top-Down Method

Physical methods that involve fragmenting bulk substances to nano-structured materials fall under this category. Some examples have been given in Table 1.

4.2 Bottom-Up Method

In these techniques, minute atoms and molecules are combined in bottom-up methods to create nano-structured particles, which include both biological and chemical approaches. The chemical methods of synthesis have been mentioned in Table 2.

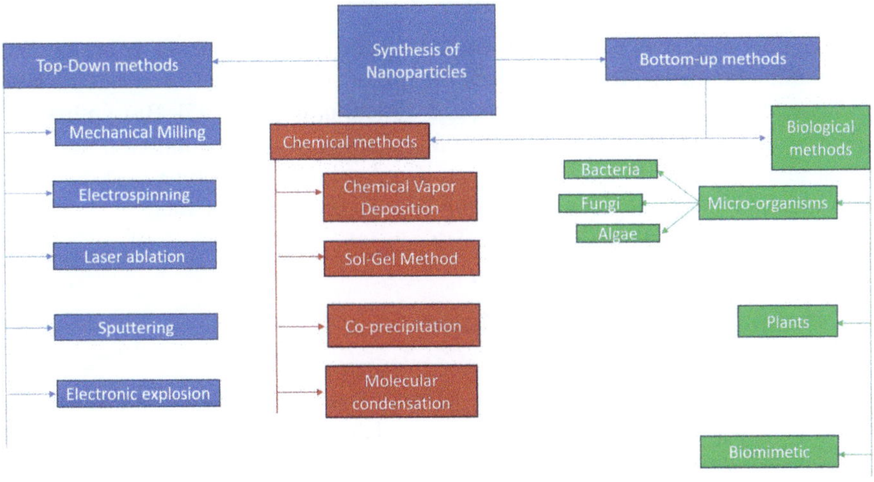

Fig. 1 Different methods of NPs synthesis

Table 1 Top-down methods of NP synthesis

Methods under the top-down approach	Technique utilized	Examples
Mechanical milling	Utilizes balls inside containers subject to a high-energy impact process to create materials at the nanoscale from larger materials [64]	1. Nanoalloys based on aluminium, nickel and copper 2. Ball-milled carbon nanomaterial, which is used in environmental remediation [162]
Electrospinning	Creation of nanofibers from polymers by drawing charged threads from polymer melts [131]. Core–shell nanoarchitectures constructed in the two coaxial capillaries using two viscous liquids in this technique [168]	Core–shell and hollow polymer, inorganic, organic and hybrid materials have been developed using this technique [93]
Laser ablation	Production of NPs by striking the precursor bulk material with an intense laser beam, leading to its vaporization following irradiation , [14]	Production of noble metal NPs along with carbon nanomaterials, oxide composites and ceramics [20, 153]
Sputtering	Irradiation of a solid material with high-energy gas particles causes expulsion of microparticles from the surface	Gold and platinum NPs can be synthesized by this method
Electron explosion	Exposure of a thin metal wire to a high current pulse causing explosion, evaporation, and vaporization of the metal and its subsequent condensation forming NPs [83]	Production of NPs from a platinum solution

Biological synthesis of NPs can be done using microorganisms, plant extract, and biomimetic means. It also involves the use of several bioactive agents such as biowastes, vegetable waste, egg shells, and agricultural waste [95]. Biosynthesized NPs have added advantage with improved scalability, biocompatibility, production of non-toxic byproducts and requirement of comparatively less time [77]. Some of the common methods of biosynthesis of NPs have been listed in Table 3.

5 Nanoparticles and Agriculture

Nanotechnology is being eyed as one of the most viable ways to transform the agricultural sector and address the current problems it is facing. Nanoparticles find widespread use as nano-fertilizers, nano-pesticides, and beneficial agents of plant growth.

Table 2 Bottom-up methods of NP synthesis with chemical methods

Methods under chemical bottom-up approach	Basic technique	Examples
Chemical vapor deposition	A thin coating of a gaseous reactant is created on the substrate surface during this process. It is beneficial in producing pure, stiff and homogenous [5]	Fullerenes, carbon nanotubules, and carbon nanofibers are widely synthesized using this technique
Sol–gel method	Metal alkoxides and metal oxides which act as metal precursors in solution, are condensed, hydrolyzed and thermally decomposed, resulting in a stable sol. By using different methods such as stirring, sonication, and shaking, the precursor is dispersed in solution and then to further create the NPs, the unstable chemical ingredients are separated [77]	Synthesis of TiO_2 NPs using precursors such as $TiCl_3$ and $Ti(OBu)_4$ [119]
Co-precipitation	This utilizes the solvent displacement technique. Initially, the polymer is mixed in solvents like ethanol and acetone and then this polymer solution is added to a non-solvent, where the interfacial stress between the two phases causes the polymers to clump together to form NPs [39]	Metal and metal oxide NPs like ZnO and TiO_2
Molecular condensation	Controlled vaporization of metals and subsequent release of vapors into a container filled with cool inert gas leads to the formation of very fine NPs due to condensation	Copper metal NP synthesis

Table 3 Bottom-up methods of NP synthesis with biological methods

Methods of biosynthesis of NPs	Basic technique	Examples
Bacteria	Bacteria can produce several inorganic compounds intracellularly and as secretory materials [74]. Bacteria secrete enzymes like NADH-dependent reductase into the surrounding environment, reducing metal ions into metal NPs, [150, 167]. The metal ions enter the cell, where enzymes and reducing agents form the NPs in vivo, which then accumulate in the cytoplasm and are extracted following cell lysis	Gold, silver, and cadmium sulfide NPs are synthesized using bacteria [152]
Fungi	Electrostatic interaction between metal ions and negatively charged carboxylate groups causes metal ions to be initially trapped on the surface of fungal cells. The enzymes in the cell wall reduce the metal ions, resulting in the formation of metal nuclei	1. Metal and metal sulfide NPs [111] 2. Chitin can be extracted from the fungal cell wall and used in the production of chitosan NPs
Algae	They have a high content of polymeric molecules, which can be reduced to accumulate heavy metal ions and convert them to malleable forms [89]	Ag and Au NP synthesis is the most common example [36]
Plants	A bioactive ingredient of desired quality can be extracted from plant tissue after an appropriate treatment [80]. Plant extracts can also be combined with metal salts at room temperature to synthesize NPs [98]	A leaf extract of *Polyalthia longifolia* can be used to create Ag NPs [94, 169]
Biomimetic	Proteins, enzymes, cells, viruses, pollen, and waste biomass are used to synthesize NPs	Nano-superstructures like dendrimers have been constructed in vitro via a protein manufacturing process [165]

5.1 Nano-Fertilizers

Nano-fertilizers are much more efficient even in smaller amounts when compared with traditional chemical fertilizers [137]. They can release nutrients as and when required by the plant, hence limiting the conversion of excess amounts of fertilizers from leaking into groundwater, a common method of pollution with traditional fertilizers [106]. These nano-fertilizers provide the plants with increased nitrogen fixation, improved seed germination, amelioration to drought stress, increased seed weight, and increased photosynthesis ability [42]. Ag [82], Zn [151], Cu, Au, Al, and Fe [92] based NPs have shown promising fertilization and plant growth-promoting properties.

5.2 *Nano-Pesticides*

Several NPs have antimicrobial and insecticidal properties making them a more sustainable and promising alternative to chemical pesticides and an economically cheaper alternative to biopesticides [44]. Some examples include TiO_2 NPs whose photocatalytic activity mediated antimicrobial properties control *Xanthomonas perforans* which is the causative agent of tomato spot disease [132]. CuO NPs show insecticidal activity against *Spodoptera littoralis* known as African cotton littworm [19]. Ag NPs show nematicidal activity against *Meloidegyne spp.,* which are known as root-knot nematodes [35]. One of the added advantages of using NPs is that it can be used for targeted delivery of pesticides and herbicides to crop plants, thus reducing the immense environmental contamination caused by the spraying of chemical pesticides. Nanotechnology thus provides a foundation for a new generation of sustainable farming practices.

6 Incorporating Nanotechnology to Improve the Agricultural Landscape Worldwide

Nanoparticles have become an excellent tool to enhance crop production under current climate conditions. Zinc oxide NPs decreased malondialdehyde (MDA) and H_2O_2 accumulation, maintained membrane stability, prevented loss of essential osmolytes and improved N, P, K, and Zn uptake [45, 52, 154]. Silica-based NPs have been found to increase photosynthesis, transpiration, relative water content, and stomatal conductance [155]. Selenium NPs enhance photosynthesis, phenolic content, antioxidant, and osmolyte levels [164].

As a more biocompatible and biodegradable alternative, CNPs have shown promising effects. Chitosan application was first studied by Allan and Hadwiger in 1979, where they demonstrated anti-microbial properties against foreign pathogens and enforced plant immune system. This would eventually lead to developing methods to improve agriculture by incorporating chitosan [87, 161]. Chitosan has been able to induce improved plant defenses in tomato [24] and cucumber [25]. Chitosan can also be used to maintain the freshness of various products like broccoli, which can get spoiled by pathogens like *Escherichia coli* and *Listeria monocytogenes* [12]. Chitosan-mediated delivery of pesticides, fertilizers for balanced and sustained nutrition, herbicides, and micronutrients can be easily carried out due to its useful structural and biological properties [138, 149]. Collectively, these few examples underline the potential of NPs in alleviating drought stress in crop plants and improving the agricultural sector. CNPs have the added advantage of biocompatibility and environmental safety over metallic and inorganic NPs. They can act both as a carrier and an active bio-stimulant, which will help in developing sustainable crop management strategies. Further in this chapter, the synthesis, characterization, and modes of action of CNPs have been described in detail.

7 Methods of Preparation of CNPs

7.1 *Emulsification and Crosslinking*

One of the first methods for the preparation of CNPs intended for drug delivery was through emulsification and crosslinking, using versatile crosslinking reagents such as glutaraldehyde. Using this method, an aqueous chitosan solution is first emulsified in the oil phase to form a water-in-oil (W/O) emulsion, after which the aqueous droplets are stabilized with a suitable surfactant such as Span 80 in toluene as a stabilizer. The two phases are mixed vigorously, followed by sonication. Thereafter, glutaraldehyde is added, which results in crosslinking of the amino groups of chitosan with the aldehyde groups of glutaraldehyde, resulting in the formation of insoluble droplets of CNPs. The CNPs can be recovered through centrifugation, followed by washing with organic solvents, viz., toluene, acetone, and methanol and then dried under vacuum. However, this method is no longer used, since glutaraldehyde was observed to cause overt toxicity and drug integrity issues [127].

7.2 *Reverse Micellization*

This is a technique similar to emulsion and crosslinking, where the droplets are now referred to as reverse micelles (microemulsion). An aqueous phase containing the chitosan solution and glutaraldehyde is taken and is mixed under constant agitation with an organic phase which contains a lipophilic surfactant (sodium bis (ethyl hexyl) sulfosuccinate or cetyl trimethyl ammonium bromide) that is dissolved in an appropriate organic solvent, such as n-hexane. Crosslinking typically occurs at the micellar interface, often leading to efficient and uniform NP synthesis. This results in the formation of reverse micelles where the core of the micelle has the CNPs formed due to crosslinking. The NPs are then purified by precipitation of the surfactant by $CaCl_2$ followed by dialysis and ultimately lyophilization. The size and properties of the NPs can be tuned by adjusting surfactant type, concentration, and the water-to-surfactant ratio. This method allows for size-controlled chitosan NP production with potential applications in drug delivery and other fields [22, 85, 129].

7.3 *Phase Inversion Precipitation*

For the synthesis of CNPs, there is a method where an aqueous solution of chitosan is first added to an organic solvent consisting of acetone and methylene chloride (a water-immiscible solvent), which contains the hydrophobic drug, in the presence of a stabilizer (poloxamer or lecithin). The organic phase is then mixed with the aqueous phase under stirring conditions using a high-pressure homogenizer to form an oil-in-water emulsion. This is followed by evaporation of the water-immiscible organic solvent (methylene chloride) from the droplets of mixed organic solution under reduced pressure using a rotavapor. This leads to acetone diffusing out of the droplets into the aqueous phase, decreasing the solubility of chitosan polymers, resulting in their solidification and subsequent precipitation as NPs. An excess quantity of water may be added in order to allow for the complete diffusion of acetone. The NPs are recovered by filtration using a 1.0 mm cellulose ester membrane filter [50].

7.4 *Emulsion-Droplet Coalescence Method*

In the emulsion-droplet coalescence method, two water-in-oil emulsions are used. A mixture of liquid paraffin oil and sorbitan sesquioleate serves as the common phase medium for two emulsions. One consists of a stable emulsion of chitosan along with the drug in the medium and the other consists of NaOH in the same medium. The two emulsions are then mixed under stirring conditions. As a result of mixing and stirring, the droplets from the two emulsions would randomly collide with each other and coalesce and the constituents would thereby come into contact with each other. NaOH serves as a precipitating agent and decreases chitosan solubility, leading to precipitation and formation of CNPs from the droplets. Nanoparticles are then recovered by centrifugation, followed by washing with toluene, ethanol, and water (in order) and then freeze drying [156].

7.5 *Desolvation*

This method is similar to the phenomenon of "salting-out." Here, first a solution of chitosan is prepared in a suitable solvent, followed by the addition of a precipitating agent of greater hydrophilicity than chitosan which competes for the solvent molecules. Upon addition into the aqueous solvent, it removes the water molecules interacting with chitosan polymers and thereby disrupts the solvation shell. As a result, the polymer becomes insoluble and precipitates. Sodium sulphate and acetone are common precipitating agents employed in the process. Stabilizers such as polysorbate-80 are used to stabilize the suspension. This process is followed by the addition of glutaraldehyde to further solidify the NPs [81].

7.6 *Ionic Gelation Method*

In this method, crosslinking occurs between oppositely charged functional groups [55]. The chitosan is dissolved in an aqueous acidic solution (example: acetic acid), which causes the amino groups to get protonated. Next, an aqueous solution of a polyanionic compound such as sodium tripolyphosphate (TPP) is added under constant agitation. The positively and negatively charged molecules interact with each other through diffusion and crosslinking occurs between the protonated amino groups of chitosan and the anionic groups of TPP due to electrostatic interactions, causing the chitosan to precipitate as spherical particles. This is followed by centrifugation to recover the nanoparticles, which are washed with water and freeze-dried or oven-dried. It is a less intensive method without involving any harmful crosslinkers or solvents and can be feasibly carried out at room temperature. The final nanoparticle size can be adjusted by controlling the chitosan/TTP ratio [102].

7.7 Ionic Gelation Combined with Radical Polymerization

It is a modified approach of the ionic gelation method where nanoparticles are prepared based on polymerizing acrylic or methacrylic acid, and added to chitosan in aqueous solution. Under stirring conditions, an aqueous solution of an acrylic acid monomer is added to an aqueous chitosan solution. Ionic gelation occurs due to electrostatic interaction between cationic groups of chitosan and anionic groups of acrylic acid monomer which undergoes polymerization in the presence of potassium persulphate under a stream of nitrogen at 6–70 °C. The NP suspension is allowed to settle overnight. The unreacted monomers of acrylic acid are removed through dialysis and subsequent washing steps [73].

7.8 Self-assembly

An emerging technique to prepare chitosan nanoparticles is through the formation of polyelectrolyte complexes (PECs). When a solution of two polyelectrolytes carrying opposite charges is mixed, it leads to the formation of polyelectrolyte complexes. A biopolymer with cationic groups, like chitosan, can react with negatively charged polyelectrolytes giving rise to the formation of polyelectrolyte complexes. Chitosan has been shown to form PECs with natural anionic polymers like alginate, pectin, carrageenan, etc. [136].

7.9 Top Down Method

This method involves the acid hydrolysis of chitin with a strong acid, like hydrochloric acid in the first step, breaking the glycosidic bonds between the sugar moieties. This yields nanocrystals of chitin after removal of the amorphous part through centrifugation and washing. Next, the nanocrystals are treated with an alkali-NaOH which deacetylates the residues (above 60%), leading to the formation of CNPs [160].

7.10 Spray Drying

This method makes use of a nanospray dryer. Here, chitosan is dissolved in aqueous acetic acid and this solution is passed through a nozzle in a hot air stream (120–150 °C). As the solution passes through the nozzle, small droplets are formed due to atomization. The solvent in the small droplets undergoes instantaneous evaporation when it comes in contact with the hot air stream, leading to the formation of CNPs [121].

7.11 Supercritical CO_2-Assisted Solubilization and Atomization

The main feature of this process is to replace dilute acetic acid with water that has been acidified by pressurized CO_2. Here, an aqueous solution of chitosan is prepared by dissolving chitosan powder in distilled water under stirring conditions. Then the vessel is filled with CO_2 under pressure. This causes acidification of the solution causing dissolution of the chitosan powder in water. The dissolution phase is fairly long-lasting, about 48 h. Pressure and temperature are tightly controlled to enhance the dissolution of CO_2 in water. It is then forced through the spraying nozzle into a fluidized bed dryer which leads to atomization and formation of droplets, followed by elimination of water from the droplets, resulting in the formation of CNPs. A filter positioned on top of the fluidized bed collects the CNPs. Long processing time and formation of relatively larger NPs are some of the drawbacks of this method [71].

8 Effect of Drought Stress on Plants

Plants encounter a plethora of biotic and abiotic stresses while growing in the natural environment. Abiotic stress factors are non-living environmental factors that can have detrimental effects on plant growth and productivity [86, 105]. Drought, temperature fluctuations, high soil salinity, toxic heavy metals, and oxidative stress are some of the key abiotic stresses faced by plants leading to hindered growth, decreased yield, altered metabolism, and mutations in the subsequent generations [140]. Abiotic stresses can also trigger a cascade of molecular responses, leading to the impaired organization and dynamics of the cytoskeleton, changes in cell wall structure, protoplasm shrinkage, and disruption of mitochondrial membrane potential.

8.1 Importance of Water Availability and Impact of Drought

Considering a substantial decline in water resources along with the increasing global population and consequently increased food demand, sustainable food production is greatly threatened by abiotic stress, primarily drought, which is becoming more unpredictable and severe due to global warming and changing climatic conditions [109, 125]. Drought stands out as a primary culprit behind crop yield reduction on a global scale among all abiotic stress factors (Begna 2020). Drought is caused due to a consistent and prolonged deficiency of freshwater available to an ecosystem. The fact that almost 80–95% of the fresh biomass of the plant body consists of water emphasizes the key role played by water availability in controlling various aspects of plant growth, development, and metabolism [1]. Drought is a multifaceted stress considering its erratic nature and unpredictable nature of spread. It depends on various factors such as uneven distribution of rainfall, evapotranspiration, and water holding capacity around rhizosphere [133].

8.2 *Types of Drought*

Drought can primarily be classified into three types, viz., (i) meteorological drought wherein precipitation deficit relative to long-term averages, frequently compared to elevated temperature, cause water scarcity, (ii) agricultural (crop) drought occurs when the meteorological deficit translates into reduced soil moisture and impaired crop growth, and (iii) hydrological drought that lower river flow, reservoir levels, and groundwater tables [99]. Apart from environmental factors, in certain cases, the physiological status of the plant can also affect drought such as in a condition called pseudo-drought, where the plant is unable to uptake water from the soil, despite the presence of sufficient moisture in the rhizosphere [37].

8.3 *Physiological Effects of Drought*

An optimal level of water availability is necessary for plant growth and development, since it plays a critical role in photosynthesis, nutrient transport, and overall plant turgor. Changes in soil moisture from the optimal affect grain yield and quality in crops. Water scarcity results in low soil moisture and low water potential in the aerial parts of the plant. When this occurs, the rate of water loss through evapotranspiration exceeds the rate of water uptake by the roots. This results in the plant adopting measures such as root expansion and closure of stomata. Drought stress in plants leads to a lower photosynthesis rate due to lower CO_2 diffusion as a result of stomatal closure, osmotic stress and water deprivation. The impact of drought on crop yield can be substantial, with global yield losses estimated at approximately 10% during the period 1964–2007 [60]. Limitation of water availability within the plant system causes severe alterations in physiological processes and metabolism, resulting in retarded or a complete cessation of growth.

One of the primary effects of drought is impaired growth and development. Water shortages during germination can delay or inhibit germination. Cell growth is dependent on turgor pressure and thus it becomes one of the most drought-sensitive processes [166]. Under conditions of water deficit, interruption of water flow from xylem to adjacent elongating cells hampers cell elongation and expansion, hindering mitosis and resulting in reduced growth. Water deficiency is also shown to reduce the number of leaves per plant and leaf size, as well as induce early senescence and abscission. Leaf rolling, aerial stunting, yellowing of leaves, leaf scorching, and permanent wilting are some of the symptoms of drought stress in plants [34]. Plant response to drought stress is widely variable and depends on the severity and duration of water deficit, species, and the developmental stage of the plant [144].

8.4 *Root Architecture and Signaling in Response to Drought*

Drought also causes massive changes in root architecture and signaling. It is known that roots can perceive water deficiency in soil and then initiate a signaling cascade to slow down the rate of water loss through stomata. Abscisic acid (ABA), cytokinins, ethylene, and malate are some of the factors implicated in mediating this root-shoot signaling. ABA promotes the efflux of K^+ ions from guard cells, resulting in loss of turgor pressure and closing of the stomata [15, 16]. ABA levels have been shown to undergo a 50-fold rise in response to loss of cell turgidity due to dehydration. Root elongation and expansion during drought give the plant access to water sequestered within deeper layers of the soil. This leads to the observation of a higher root:shoot ratio in plants exposed to water stress [57]. The plant allocates more resources to drive root proliferation while limiting the growth of aerial parts. However, extended periods of drought have been shown to decrease respiration in roots, resulting in an inadequate supply of energy in the form of ATP and can negatively affect root growth [91].

8.5 *Oxidative Stress Induced by Drought*

Environmental constraints, such as drought, cause a perturbation in the balance maintained between the production of reactive oxygen species (ROS) and their breakdown in cells in a steady state. This results in an elevation in the levels of ROS produced as a result of oxygen metabolism. This can lead to oxidative degradation of cellular macromolecules like proteins, lipids, and DNA, which can negatively impact cellular processes [145]. ROS such as superoxide, hydroxyl, and H_2O_2 directly attack lipid membranes, thereby increasing lipid peroxidation. Drought-induced overproduction of ROS increases the amount of malondialdehyde which is the end product of lipid peroxidation of unsaturated fatty acid chains in phospholipids and serves as a marker for oxidative damage [112]. Stomatal closure leading to decreased CO_2 fixation by Rubisco results in the accumulation of ATP and NADPH as they are no longer utilized in the Calvin-Benson Cycle. Thus, the photosynthetic electron transport chain becomes saturated, leading to an accumulation of excess excitation energy in chloroplasts. This causes the reaction centers of PS I and PS II in the thylakoid membrane of chloroplasts to produce reactive oxygen species [158].

8.6 Impact on Photosynthesis and Pigments

Drought stress has a direct and severe impact on photosynthesis by affecting all major constituents of the photosynthetic apparatus, such as membrane electron transport, carbon dioxide fixation, and stomatal control of CO_2 levels. Stomatal closure, which is one of the first responses to water stress, limits the entry of CO_2 into the leaves, thereby reducing the rate of photosynthesis [56]. Photosynthetic assimilation of carbon is decreased and photorespiration is increased. Gaseous exchange is further hampered due to a reduction in leaf size and premature leaf senescence [54]. Drought stress results in the generation of ROS which can hinder chlorophyll synthesis and promote chlorophyll degradation as well as that of other photosynthetic pigments like carotenoids. The changes in levels of plant pigments under water stress cause a change in the color of leaves. It is found that when plants are exposed to drought stress, there are changes in chlorophyll pigments and disruption of the chlorophyll apparatus. The activity of Rubisco is also negatively impacted during water stress [32]. Reduction in cell volume can cause an increase in cytoplasmic viscosity and protein aggregation caused by a rise in solute concentration [108]. Reduced water potential also affects enzyme activity. Thus, the activity of enzymes involved in photosynthesis is also reduced [27]. Plants employ a wide array of defense mechanisms to overcome drought stress and improve their survival. These responses can be clubbed under three major categories: drought escape, drought avoidance, and drought tolerance. Plants undergo considerable physiological and biochemical alterations when they are subjected to water stress.

8.7 Drought Escape

In this mechanism, plants combat drought conditions by avoiding exposure to desiccating conditions. This is achieved through completion of their life cycle in a period of adequate availability of moisture before drought hits them. This includes early flowering, accelerated plant development, quick germination, and seed production before the dry spells [130]. Certain crop plants have been shown to adjust their development cycles using phenological strategies in such a way that the most critical phases of plant development sensitive to the availability of water aligns with periods of reliable water supply [43], for example, wheat may undergo early blooming and accelerated grain filling to complete its life cycle before establishment of drought. Similar strategies are also used by rice and maize, which involve altering the timescale of their growth and maturation phases to coincide with the availability of water. Thus, by synchronizing their reproductive and developmental cycles with water availability, plants are able to mitigate the negative impacts of drought.

8.8 Drought Avoidance

Avoidance is a versatile approach employed by plants growing under conditions of water scarcity to acclimate and adapt to the prevailing conditions. Plants adopt several methods so that water potential is maintained during drought conditions. This is mainly achieved by reducing transpirational water losses through stomatal closure and increasing the rate of water uptake through roots by promoting root growth and reducing shoot growth. Morphological adaptations to avoid desiccation include developing a thick cuticle, reduction in leaf size and number, changes in leaf orientation and trichome production [75].

8.9 Drought Tolerance

At the biochemical level, plants synthesize certain key macromolecules, hormones, and secondary metabolites that participate in the stress response cascade and promote drought tolerance (Table 4).

9 Methods of Application of CNPs

9.1 Foliar Spray

CNPs are most commonly applied as foliar sprays, directly onto plant leaves. CNPs can easily be absorbed by leaves, enter through the stomata and traverse further into the plant body through conducting tissues [2]. This allows rapid nutrient uptake and induction of stress tolerance mechanisms [120]. Foliar applications have been shown to improve relative water content, chlorophyll levels, antioxidant enzyme activity, and osmolyte accumulation (such as proline and soluble sugars), which collectively mitigate drought stress effects in crops like *Salvia abrotanoides* and lettuce [18, 124].

9.2 Soil or Root Drench

CNPs can also be applied to the soil or as a root drench. This method delivers CNPs to the rhizosphere for root absorption to enhance root growth and nutrient uptake.

9.3 Seed Priming

Priming seeds with CNPs prior to planting enhances initial stress tolerance, improving germination rates and vigor in early developmental stages under drought conditions. It can be applied as a thin film covering the seeds. It is a more optimized technique compared to soil treatment as it uses a targeted and controlled delivery approach. A few examples of the various methods of application of CNPs are tabulated (Table 5).

Table 4 Different mechanisms of drought tolerance in plants

Response	Physiology	Mechanism	References
Oxidative adjustment	Production of ROS	Act as secondary messengers in stress signaling	Sharma et al. [145]
Osmotic adjustment	Accumulation and synthesis of osmolytes, ion transport and compartmentalization, water uptake and retention	Maintain turgor pressure and water holding capacity of the cell without hindering normal cell function, preserve protein activity and cell membrane integrity	Wach and Skowron [158]
Antioxidant enzyme production	Synthesis of antioxidant enzymes like peroxidase (POD), superoxide dismutase (SOD), ascorbate peroxidase (APX), catalase (CAT), guaiacol peroxidase (GPX), monodehydroascorbate reductase (MDHAR), glutathione reductase (GR), and dehydroascorbate reductase (DHAR)	Involved in directly scavenging ROS or activating a nonenzymatic defense system that upregulates the levels of antioxidants within cellular compartments	Foyer and Noctor [58]
Phytohormone production	Produced in response to the cues of environmental stress	Participate in signaling cascades, regulate gene expression to produce physiological and biochemical changes within the plant to help it combat stress conditions. Altered root architecture, increased root: shoot ratio, leaf senescence, stomatal closure, change in growth dynamics, ROS amelioration, and adaptation mechanisms	Iqbal et al. [78]

10 Mode of Action of CNPs

Chitosan, a biodegradable and eco-friendly biopolymer has shown potential in improving plant growth and resistance to various abiotic stresses, including drought. CNPs with their small size and large surface area are more effective than bulk chitosan in enhancing plant metabolism and improving stress tolerance [18]. The precise mechanism of action of chitosan in plants is still being elucidated, but evidence suggests that it acts as an elicitor and activates defense responses such as antioxidant enzyme production, phytoalexin synthesis, and cell wall reinforcement.

Table 5 Different methods of application of CNPs

Sl. no	Crop used	Type of CNPs formulation applied	Type of application	Effects observed	References
1	*Hordeum vulgare* L.	CNPs	Foliar and soil application	Improved relative water content, proline levels, and upregulated activity of CAT and SOD. Grain yield enhanced	Behboudi et al. [23]
2	Pearl millet (*Pennisetum glaucum* L.)	Chitosan nanoemulsion	Foliar application	Reduction in stomatal conductance, thereby limiting the transpiration rate and rise in leaf temperature than unsprayed plants	Priyaadharshini et al. [170]
3	Sugarcane	CNPs: S-nitrosoglutathione into CNPs	Foliar application	Mitigated effects of water stress, increased photosynthetic efficiency, increase in root/shoot ratio	Silveira et al. [148]
4	*Catharanthus roseus* (L.)	CNPs DDA: 80% deacetylation	Foliar application	Enhanced growth of water-stressed *C. roseus*, accumulation of H_2O_2 and MDA lowered, activities of CAT and APX enzymes upregulated, an increase in alkaloid content, and upregulation of drought resistance genes	Ali et al. [144]

(continued)

Table 5 (continued)

Sl. no	Crop used	Type of CNPs formulation applied	Type of application	Effects observed	References
5	*Salvia abrotanoides* (Kar.)	CNPs	Soil treatment	Improved relative water content, total chlorophyll, carotenoids, phenol, flavonoids, and compatible solutes. Increase in activity of antioxidant enzymes, decrease in the size of stomatal aperture	Attaran Dowom et al. [18]
6	*Lupinus termis* L.	Chitosan and CNPs	Seed priming	Enhanced drought tolerance response observed, increased photosynthetic efficiency, pigment content, and hormone levels, resulting in improved growth and yield	Bakhoum et al. [21]

Chitosan also enhances resistance by cross-linking cell wall proteins and upregulating pathogenesis-related genes like *β-glucanase and chitinase.* However, further research is needed due to variability across species and conditions [147]. Chitosan effects are mediated by chitin-specific receptors including chitosan-binding proteins and CEBiP, which recognize it as a PAMP to trigger defenses [79]. In *Arabidopsis thaliana*, it activates CERK1, MAP kinases, and receptor-like kinases to induce defense gene expression through CERK1-dependent and independent pathways [135]. Its cationic nature allows interaction with chromatin, potentially influencing gene expression. Chitosan also scavenges ROS through its amino and hydroxyl groups and promotes NO-mediated signaling that enhances defense enzymes such as SOD, CAT, POD, phenylalanine ammonia-lyase (PAL), and polyphenol oxidase (PPO). By activating enzymes like chitinases and β-glucanases and various signaling molecules, chitosan orchestrates complex defense pathways [115, 147]. Thus, its mechanism of action involves interaction of chitosan with chitin-specific receptors such as chitin elicitor binding proteins (CEBiP) and acts as a pathogen-associated molecular pattern (PAMP), triggering pattern recognition receptors (PRRs). This activation stimulates signaling cascades involving MAP kinases, receptor-like kinases (e.g., CERK1), ROS, nitric oxide (NO), hydrogen peroxide (H_2O_2), salicylic acid (SA), jasmonic acid (JA), ABA, and calcium pathways. Chitosan enhances defense by upregulating genes encoding pathogenesis-related proteins and antioxidant enzymes [147]. Collectively, these actions improve stress tolerance and physiological resilience in plants, as shown in Fig. 2.

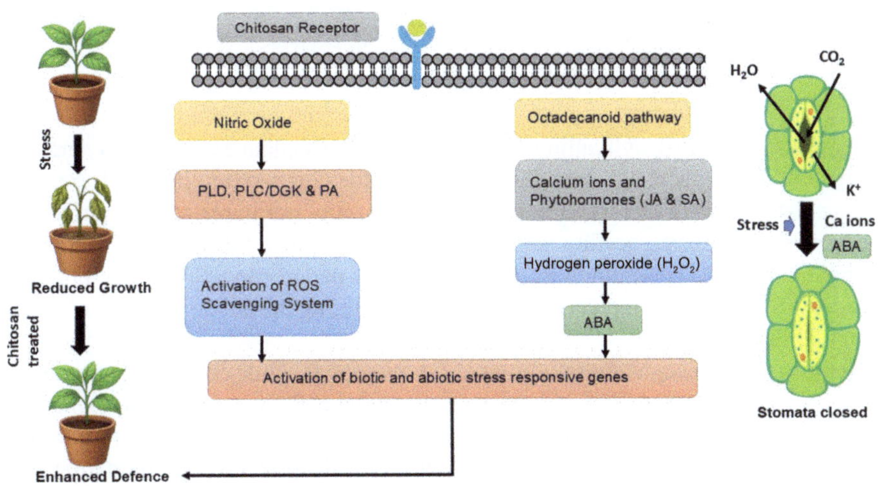

Fig. 2 General mechanism of action of chitosan in alleviating plant stress

10.1 Characterization of Nano-Chitosan/tripolyphosphate (TPP) NPs

A study characterized and evaluated nano-chitosan/TPP NPs and the in vitro biological effects. The NPs exhibited a positive zeta potential, good stability, narrow size distribution, and spherical morphology and crystalline nature. These properties suggested that improved charge, size, and structure of CNPs would result in stronger biological interactions compared to bulk chitosan.

10.2 Dose-Dependent Biological Effects of CNPs

Several biological assays revealed concentration-dependent effects. At lower concentrations, CNPs promoted root and shoot growth, biomass and organogenesis while at higher doses, there was inhibited growth, particularly in the roots. Growth-promoting doses also increased antioxidant enzymes, secondary metabolites like proline, phenols, alkaloids, etc., as well as PAL activity, indicating enhanced defense and elicitor effects. However, the activity of root antioxidant enzymes decreased at high doses, suggesting stress-induced H_2O_2 accumulation. These NPs also demonstrated strong hormone-like activity enhancing organogenesis and biomass in micropropagation at lower doses than bulk chitosan [17].

10.3 Mitigation of Drought Stress by CNPs

Encapsulating S-nitrosoglutathione (GSNO) into CNPs enhances its effectiveness in mitigating the impacts of water deficit in sugarcane compared to its free form. The encapsulated GSNO increased the root-to-shoot ratio and improved the photosynthetic performance, suggesting that the controlled sustained release of NO enhances drought tolerance. These findings demonstrate that nanotechnology can amplify the benefits of NO donors in plants under stress, alleviating the adverse effects of water deficit [148].

In *Vicia faba*, drought stress suppressed shoot growth while enhancing root development—a typical adaptation to improve water uptake. Photosynthetic pigments and IAA levels decreased, there was an increase in oxidative damage due to elevated H_2O_2, MDA, and lipoxygenase (LOX) activity reflecting lipid peroxidation and ROS accumulation. Plants naturally respond by enhancing antioxidant enzymes and accumulating secondary metabolites like glutathione and α-tocopherol, along with osmoprotectants like sugars and proline to mitigate stress damage. The application of chitosan and its NPs alleviated drought stress by improving antioxidant defenses, reducing ROS, and preserving membrane integrity with enhanced osmotic adjustment. CNPs were particularly effective due to greater bioactivity, better cellular uptake, and controlled nutrient release. Chlorophyll and carotenoids were restored, possibly by the activation of genes responsible for pigment biosynthesis and maintaining hormonal balance, particularly IAA levels which support growth under stress. It also reduced H_2O_2 and MDA and suppressed LOX activity along with the enhancement of antioxidants, suggesting that it improved the ROS-scavenging system. Enhanced phenolics, glutathione, and α-tocopherol contributed to stabilizing membranes and quenching of free radicals, while increased soluble sugars and proline helped maintain osmotic balance and protect cellular structures [41].

A study on *Salvia abrotanoides* and *Salvia yangii* demonstrated how drought stress and chitosan application affected the accumulation of phenolic and flavonoid compounds, secondary metabolites and gene expression. Both drought and chitosan individually and interactively influenced the total phenolic and flavonoid contents, with unique responses observed between the species and various plant organs. Phenolic compounds crucial for antioxidant defense, generally increased under mild drought and moderate chitosan application, particularly in *S. abrotanoides.* Chitosan also enhanced phenolic and flavonoid levels by inducing biosynthetic pathways, supported by the upregulation of key genes such as 4-coumarate: CoA ligase (4CL), 1-deoxy-D-xylulose-5-phosphate synthase 2 (DXS2), and tyrosine aminotransferase (TAT), which are involved in the methyl erythritol phosphate (MEP) and phenylpropanoid pathways. Likewise, drought and chitosan increased tanshinone levels underscoring their role in stress adaptation. At the molecular level, chitosan and drought modulated the expression of key genes involved in tanshinone biosynthesis in both *S. abrotanoides* and *S. yangii,* with the MEP pathway emerging as particularly important. Further analysis revealed that optimal chitosan concentrations improved phenolic and tanshinone accumulation while regulating gene expression. However, high doses of chitosan under severe drought stress reduced phenolic content in some tissues, suggesting a trade-off in stress adaptation. These findings highlight the ability of chitosan to act as a stress modulator, increasing antioxidant defenses and secondary metabolism in *Salvia* species while interacting with genetic and environmental factors. This species-specific responses highlight the complex interplay between environmental stress, elicitor treatments and plant metabolic potential [90].

The bioactivity of chitosan oligosaccharides (COS) is strongly influenced by their chemical structure, particularly their degree of polymerization (DP) and degree of acetylation (DA). COS with lower molecular weight are more effective in promoting plant growth and enhancing stress tolerance compared to high-molecular-weight chitosan. N-acetylation influences defense elicitation and growth promotion, with fully acetylated chitin oligosaccharides enhancing defense and fully deacetylated COS hexamers boosting growth through distinct pathways. Chitosan hexamers have been shown to improve photosynthetic parameters, chlorophyll fluorescence, carbon fixation, and carbohydrate accumulation, contributing to higher biomass. While N-acetylation is important for elicitor recognition, growth-promoting effects appear to be stronger with deacetylated COS, possibly due to their positive charge interacting with negatively charged plant cell walls. The exact molecular mechanisms behind these growth-promoting effects remain unclear, though evidence suggests that they involve broad modulation of carbon and nitrogen metabolism [97].

Drought stress reduced sugarcane growth, photosynthesis, and chlorophyll content, while increasing oxidative and osmotic stress indicators such as MDA, proline, and soluble sugars. COS alleviated these effects by reducing stress markers and improving chlorophyll levels, photosynthetic efficiency, and osmotic adjustment. Transcriptome and metabolome analyses revealed that COS modulated key pathways, particularly phenylalanine metabolism, carbon assimilation and also upregulated drought-responsive genes such as pyruvate kinase and PAL. Overall, COS could enhance drought tolerance in sugarcane by improving photosynthesis, regulating stress responses and modulating metabolic and gene expression pathways [163].

Oligo-chitosan prepared from chitosan using γ-irradiation retained the typical structural properties of chitosan with minor variations. Both maintained a similar degree of deacetylation, but oligo-chitosan showed improved antioxidant activity. Foliar application of chitosan and oligo-chitosan on potato plants enhanced growth traits such as shoot height and node number, better membrane stability, improved chlorophyll, carotenoid, proline, and sugar levels with reduced expression of markers of oxidative stress, while antioxidant and defense enzyme activities were elevated. Chitosan and oligo-chitosan NPs could effectively promote plant growth and improve drought stress tolerance by enhancing physiological and defense responses [116].

Foliar-applied chitosan NPs in drought-stressed maize hybrids with contrasting tolerance levels showed maintained water status, chlorophyll content, gas exchange, and photochemical efficiency close to irrigated plants. CNPs also enhanced the activity of antioxidant enzyme like SOD, CAT, APX, and PAL that helped to reduce oxidative damage and improve yield-related parameters by promoting water retention, osmotic adjustment, and efficient allocation of photoassimilates. These effects were possibly due to its positive charge, which caused its interaction with cell components, reducing stomatal conductance and water loss while maintaining photosynthetic activity. Thus, CNPs improved drought resilience, suggesting their potential as biostimulant under water-limited conditions [10].

Applying high molecular weight (MW) CNPs through seed priming and foliar spray to rice seedlings (*Oryza sativa* L.) improved shoot and root growth, maintained water status and reduced oxidative damage by boosting antioxidant enzymes like GPX and APX. It also sustained higher photosynthetic rates and chlorophyll content, likely due to better membrane interaction and signaling. High MW chitosan showed longer-lasting effects than low MW making it an effective strategy to improve drought resilience in rice [113]. Its application improved the physiological parameters like plant growth, relative water content, stomatal conductance, etc. It even enhanced the proline accumulation and increased the activity of antioxidant enzymes like CAT and APX, mitigating the drought-induced oxidative stress. This treatment reduced H_2O_2 and MDA accumulation preserving membrane integrity. Furthermore, CNPs boosted alkaloid content with increased expression of key biosynthesis genes such as strictosidine synthase (STR) and deacetylvindoline-4-O-acetyltransferase (DAT) [144].

Impact of zinc-chitosan-salicylic acid (ZCS NPs) on wheat plants under drought stress significantly improved growth, physiological traits and yield, particularly in drought-sensitive wheat varieties. The nanoparticles enhanced chlorophyll and carotenoid levels, boosted antioxidant enzyme activities, and maintained relative water content, thereby mitigating the negative effects of water stress. ZCS NPs offer an effective and eco-friendly approach to improving drought tolerance, crop yield, and overall plant health, making them a promising tool for sustainable agriculture [38].

Chitosan-coated iron oxide nanoparticles (Fe-CTs NPs) and Kitoplus® growth stimulant were applied on peppermint plants under drought stress. The plants showed improvements in chlorophyll content, photosynthetic activity, and overall growth. The combination of these treatments also enhanced the production of essential oil with certain concentrations, improving the levels of key components like menthone and menthol. These findings highlight the potential of combining biostimulants and nanomaterials to improve plant growth and essential oil production under drought stress [61].

Salvia abrotanoides, a species native to arid regions, has been shown to tolerate water stress due to its ability to withstand harsh climate. The study found that CNPs, particularly at concentrations of 60 and 90 ppm, significantly improved drought tolerance by enhancing physiological traits such as relative water content, carotenoid, phenol, and flavonoid content as well as proline, sugars, and antioxidant enzyme activities. Additionally, there was an improvement in stomatal morphology and chlorophyll content indicating enhanced water use efficiency. These findings suggest that CNPs could be a sustainable and effective alternative to agrochemicals in mitigating drought stress in plants, especially in ornamental-medicinal species like *S. abrotanoides*. Future research should explore the molecular mechanisms and gene expression patterns associated with NP-induced drought tolerance [18]. The various applications of CNPs in drought stress management have been listed in Table 6.

11 Biotic Stressors

The immunomodulatory effects of CNPs in plants were seen to enhance the plant innate immune response with increased defense enzyme activity, upregulation of defense-related genes (including antioxidant enzymes) and an elevation in total phenolic levels. The extracellular localization of CNPs is also suggested to play a role in the upregulation of the defense response. Furthermore, an increase in NO, a key signaling molecule in plant defense was observed following CNP treatment. The immune response was significantly reduced when NO production was inhibited, indicating the involvement of NO in the immune stimulation by CNPs. CNPs can be used as a more effective tool in providing a sustainable approach for disease control in organic farming [29].

Fusarium wilt is a major threat to tomato production, and traditional control methods often rely on chemical treatments, which can have detrimental effects on the environment and human health. Chitosan, a natural biodegradable biopolymer, has shown promise as a plant defense elicitor. However, there has been limited research on its ability to combat fungal pathogens, particularly in the form of nanoparticles. CNPs, due to their small size and high surface area, have greater efficacy compared to bulk chitosan in enhancing plant defense mechanisms. This study focused on evaluating the potential of chitosan and CNPs as biocontrol agents against *Fusarium*

Table 6 Different types of CNPs and their methods of drought tolerance

Name of plant	Type of CNPs used	Mechanism of drought tolerance	References
Sugarcane	CNPs encapsulating S-nitrosoglutathione (GSNO)	Increased root-to-shoot ratio, improved photosynthetic performance, enhanced drought tolerance via controlled NO release	Silveira et al. [148]
Sugarcane	Chitosan Oligosaccharides (COS)	Reduced stress markers, improved chlorophyll levels, photosynthetic efficiency, osmotic adjustment, upregulated drought-responsive genes (pyruvate kinase, PAL)	Yang et al. [163]
Vicia faba	Chitosan and chitosan NPs	Improved antioxidant defenses, reduced ROS, preserved membrane integrity, enhanced osmotic adjustment, restored chlorophyll, carotenoid and IAA levels, reduced H_2O_2 and MDA	Dawood et al. [41]
Salvia abrotanoides and *S. yangii*	Chitosan, chitosan NPs (60 and 90 ppm)	Enhanced tanshinone accumulation, upregulated biosynthetic genes (4CL, DXS2, TAT), stress modulation, enhanced relative water content, carotenoid, phenol and flavonoid content, proline, sugars, antioxidant enzyme activities, improved stomatal morphology, and chlorophyll content, increased water use efficiency	Khodadadi et al., [90] and Attaran Dowom et al. [18]
Potato	Chitosan and oligo-chitosan (γ-irradiated)	Enhanced shoot height, node number, chlorophyll, carotenoids, proline, sugar levels, improved membrane stability, increased antioxidant, and defense enzyme activities	Muley et al. [116]
Maize	Foliar applied CNPs	Maintained water status, chlorophyll content, gas exchange, photochemical efficiency, enhanced antioxidant enzyme activities (SOD, CAT, APX, PAL), improved drought resilience	Almeida et al. [10]

(continued)

Table 6 (continued)

Name of plant	Type of CNPs used	Mechanism of drought tolerance	References
Rice	High MW chitosan NPs (seed priming and foliar spray) and CNPs	Improved shoot and root growth, maintained water status, reduced oxidative damage via GPX and APX, sustained photosynthesis and chlorophyll, increased proline, maintained membrane integrity, increased alkaloid content, upregulated biosynthesis genes (*STR*, *DAT*)	Ali et al. [9] and Moolphuerk et al. [113]
Wheat	Zinc-chitosan-salicylic acid NPs (ZCh NPs)	Improved growth, physiological traits, yield, chlorophyll and carotenoid levels, boosted antioxidant enzyme activities, maintained relative water content, enhanced drought tolerance	Das et al. [38]
Peppermint	Chitosan-coated iron oxide NPs (Fe-CTs NPs) + Kitoplus®	Improved chlorophyll content, photosynthetic activity, essential oil production (menthone, menthol), enhanced drought tolerance	Giglou et al. [61]

wilt disease in tomato caused by *Fusarium andiyazi*. In this study, the researchers applied chitosan and CNPs to tomato plants and analyzed them for the expression of key defense-related genes like the pathogenesis-related (PR) proteins. These proteins are part of the plant immune response and play an essential role in defending against pathogen invasion. The study found that both chitosan and CNP treatments resulted in the upregulation of these defense genes, indicating that the nanoparticles effectively trigger the plant defense system. One of the key findings of this research was that CNPs not only enhanced the expression of these defense genes, but also boosted the activity of antioxidant enzymes which helped reduce oxidative stress in plants during disease conditions. This ability to enhance systemic resistance and mitigate oxidative damage is critical for improving plant health and resilience to biotic stressors. Furthermore, the application of CNPs showed a positive effect on the overall growth and health of the plants. This research paves the way for future studies on the use of CNPs in integrated pest management and plant disease control strategies [33].

Wheat leaf rust, caused by *Puccinia triticina,* is a major global crop disease. Both CNPs and salicylic acid significantly reduced the disease incidence with visible effects such as abnormalities in urediniospore germination, including collapse and shrinkage. The treatments also increased the activities of POD and CAT enzymes and caused more discoloration compared to untreated plants. Anatomical analysis revealed enhanced tissue structure in plants treated with CNPs such as increased mesophyll thickness and epidermis integrity. Transcription of defense-related genes was activated in CNP-treated plants after inoculation, suggesting enhanced plant defense. CNPs could be a promising, eco-friendly approach to mitigate leaf rust in wheat and improve plant health [51].

CNPs can protect chilli plants from *Alternaria* leaf spot disease by enhancing innate immunity, including defense enzyme activity, phenol, and flavonoid levels and NO production [141]. Chitosan/dextran nanoparticles (CDNPs) are known to possess antiviral properties against alfalfa mosaic virus (AMV) in *Nicotiana glutinosa* plants. CDNPs significantly reduced viral accumulation and also enhanced plant immune responses. Thus, CDNPs act as an effective, eco-friendly approach for managing viral diseases in plants [4].

Chitosan-based nanoparticles like CSAg and CSCu improved physiological traits, enhanced secondary metabolite production and increased disease resistance in *Capsicum annuum* plants, offering an eco-friendly approach to enhance plant growth, yield, and protection against pests for sustainable agriculture [107].

Thiamine-loaded CNPs enhance seed germination, growth, indole acetic acid production, and protection against wilt disease in chickpea plants. The nanoparticles also activated defense enzymes and reduced cell death in roots, highlighting their potential as both growth stimulators and defense activators [118]. Copper-chitosan nanoparticles applied to finger millet to control blast disease showed improved plant growth and an increase in yield with the combined seed coat and foliar spray method. The treatment provided protection against blast disease and enhanced defense enzyme activity, which correlated with the suppression of disease development [142].

12 Conclusion

The use of conventional chemical fertilizers poses negative impacts on the environment and human health. Nanoparticles are emerging as a transformative agent in agriculture and are considered to enhance plant growth, nutrient uptake, and confer stress resilience, especially under drought conditions. CNPs are biodegradable and eco-friendly and hence has no toxic effects on soil and crops. They offer biocompatibility, immune system activation, and sustainable crop management potential. CNPs can be prepared using diverse methods, including emulsification, reverse micellization, phase inversion, desolvation, ionic gelation, etc. The various methods vary in their use of crosslinkers, solvents, electrostatic interactions, or polymerization to synthesize the nanoparticles. Techniques also differ in complexity, particle size control, and suitability for drug delivery or agricultural applications. Ionic gelation and self-assembly are safer and environmentally friendly, while some older methods involve toxic reagents or longer processing times. Drought stress adversely affects plant growth, photosynthesis, root-shoot balance, and induces oxidative damage through ROS accumulation. Plants can respond via drought escape, avoidance, and several tolerance mechanisms, including stomatal closure, root proliferation, osmolyte accumulation, and antioxidant enzyme production. CNPs help to mitigate drought effects by enhancing water retention, photosynthetic pigments, antioxidant activity and upregulation of stress-responsive gene expression. CNPs can be applied via various and effectively improve plant resilience and yield under water deficit conditions. Their small size, positive charge and controlled release improve bioavailability, photosynthesis, and hormone balance, promoting growth under water deficit. CNPs and COS regulate secondary metabolism, increasing the production of phenolics, flavonoids, and osmolytes while modulating gene expression in key biosynthetic pathways. Applications across crops like maize, rice, wheat, sugarcane, peppermint, and *Salvia* improve water retention, chlorophyll content, antioxidant activity, and yield. Combined nanomaterial and biostimulant strategies offer sustainable eco-friendly approaches for mitigating drought stress and enhancing plant resilience. Though a substantial knowledge on the mode of action of CNPs has been derived, further transcriptomic and proteomic studies are needed for better understanding and use of CNPs in abiotic and biotic stress management.

References

1. Abbasi T, Abbasi SA (2010) Biomass energy and the environmental impacts associated with its production and utilization. In: Renewable and sustainable energy reviews, vol 14, Issue 3. https://doi.org/10.1016/j.rser.2009.11.006
2. Abdel-Aziz H, Hasaneen MN, Omar A (2018) Effect of foliar application of nano chitosan NPK fertilizer on the chemical composition of wheat grains. Egypt J Bot 0(0). https://doi.org/10.21608/ejbo.2018.1907.1137
3. Abdel Latef AAH, Zayed EM, Omar AA (2025) Bioremediation and nanotechnology for climate change mitigation. Springer Nature, Singapore Pte Ltd. pp 1–652. https://doi.org/10.1007/978-981-96-3069-1
4. Abdelkhalek A, Qari SH, Abu-Saied MAA-R, Khalil AM, Younes HA, Nehela Y, Behiry SI (2021) Chitosan nanoparticles inactivate alfalfa mosaic virus replication and boost innate immunity in *Nicotiana glutinosa* plants. Plants 10(12):2701. https://doi.org/10.3390/plants10122701
5. Ago H (2015) CVD growth of high-quality single-layer grapheme. In: Matsumoto K (ed) Frontiers of graphene and carbon nanotubes: Devices and applications. Springer Japan, pp 3–20
6. Ahlawat J, Masoudi Asil S, Guillama Barroso G, Nurunnabi M, Narayan M (2021) Application of carbon nano onions in the biomedical field: Recent advances and challenges. Biomaterials Science 9(3):626–644. https://doi.org/10.1039/d0bm01476a
7. Alabdallah NM, Hasan MM, Hammami I, Alghamdi AI, Alshehri D, Alatawi HA (2021) Green synthesized metal oxide nanoparticles mediate growth regulation and physiology of crop plants under drought stress. Plants 10(8):1730. https://doi.org/10.3390/plants10081730
8. Ali EF, El-Shehawi AM, Ibrahim OHM, Abdul-Hafeez EY, Moussa MM, Hassan FAS (2021) A vital role of chitosan nanoparticles in improvisation the drought stress tolerance in *Catharanthus roseus* (L.) through biochemical and gene expression modulation. Plant Physiol Biochem 161:166–175. https://doi.org/10.1016/j.plaphy.2021.02.008
9. Ali EF, El-Shehawi AM, Ibrahim OHM, Abdul-Hafeez EY, Moussa MM, Hassan FAS (2021) A vital role of chitosan nanoparticles in improvisation the drought stress tolerance in Catharanthus roseus (L.) through biochemical and gene expression modulation. Plant Physiol Biochem 161. https://doi.org/10.1016/j.plaphy.2021.02.008
10. Almeida LG, Magalhães PC, Karam D, Silva EMD, Alvarenga AA (2019) Chitosan application in the induction of water deficit tolerance in maize plants. Acta Sci Agron 42:e42463. https://doi.org/10.4025/actasciagron.v42i1.42463
11. Altammar KA (2023) A review on nanoparticles: characteristics, synthesis, applications, and challenges. Front Microbiol 17(14):1155622. https://doi.org/10.3389/fmicb.2023.1155622. PMID:37180257;PMCID:PMC10168541

12. Álvarez MV, Ponce AG, Moreira MR (2013) Antimicrobial efficiency of chitosan coating enriched with bioactive compounds to improve the safety of fresh-cut broccoli. LWT—Food Sci Technol 50(1):78–87. https://doi.org/10.1016/j.lwt.2012.06.021
13. Alvarez MV, Ponce AG, Moreira MD (2013) Antimicrobial efficiency of chitosan film enriched with bioactive compounds to improve the safety of fresh cut broccoli. Leb-Wiss und-Technologie 50:78–87. https://doi.org/10.1016/j.lwt.2012.06.021
14. Amendola V, Meneghetti M (2009) Laser ablation synthesis in solution and size manipulation of noble metal nanoparticles. Phys Chem Chem Phys 11(20):3805–3821. https://doi.org/10.1039/b900654k
15. Anjum SA, Wang LC, Farooq M, Hussain M, Xue LL, Zou CM (2011a) Brassinolide application improves the drought tolerance in maize through modulation of enzymatic antioxidants and leaf gas exchange. J Agron Crop Sci 197(3). https://doi.org/10.1111/j.1439-037X.2010.00459.x
16. Anjum SA, Xie X, Wang LC, Saleem MF, Man C, Lei W (2011b) Morphological, physiological and biochemical responses of plants to drought stress. Afr J Agric Res 6:9. https://doi.org/10.21921/jas.5.3.7
17. Asgari-Targhi G, Iranbakhsh A, Ardebili ZO (2018) Potential benefits and phytotoxicity of bulk and nano-chitosan on the growth, morphogenesis, physiology, and micropropagation of *Capsicum annuum*. Plant Physiol Biochem 127:393–402. https://doi.org/10.1016/j.plaphy.2018.04.013
18. Attaran Dowom S, Karimian Z, Mostafaei Dehnavi M, Samiei L (2022) Chitosan nanoparticles improve physiological and biochemical responses of *Salvia abrotanoides* (Kar.) under drought stress. BMC Plant Biol 22(1):364. https://doi.org/10.1186/s12870-022-03689-4
19. Ayoub HA, Khairy M, Elsaid S, Rashwan FA, Abdel-Hafez HF (2018) Pesticidal activity of nanostructured metal oxides for generation of alternative pesticide formulations. J Agric Food Chem 66(22):5491–5498. https://doi.org/10.1021/acs.jafc.8b01726
20. Baig N, Kammakakam I, Falath W (2021) Nanomaterials: A review of synthesis methods, properties, recent progress, and challenges. Mater Adv 2:1821–1871. https://doi.org/10.1039/D0MA00807A
21. Bakhoum GS, Sadak MS, Tawfik MM (2022) Chitosan and chitosan nanoparticle effect on growth, productivity and some biochemical aspects of *Lupinustermis* L plant under drought conditions. Egypt J Chem 65(5). https://doi.org/10.21608/ejchem.2021.97832.4563
22. Banerjee T, Mitra S, Kumar Singh A, Kumar Sharma R, Maitra A (2002) Preparation, characterization and biodistribution of ultrafine chitosan nanoparticles. Int J Pharm 243(1–2). https://doi.org/10.1016/S0378-5173(02)00267-3
23. Behboudi F, Tahmasebi Sarvestani Z, Kassaee MZ, Modares Sanavi SAM, Sorooshzadeh A, Ahmadi SB (2018) Evaluation of chitosan nanoparticles effects on yield and yield components of barley (*Hordeum vulgare* L.) under late season drought stress. J Water Environ Nanotechnol 3(1)
24. Benhamou N, Kloepper J, Tuzun S (1998) Induction of resistance against *Fusarium* wilt of tomato by combination of chitosan with an endophytic bacterial strain: Ultrastructure and cytochemistry of the host response. Planta 204:153–168. https://doi.org/10.1007/s004250050242
25. Ben-Shalom N, Ardi R, Pinto R, Aki C, Fallik E (2003) Controlling gray mould caused by *Botytis cinerea* in cucumber plants by means of chitosan. Crop Pro 22:285–290. https://doi.org/10.1016/S0261-2194(02)00149-7
26. Brunelle T, Chakir R, Carpentier A et al (2024) Reducing chemical inputs in agriculture requires a system change. Commun Earth Environ 5:369. https://doi.org/10.1038/s43247-024-01533-1
27. Cai F, Zhang Y, Mi N, Ming H, Zhang S, Zhang H, Zhao X (2020) Maize (*Zea mays* L.) physiological responses to drought and rewatering, and the associations with water stress degree. Agric Water Manag 241. https://doi.org/10.1016/j.agwat.2020.106379
28. Cao Y, Turk K, Bibi N, Ghafoor A, Ahmed N, Azmat M, Ahmed R, Ghani MI, Ahanger MA (2025) Nanoparticles as catalysts of agricultural revolution: enhancing crop tolerance to

abiotic stress: a review. Front Plant Sci 17(15):1510482. https://doi.org/10.3389/fpls.2024.1510482.PMID:39898270;PMCID:PMC11782286
29. Chandra S, Chakraborty N, Dasgupta A, Sarkar J, Panda K, Acharya K (2015) Chitosan nanoparticles: A positive modulator of innate immune responses in plants. Sci Rep 5(1):15195. https://doi.org/10.1038/srep15195
30. Chandra S, Das P, Bag S, Laha D, Pramanik P (2011) Synthesis, functionalization and bioimaging applications of highly fluorescent carbon nanoparticles. Nanoscale 3(4):1533–1540. https://doi.org/10.1039/c0nr00908a
31. Chandrashekar HK, Singh G, Kaniyassery A, Thorat SA, Nayak R, Murali TS, Muthusamy A (2023) Nanoparticle-mediated amelioration of drought stress in plants: a systematic review. 3 Biotech 13(10):336. https://doi.org/10.1007/s13205-023-03751-4. Epub 2023 Sep 8. PMID: 37693636; PMCID: PMC10491566
32. Chen JQ, Meng XP, Zhang Y, Xia M, Wang XP (2008) Over-expression of *OsDREB* genes lead to enhanced drought tolerance in rice. Biotechnol Lett 30(12). https://doi.org/10.1007/s10529-008-9811-5
33. Chun S-C, Chandrasekaran M (2019) Chitosan and chitosan nanoparticles induced expression of pathogenesis-related proteins genes enhances biotic stress tolerance in tomato. Int J Biol Macromol 125:948–954. https://doi.org/10.1016/j.ijbiomac.2018.12.167
34. Corso D, Delzon S, Lamarque LJ, Cochard H, Torres-Ruiz JM, King A, Brodribb T (2020) Neither xylem collapse, cavitation, or changing leaf conductance drive stomatal closure in wheat. Plant Cell Environ 43(4). https://doi.org/10.1111/pce.13722
35. Cromwell WA, Yang J, Starr JL, Jo Y-K (2014) Nematicidal effects of silver nanoparticles on root-knot nematode in bermudagrass. J Nematol 46(3):261
36. Dahoumane SA, Mechouet M, Wijesekera K, Filipe CD, Sicard C, Bazylinski DA, Jeffryes C (2017) Algae-mediated biosynthesis of inorganic nanomaterials as a promising route in nanobiotechnology–a review. Green Chem 19(3):552–587. https://doi.org/10.1039/C6GC02346K
37. Daryanto S, Wang L, Jacinthe PA (2017) Global synthesis of drought effects on cereal, legume, tuber and root crops production: A review. Agric Water Manag 179. https://doi.org/10.1016/j.agwat.2016.04.022
38. Das D, Bisht K, Chauhan A, Gautam S, Jaiswal JP, Salvi P, Lohani P (2023) Morpho-physiological and biochemical responses in wheat foliar sprayed with zinc-chitosan-salicylic acid nanoparticles during drought stress. Plant Nano Biology 4:100034. https://doi.org/10.1016/j.plana.2023.100034
39. Das S, Srivastava VC (2016) Synthesis and characterization of ZnO–MgO nanocomposite by co-precipitation method. Smart Science 4:190–195
40. Dawood MFA, Tahjib-Ul-Arif M, Shirazy BJ, Abdel Latef AHA. (2025) Unraveling the role of κ-carrageenan on the combined effect of drought and chromium stress in wheat (Triticum aestivium L.). Plant Physiol Biochem 223:J109781
41. Dawood MG, El-Awadi ME, Sadak MS (2024) Chitosan and its nanoform regulates physiological processes and antioxidant mechanisms to improve drought stress tolerance of *Vicia faba* Plant. J Soil Sci Plant Nutr 24(3):5696–5709. https://doi.org/10.1007/s42729-024-01934-3
42. Delfani M, Baradarn Firouzabadi M, Farrokhi N, Makarian H (2014) Some physiological responses of black-eyed pea to iron and magnesium nanofertilizers. Commun Soil Sci Plant Anal 45(4):530–540. https://doi.org/10.1080/00103624.2013.869307
43. Dietz KJ, Zörb C, Geilfus CM (2021) Drought and crop yield. In: Plant Biology, vol 23, Issue 6. https://doi.org/10.1111/plb.13304
44. Dikshit PK, Kumar J, Das AK, Sadhu S, Sharma S, Singh S et al (2021) Green synthesis of metallic nanoparticles: applications and limitations. Catalysts 11(8):902. https://doi.org/10.3390/catal11080902
45. Dimkpa CO, Singh U, Bindraban PS, Elmer WH, Gardea-Torresdey JL, White JC (2019) Zinc oxide nanoparticles alleviate drought-induced alterations in sorghum performance, nutrient acquisition, and grain fortification. Sci Total Environ 688:926–934. https://doi.org/10.1016/j.scitotenv.2019.06.392

46. Downer R, Manteuffel-Ross T, Tymon L, Aghai M (2024) Direct seeding for conifer regeneration in the Pacific Northwest: A past, present, and future of operational practices. In: Journal of Forestry, vol 122, Issue 2. https://doi.org/10.1093/jofore/fvad050
47. Ealia SAM, Saravanakumar MP (2017) A review on the classification, characterisation, synthesis of nanoparticles and their application. IOP Conf Ser: Mater Sci Eng 263:032019. https://doi.org/10.1088/1757-899X/263/3/032019
48. Easwaran C, Moorthy G, Christopher SR, Mohan P, Marimuthu R, Koothan V et al (2024) Nano hybrid fertilizers: A review on the state of the art in sustainable agriculture. Sci Total Environ:172533. https://doi.org/10.1016/j.scitotenv.2024.172533
49. Effects of Drought Stress on Crop Production and Productivity (2020) Int J Res Stud Agric Sci 6(9). https://doi.org/10.20431/2454-6224.0609005
50. El-Shabouri MH (2002) Positively charged nanoparticles for improving the oral bioavailability of cyclosporin-A. Int J Pharm 249(1–2). https://doi.org/10.1016/S0378-5173(02)00461-1
51. Elsharkawy M, Omara R, Mostafa Y, Alamri S, Hashem M, Alrumman S, Ahmad A (2022) Mechanism of wheat leaf rust control using chitosan nanoparticles and salicylic acid. JoF 8(3):304. https://doi.org/10.3390/jof8030304
52. El-Zohri M, Al-Wadaani NA, Bafeel SO (2021) Foliar sprayed green zinc oxide nanoparticles mitigate drought-induced oxidative stress in tomato. Plants 10(11):2400. https://doi.org/10.3390/plants10112400
53. Encyclopedia of Life Sciences (2005) In: Encyclopedia of life sciences. https://doi.org/10.1002/047001590x
54. Faisal S, Mujtaba SM, Asma, Mahboob W (2019) Polyethylene glycol mediated osmotic stress impacts on growth and biochemical aspects of wheat (*Triticum aestivum* L.). J Crop Sci Biotechnol 22(3). https://doi.org/10.1007/s12892-018-0166-0
55. Fan W, Yan W, Xu Z, Ni H (2012) Formation mechanism of monodisperse, low molecular weight chitosan nanoparticles by ionic gelation technique. Colloids Surf B: Biointerfaces 90(1). https://doi.org/10.1016/j.colsurfb.2011.09.042
56. Farooq M, Wahid A, Kobayashi N, Fujita D, Basra SMA (2009) Plant drought stress: Effects, mechanisms and management. In: Agronomy for sustainable development, vol 29, Issue 1. https://doi.org/10.1051/agro:2008021
57. Figueroa-Bustos V, Palta JA, Chen Y, Stefanova K, Siddique KHM (2020) Wheat cultivars with contrasting root system size responded differently to terminal drought. Front Plant Sci 11. https://doi.org/10.3389/fpls.2020.01285
58. Foyer CH, Noctor G (2005) Redox homeostasis and antioxidant signaling: A metabolic interface between stress perception and physiological responses. In: Plant cell, vol 17, Issue 7. https://doi.org/10.1105/tpc.105.033589
59. Gajanan K, Tijare SN (2018) Applications of nanomaterials. Materials Today. Proceedings 5(1):1093–1096
60. Ghazaryan G, König S, Rezaei EE, Siebert S, Dubovyk O (2020) Analysis of drought impact on croplands from global to regional scale: A remote sensing approach. Remote Sens 12(24). https://doi.org/10.3390/rs12244030
61. Giglou MT, Giglou RH, Esmaeilpour B, Azarmi R, Padash A, Falakian M, Śliwka J, Gohari G, Lajayer HM (2022) A new method in mitigation of drought stress by chitosan-coated iron oxide nanoparticles and growth stimulant in peppermint. Ind Crops Prod 187:115286. https://doi.org/10.1016/j.indcrop.2022.115286
62. Gohari G, Farhadi H, Panahirad S, Zareei E, Labib P, Jafari H, Mahdavinia G, Hassanpouraghdam MB, Ioannou A, Kulak M, Fotopoulos V (2023) Mitigation of salinity impact in spearmint plants through the application of engineered chitosan-melatonin nanoparticles. Int J Biol Macromol 224:893–907. https://doi.org/10.1016/j.ijbiomac.2022.10.175
63. Gomes DG, Debiasi TV, Pelegrino MT, Pereira RM, Ondrasek G, Batista BL, Seabra AB, Oliveira HC (2022) Soil treatment with nitric oxide-releasing chitosan nanoparticles protects the root system and promotes the growth of soybean plants under copper stress. Plants 11(23):3245. https://doi.org/10.3390/plants11233245

64. Gorrasi G, Sorrentino A (2015) Mechanical milling as a technology to produce structural and functional bio-nanocomposites. Green Chem 17:2610–2625
65. Gujrati M, Malamas A, Shin T, Jin E, Sun Y, Lu Z-R (2014) Multifunctional cationic lipid-based nanoparticles facilitate endosomal escape and reduction-triggered cytosolic siRNA release. Mol Pharm 11(8):2734–2744. https://doi.org/10.1021/mp500174w
66. Gupta SM, Tripathi M (2012) An overview of commonly used semiconductor nanoparticles in photocatalysis. High Energy Chem 46:1–9. https://doi.org/10.1134/S0018143912010134
67. Hashem AH, Abdelaziz AM, Askar AA, Fouda HM, Khalil AM, Abd-Elsalam KA et al (2021) *Bacillus megaterium*-mediated synthesis of selenium nanoparticles and their antifungal activity against *Rhizoctonia solani* in faba bean plants. J Fungi 7:195. https://doi.org/10.3390/jof7030195
68. Hayat F, Khanum F, Li J, Iqbal S, Khan U, Javed HU, Razzaq MK, Altaf MA, Peng Y, Ma X, Li C (2023) Nanoparticles and their potential role in plant adaptation to abiotic stress in horticultural crops: A review. Sci Hortic 321, Article 112285. https://doi.org/10.1016/j.scienta.2023.112285
69. Heiligtag F, Niederberger M (2013) The fascinating world of nanoparticle research. Mater Today 16:262–271. https://doi.org/10.1016/j.mattod.2013.07.004
70. Hidangmayum A, Dwivedi P, Katiyar D, Hemantaranjan A (2019) Application of chitosan on plant responses with special reference to abiotic stress. Physiol Mol Biol Plants 25(2):313–326. https://doi.org/10.1007/s12298-018-0633-1
71. Hijazi N, Rodier E, Letourneau JJ, Louati H, Sauceau M, le Moigne N, Benezet JC, Fages J (2014) Chitosan nanoparticles generation using CO_2 assisted processes. J Supercrit Fluids 95. https://doi.org/10.1016/j.supflu.2014.08.003
72. Horikoshi S, Serpone N (2013) Introduction to nanoparticles. In: Microwaves in nanoparticle synthesis: Fundamentals and applications. Wiley-VCH, Berlin, pp 1–24. https://doi.org/10.1002/9783527648122.ch1
73. Hu Y, Jiang X, Ding Y, Ge H, Yuan Y, Yang C (2002) Synthesis and characterization of Chitosan-poly(acrylic acid) nanoparticles. Biomaterials 23(15). https://doi.org/10.1016/S0142-9612(02)00071-6
74. Hulkoti NI, Taranath TC (2014) Biosynthesis of nanoparticles using microbes—a review. Colloids Surf, B 121:474–483. https://doi.org/10.1016/j.colsurfb.2014.05.027
75. Hussain S, Hussain S, Qadir T, Khaliq A, Ashraf U, Parveen A, Saqib M, Rafiq M (2019) Drought stress in plants: An overview on implications, tolerance mechanisms and agronomic mitigation strategies. In: Plant science today, vol 6, Issue 4. https://doi.org/10.14719/pst.2019.6.4.578
76. Huy VN, Methela NJ, Al-Azawi TNI, Khan M, Faluku M, Brown A, Lee D, Das AK, Amir R, Lay L, Mun B, Kim Y, Hussian A, Yun B (2025) Fulvic acid-releasing chitosan nanoparticles promote the growth and salt stress tolerance of soybean plants. Physiol Plant 177(3):e70254. https://doi.org/10.1111/ppl.70254
77. Ijaz M, Khan F, Ahmed T, Noman M, Zulfiqar F, Rizwan M, Chen J, Siddique KHM, Li B (2023) Nanobiotechnology to advance stress resilience in plants: Current opportunities and challenges. Materials Today: Bio 22:100759. https://doi.org/10.1016/j.mtbio.2023.100759
78. Iqbal S, Wang X, Mubeen I, Kamran M, Kanwal I, Díaz GA, Abbas A, Parveen A, Atiq MN, Alshaya H, Zin El-Abedin TK, Fahad S (2022) Phytohormones trigger drought tolerance in crop plants: outlook and future perspectives. In: Frontiers in plant science, vol 12. https://doi.org/10.3389/fpls.2021.799318
79. Iriti M, Faoro F (2009) Chitosan as a MAMP, searching for a PRR. Plant Signal Behav 4(1):66–68. https://doi.org/10.4161/psb.4.1.7408
80. Jadoun S, Arif R, Jangid NK, Meena RK (2021) Green synthesis of nanoparticles using plant extracts: A review. Environ Chem Lett 19(1):355–374. https://doi.org/10.1007/s10311-020-01074-x
81. Janes KA, Calvo P, Alonso MJ (2001) Polysaccharide colloidal particles as delivery systems for macromolecules. Adv Drug Deliv Rev 47(1). https://doi.org/10.1016/S0169-409X(00)00123-X

82. Jaskulski D, Jaskulska I, Majewska J, Radziemska M, Bilgin A, Brtnicky M (2022) Silver nanoparticles (AgNPs) in urea solution in laboratory tests and field experiments with crops and vegetables. Materials 15(3):870. https://doi.org/10.3390/ma15030870
83. Joh D-W, Jung T-K, Lee H-S, Kim D-H (2013) Synthesis of nanoparticles using electrical explosion of Ni wire in Pt solution. J Nanosci Nanotechnol 13(9):6092–6094. https://doi.org/10.1166/jnn.2013.7677
84. Joudeh N, Linke D (2022) Nanoparticle classification, physicochemical properties, characterization, and applications: a comprehensive review for biologists. J Nanobiotechnology 20:262. https://doi.org/10.1186/s12951-022-01477-8
85. Kafshgari MH, Khorram M, Mansouri M, Samimi A, Osfouri S (2012) Preparation of alginate and chitosan nanoparticles using a new reverse micellar system. Iran Polym J (Engl Ed) 21(2). https://doi.org/10.1007/s13726-011-0010-1
86. Kappachery S, AlHosani M, Khan TA, AlKharoossi SN, AlMansoori N, AlShehhi SAS, AlMansoori H, AlKarbi M, Sasi S, Karumannil S, Elangovan SK, Shah I, Gururani MA (2024) Modulation of antioxidant defense and PSII components by exogenously applied acetate mitigates salinity stress in *Avena sativa*. Sci Rep 14(1). https://doi.org/10.1038/s41598-024-51302-5
87. Katiyar D, Hemantaranjan A, Singh B, Bhanu AN (2014) A future perspective in crop protection: chitosan and its oligosaccharides. Adv Plants & Agric Res 1:6. https://doi.org/10.15406/apar.2014.01.00006
88. Katiyar D, Hemantaranjan A, Singh B (2015) Chitosan as a promising natural compound to enhance potential physiological responses in plant: a review. Indian J Plant Physiol 20. https://doi.org/10.1007/s40502-015-0139-6
89. Khanna P, Kaur A, Goyal D (2019) Algae-based metallic nanoparticles: Synthesis, characterization and applications. J Microbiol Methods 163:105656. https://doi.org/10.1016/j.mimet.2019.105656
90. Khodadadi F, Ahmadi FS, Talebi M, Matkowski A, Szumny A, Afshari M, Rahimmalek M (2023) Metabolic and transcriptomic approaches of chitosan and water stress on polyphenolic and terpenoid components and gene expression in *Salvia abrotanoides* (Karl.) and S. yangii. IJMS 24(20):15426. https://doi.org/10.3390/ijms242015426
91. Kim I, Chae HM, Choi B (2020) Effects of experimental throughfall exclusion on soil respiration in a continental coniferous stand, South Korea. Forests 11(9). https://doi.org/10.3390/f11090972
92. Kopittke PM, Menzies NW, Wang P, McKenna BA, Lombi E (2019) Soil and the intensification of agriculture for global food security. Environ Int 132:105078. https://doi.org/10.1016/j.envint.2019.105078
93. Kumar R, Singh RK, Dubey PK, Kumar P, Tiwari RS, Oh I-K (2013) Pressure-dependent synthesis of high-quality few-layer graphene by plasma-enhanced arc discharge and their thermal stability. J Nanopart Res 15(4):1556. https://doi.org/10.1007/s11051-013-1556-8
94. Kumar V, Bano D, Mohan S, Singh DK, Hasan SH (2016) Sunlight-induced green synthesis of silver nanoparticles using aqueous leaf extract of *Polyalthia longifolia* and its antioxidant activity. Mater Lett 181:371–377. https://doi.org/10.1016/j.matlet.2016.06.031
95. Kumari SC, Dhand V, Padma PN (2021) Green synthesis of metallic nanoparticles: a review. Nanomaterials:259–281. https://doi.org/10.1016/B978-0-12-822401-4.00022-2
96. Landi L, De Miccolis Angelini RM, Pollastro S, Feliziani E, Faretra F, Romanazzi G (2017) Global transcriptome analysis and identification of differentially expressed genes in strawberry after preharvest application of benzothiadiazole and chitosan. Front Plant Sci 8:235. https://doi.org/10.3389/fpls.2017.00235
97. Li KC, Zhang XQ, Yu Y, Xing RE, Liu S, Li PC (2020) Effect of chitin and chitosan hexamers on growth and photosynthetic characteristics of wheat seedlings. Photosynt 58(3):819–826. https://doi.org/10.32615/ps.2020.021
98. Li X, Xu H, Chen Z-S, Chen G (2011) Biosynthesis of nanoparticles by microorganisms and their applications. J Nanomater 2011:270974

99. Lima RPC, da Silva DD., Pereira SB, Moreira MC, Passos JBMC, Coelho CD, Elesbon AAA (2019) Development of an annual drought classification system based on drought severity indexes. An Da Acad Bras Cienc 91(1). https://doi.org/10.1590/0001-3765201920180188
100. Linh TM, Mai NC, Hoe PT, Lien LQ, Ban NK, Hien LTT, Chau NH, Van NT (2020) Metal-based nanoparticles enhance drought tolerance in soybean. J Nanomater, Article 4056563. https://doi.org/10.1155/2020/4056563
101. Liu G, Shao L, Ge F, Chen J (2007) Preparation of ultrafine chitosan particles by reverse microemulsion. China Particuology 5(6). https://doi.org/10.1016/j.cpart.2007.08.002
102. Liu H, Gao C (2009) Preparation and properties of ionically cross-linked chitosan nanoparticles. Polym Adv Technol 20(7). https://doi.org/10.1002/pat.1306
103. Machac P, Cichon S, Lapcak L, Fekete L (2020) Graphene prepared by chemical vapour deposition process. Graphene Tech 5:9–17
104. Malerba M, Crosti P, Cerana R (2012) Defense/stress responses activated by chitosan in sycamore cultured cells. Protoplasma 249:89–98. https://doi.org/10.1007/s00709-011-0264-7
105. Mareri L, Parrotta L, Cai G (2022). Environmental stress and plants. In: International journal of molecular sciences, vol 23, Issue 10. https://doi.org/10.3390/ijms23105416
106. Mastronardi E, Tsae P, Zhang X, Monreal C, DeRosa MC (2015) Strategic role of nanotechnology in fertilizers: potential and limitations, pp 25–67. https://doi.org/10.1007/978-3-319-14024-7_2
107. Mawale KS, Giridhar P (2024) Chitosan nanoparticles modulate plant growth, and yield, as well as thrips infestation in *Capsicum* spp. Int J Biol Macromol 254:127682. https://doi.org/10.1016/j.ijbiomac.2023.127682
108. McDowell NG, Sapes G, Pivovaroff A, Adams HD, Allen CD, Anderegg WRL, Arend M, Breshears DD, Brodribb T, Choat B, Cochard H, de Cáceres M, de Kauwe MG, Grossiord C, Hammond WM, Hartmann H, Hoch G, Kahmen A, Klein T, Xu C et al (2022) Mechanisms of woody-plant mortality under rising drought, CO_2 and vapour pressure deficit. In: Nature reviews earth and environment, vol 3, Issue 5. https://doi.org/10.1038/s43017-022-00272-1
109. Mishra N, Jiang C, Chen L, Paul A, Chatterjee A, Shen G (2023) Achieving abiotic stress tolerance in plants through antioxidative defense mechanisms. In: Front Plant Sci, vol 14. https://doi.org/10.3389/fpls.2023.1110622
110. Mohamed NG, Abdel-Hakeem MA (2023) Chitosan nanoparticles enhance drought tolerance in tomatoes (*Solanum lycopersicum* L) via gene expression modulation. Plant Gene, 34. https://doi.org/10.1016/j.plgene.2023.100406
111. Mohanpuria P, Rana NK, Yadav SK (2008) Biosynthesis of nanoparticles: technological concepts and future applications. J Nanopart Res 10(3):507–517. https://doi.org/10.1007/s11051-007-9275-x
112. Møller IM, Jensen PE, Hansson A (2007) Oxidative modifications to cellular components in plants. In: Annual review of plant biology, vol 58. https://doi.org/10.1146/annurev.arplant.58.032806.103946
113. Moolphuerk N, Lawson T, Pattanagul W (2022) Chitosan mitigates the adverse effects and improves photosynthetic activity in rice (*Oryza sativa* L.) seedlings under drought condition. J Crop Improv 36(5):638–655. https://doi.org/10.1080/15427528.2021.2000544
114. Moreno-Vega A-I, Gómez-Quintero T, Nuñez-Anita R-E, Acosta-Torres L-S, Castaño V (2012) Polymeric and ceramic nanoparticles in biomedical applications. J Nanotechnol, Article ID 936041. https://doi.org/10.1155/2012/936041
115. Mukhtar Ahmed KB, Khan MMA, Siddiqui H, Jahan A (2020) Chitosan and its oligosaccharides, a promising option for sustainable crop production- a review. Carbohyd Polym 227:115331. https://doi.org/10.1016/j.carbpol.2019.115331
116. Muley AB, Shingote PR, Patil AP, Dalvi SG, Suprasanna P (2019) Gamma radiation degradation of chitosan for application in growth promotion and induction of stress tolerance in potato (*Solanum tuberosum* L.). Carbohyd Polym 210:289–301. https://doi.org/10.1016/j.carbpol.2019.01.056
117. Muñoz-García J, Vázquez L, Cuerno R, Sánchez-García JA, Castro M, Gago R (2009) Self-organized surface nanopatterning by ion beam sputtering. In: Wang ZM (ed) Toward Functional Nanomaterials. Springer, Berlin, pp 323–398

118. Muthukrishnan S, Murugan I, Selvaraj M (2019) Chitosan nanoparticles loaded with thiamine stimulate growth and enhances protection against wilt disease in Chickpea. Carbohyd Polym 212:169–177. https://doi.org/10.1016/j.carbpol.2019.02.037
119. Nagarjuna R, Roy S, Ganesan R (2015) Polymerizable sol–gel precursor mediated synthesis of TiO_2 supported zeolite-4A and its photodegradation of methylene blue. Microporous Mesoporous Mater 211. https://doi.org/10.1016/j.micromeso.2015.02.044.
120. Nguyen Van S, Dinh Minh H, Nguyen Anh D (2013) Study on chitosan nanoparticles on biophysical characteristics and growth of Robusta coffee in green house. Biocatal Agric Biotechnol 2(4). https://doi.org/10.1016/j.bcab.2013.06.001
121. Nguyen TV, Nguyen TTH, Wang SL, Vo TPK, Nguyen AD (2017) Preparation of chitosan nanoparticles by TPP ionic gelation combined with spray drying, and the antibacterial activity of chitosan nanoparticles and a chitosan nanoparticle–amoxicillin complex. Res Chem Intermed 43(6). https://doi.org/10.1007/s11164-016-2428-8
122. Nile SH, Thiruvengadam M, Wang Y, Samynathan R, Shariati MA, Rebezov M et al (2022) Nano-priming as emerging seed priming technology for sustainable agriculture—recent developments and future perspectives. J Nanobiotechnology 20:254. https://doi.org/10.1186/s12951-022-01423-8
123. Niwa T, Takeuchi H, Hino T, Kunou N, Kawashima Y (1993) Preparations of biodegradable nanospheres of water-soluble and insoluble drugs with D,L-lactide/glycolide copolymer by a novel spontaneous emulsification solvent diffusion method, and the drug release behavior. J Control Release 25(1–2). https://doi.org/10.1016/0168-3659(93)90097-O
124. Nurliana S, Fachriza S, Hemelda NM, Yuniati R (2022) Chitosan application for maintaining the growth of lettuce (*Lactuca sativa*) under drought condition. IOP Conf Ser: Earth Environ Sci 980(1). https://doi.org/10.1088/1755-1315/980/1/012013
125. O'Connell E (2017) Towards adaptation of water resource systems to climatic and socio-economic change. Water Resour Manag 31(10). https://doi.org/10.1007/s11269-017-1734-2
126. Oh W-K, Yoon H, Jang J (2010) Size control of magnetic carbon nanoparticles for drug delivery. Biomaterials 31(6):1342–1348. https://doi.org/10.1016/j.biomaterials.2009.11.030
127. Ohya Y, Shiratani M, Kobayashi H, Ouchi T (1994) Release behavior of 5-fluorouracil from chitosan-gel nanospheres immobilizing 5-fluorouracil coated with polysaccharides and their cell specific cytotoxicity. J Macromol Sci, Part A 31 (5). https://doi.org/10.1080/10601329409349743
128. Oliveira HC, Gomes BCR, Pelegrino MT, Seabra AB (2016) Nitric oxide-releasing chitosan nanoparticles alleviate the effects of salt stress in maize plants. Nitric Oxide 61:10–19. https://doi.org/10.1016/j.niox.2016.09.010
129. Orellano MS, Longo GS, Porporatto C, Correa NM, Falcone RD (2020) Role of micellar interface in the synthesis of chitosan nanoparticles formulated by reverse micellar method. Colloids Surf A: PhysChemical Eng Asp 599. https://doi.org/10.1016/j.colsurfa.2020.124876
130. Ortiz J, Hernández-Fuentes C, Sáez PL, Fernández Del Saz N, Cavieres LA (2023) Compound and simple leaf woody species of the Chilean matorral are equally affected by extreme drought. Plant Ecology 224(1). https://doi.org/10.1007/s11258-022-01277-y
131. Ostermann R, Cravillon J, Weidmann C, Wiebcke M, Smarsly BM (2011) Metal–organic framework nanofibers via electrospinning. Chem Commun 47:442–444. https://doi.org/10.1039/C0CC02271C
132. Paret ML, Vallad GE, Averett DR, Jones JB, Olson SM (2013) Photocatalysis: effect of light-activated nanoscale formulations of TiO_2 on *Xanthomonas perforans* and control of bacterial spot of tomato. Phytopathology 103(3):228–236
133. Passioura JB, Angus JF (2010) Improving productivity of crops in water-limited environments. In: Advances in agronomy, vol 106, Issue C. https://doi.org/10.1016/S0065-2113(10)06002-5
134. Patil N, Bhaskar R, Vyavhare V, Dhadge R, Khaire V, Patil Y (2021) Overview on methods of synthesis of nanoparticles. Int J Curr Pharm Res 13:11–16
135. Povero G, Loreti E, Pucciariello C, Santaniello A, Di Tommaso D, Di Tommaso G, Kapetis D, Zolezzi F, Piaggesi A, Perata P (2011) Transcript profiling of chitosan-treated *Arabidopsis* seedlings. J Plant Res 124(5):619–629. https://doi.org/10.1007/s10265-010-0399-1

136. Quiñones JP, Peniche H, Peniche C (2018) Chitosan based self-assembled nanoparticles in drug delivery. In: Polymers, vol 10, Issue 3. https://doi.org/10.3390/polym10030235
137. Rameshaiah GN, Pallavi J, Shabnam S (2015) Nano fertilizers and nano sensors—an attempt for developing smart agriculture. Int J Eng Res Gen Sci 3(1):314–320
138. Rodrigues S, Dionísio M, López CR, Grenha A (2012) Biocompatibility of chitosan carriers with application in drug delivery. J Funct Biomater. 3(3):615–641. https://doi.org/10.3390/jfb3030615
139. Sadak MS, Tawfik MM, Bakhoum, GSH (2022) Role of chitosan and chitosan-based nanoparticles on drought tolerance in plants: probabilities and prospects. In: Santosh Kumar, Madihally SV (eds) Nanomaterial-plant interactions role of chitosan and chitosan-based nanomaterials in plant sciences. https://doi.org/10.1016/B978-0-323-85391-0.00013-7
140. Saharan BS, Brar B, Duhan JS, Kumar R, Marwaha S, Rajput VD, Minkina T (2022) Molecular and physiological mechanisms to mitigate abiotic stress conditions in plants. In: Life, vol 12, Issue 10. https://doi.org/10.3390/life12101634
141. Sarkar A, Chakraborty N, Acharya K (2022) Chitosan nanoparticles mitigate *Alternaria* leaf spot disease of chilli in nitric oxide dependent way. Plant Physiol Biochem 180:64–73. https://doi.org/10.1016/j.plaphy.2022.03.038
142. Sathiyabama M, Manikandan A (2018) Application of copper-chitosan nanoparticles stimulate growth and induce resistance in finger millet (*Eleusine coracana* Gaertn.) Plants against Blast Disease. J Agric Food Chem 66(8):1784–1790. https://doi.org/10.1021/acs.jafc.7b05921
143. Schmidt P, Poels R, Ketner P, Jonkers W (2011) Impacts on biomass, nutrients and water. In: Sustainable management of tropical rainforests
144. Seleiman MF, Al-Suhaibani N, Ali N, Akmal M, Alotaibi M, Refay Y, Dindaroglu T, Abdul-Wajid HH, Battaglia ML (2021) Drought stress impacts on plants and different approaches to alleviate its adverse effects. In: Plants, vol 10, Issue 2. https://doi.org/10.3390/plants10020259
145. Sharma P, Jha AB, Dubey RS, Pessarakli M (2012) Reactive oxygen species, oxidative damage, and antioxidative defense mechanism in plants under stressful conditions. J Bot 2012. https://doi.org/10.1155/2012/217037
146. Shekhawat GS, Mahawar L, Rajput P, Rajput VD, Minkina T, Singh RK (2021) Role of engineered carbon nanoparticles (CNPs) in promoting growth and metabolism of *Vigna radiata* (L.) Wilczek: Insights into the biochemical and physiological responses. Plants 10(7):1317. https://doi.org/10.3390/plants10071317
147. Shinde NA, Kawar PG, Dalvi SG (2024) Chitosan-based nanoconjugates: A promising solution for enhancing crops drought-stress resilience and sustainable yield in the face of climate change. Plant Nano Biology 7:100059. https://doi.org/10.1016/j.plana.2024.100059
148. Silveira NM, Seabra AB, Marcos FCC, Pelegrino MT, Machado EC, Ribeiro RV (2019) Encapsulation of S-nitrosoglutathione into chitosan nanoparticles improves drought tolerance of sugarcane plants. Nitric Oxide-Biol Chem 84:38–44. https://doi.org/10.1016/j.niox.2019.01.004
149. Sinha VR et al (2004) Chitosan microspheres as a potential carrier for drugs. Int J Pharm 274(1–2):1–33. https://doi.org/10.1016/j.ijpharm.2003.12.026
150. Slavin YN, Asnis J, Häfeli UO, Bach H (2017) Metal nanoparticles: understanding the mechanisms behind antibacterial activity. Journal of Nanobiotechnology 15:1–20
151. Song U, Kim J (2020) Zinc oxide nanoparticles: a potential micronutrient fertilizer for horticultural crops with little toxicity. Hortic Environ Biotechnol 61(3):625–631. https://doi.org/10.1007/s13580-020-00244-8
152. Sriram MI, Kalishwaralal K, Barathmanikanth S, Gurunathani S (2012) Size-based cytotoxicity of silver nanoparticles in bovine retinal endothelial cells. Nanosci. Methods 1:56–77
153. Su SS, Chang I (2018) Review of production routes of nanomaterials. In: Brabazon D, Pellicer E, Zivic F, Sort J, Baró MD, Grujovic N, Choy KL (eds) Commercialization of nanotechnologies–a case study approach. Springer International Publishing, Berlin, pp 15–29. https://doi.org/10.1007/978-3-319-56979-6_2

154. Sun L, Song F, Zhu X, Liu S, Liu F, Wang Y, Li X (2021) Nano-ZnO alleviates drought stress via modulating the plant water use and carbohydrate metabolism in maize. Arch Agron Soil Sci 67(2):245–259. https://doi.org/10.1080/03650340.2020.1723003
155. Sutulienė R, Ragelienė L, Samuolienė G, Brazaitytė A, Urbutis M, Miliauskienė J (2022) The response of antioxidant system of drought-stressed green pea (*Pisum sativum* L.) affected by watering and foliar spray with silica nanoparticles. Horticulturae 8(1):35. https://doi.org/10.3390/horticulturae8010035
156. Tokumitsu H, Ichikawa H, Fukumori Y (1999) Chitosan-gadopentetic acid complex nanoparticles for gadolinium neutron- capture therapy of cancer: Preparation by novel emulsion-droplet coalescence technique and characterization. Pharm Res 16(12). https://doi.org/10.1023/A:1018995124527
157. Vares G, Jallet V, Matsumoto Y, Rentier C, Takayama K, Sasaki T, Hayashi Y, Kumada H, Sugawara H (2020) Functionalized mesoporous silica nanoparticles for innovative boron-neutron capture therapy of resistant cancers. Nanomedicine: Nanotechnol, Biol, Med 27. https://doi.org/10.1016/j.nano.2020.102195
158. Wach, D., & Skowron, P. (2022). An overview of plant responses to the drought stress at morphological, physiological and biochemical levels. Pol J Agron 50
159. Wang A, Li J, AL-Huqail AA, AL-Harbi MS, Ali EF, Wang J, Ding Z, Rekaby SA, Ghoneim AM, Eissa MA (2021) Mechanisms of chitosan nanoparticles in the regulation of cold stress resistance in banana plants. Nanomaterials 11(10):2670. https://doi.org/10.3390/nano11102670
160. Wijesena RN, Tissera N, Kannangara YY, Lin Y, Amaratunga GAJ, de Silva KMN (2015) A method for top down preparation of chitosan nanoparticles and nanofibers. Carbohydr Polym 117. https://doi.org/10.1016/j.carbpol.2014.10.055
161. Xing K, Zhu X, Peng X, Qin S (2014) Chitosan antimicrobial and eliciting properties for pest control in agriculture: A review. Agron Sustain Dev 35:569–588. https://doi.org/10.1007/s13593-014-0252-3
162. Yadav TP, Yadav RM, Singh DP (2012) Mechanical milling: a top down approach for the synthesis of nanomaterials and nanocomposites. Nanosci Nanotechnol 2:22–48
163. Yang S, Chu N, Zhou H, Li J, Feng N, Su J, Deng Z, Shen X, Zheng D (2022) Integrated Analysis of Transcriptome and Metabolome Reveals the Regulation of Chitooligosaccharide on Drought Tolerance in Sugarcane (*Saccharum* spp. Hybrid) under Drought Stress. IJMS 23(17):9737. https://doi.org/10.3390/ijms23179737
164. Zahedi SM, Hosseini MS, Daneshvar Hakimi Meybodi N, Peijnenburg W (2021) Mitigation of the effect of drought on growth and yield of pomegranates by foliar spraying of different sizes of selenium nanoparticles. J Sci Food Agric 101(12):5202–5213. https://doi.org/10.1002/jsfa.11167
165. Zan G, Wu Q (2016) Biomimetic and bioinspired synthesis of nanomaterials/nanostructures. Adv Mater 28:2099–2147
166. Zeid IM, Shedeed ZA (2006) Response of alfalfa to putrescine treatment under drought stress. Biologia Plantarum 50(4). https://doi.org/10.1007/s10535-006-0099-9
167. Zhang X, Yan S, Tyagi R, Surampalli R (2011) Synthesis of nanoparticles by microorganisms and their application in enhancing microbiological reaction rates. Chemosphere 82:489–494
168. Taokaew S and Chuenkaek T (2024) Developments of Core/Shell Chitosan-Based Nanofibers by Electrospinning Techniques: A Review. Fibers, 12(3), 26.https://doi.org/10.3390/fib12030026
169. Kumar V and Yadav SK (2009) Plant-Mediated Synthesis of Silver and Gold Nanoparticles and Their Applications. Journal of Chemical Technology & Biotechnology, 84, 151-157.http://dx.doi.org/10.1002/jctb.2023
170. Priyaadharshini M, Sritharan N, Senthil A, Marimuthu S (2019) Physiological studies on effect of chitosan nanoemulsion in pearl millet under drought condition. Journal of Pharmacognosy and Phytochemistry. 8(3):3304–3307.

Chitosan Nanoparticles Under Temperature Extremes

Nayab Ahmad, Sidra Ahmad, Md. Shahinoor Islam, Muhammad Nauman Khan, and Arafat Abdel Hamed Abdel Latef

Abstract Chitosan nanoparticles (CNPs) have been reported to have a multidirectional role in agriculture, particularly increasing the resistance of plants to a variety of stress conditions. This chapter focuses on the behavior, stability, and efficacy of CNPs at both high and low temperatures, which significantly affect physicochemical properties as well as biological activities of CNPs. It looks into the impact of temperature variation on structural stability, kinetics of release, and interaction of CNPs with plant tissues. The chapter also delves into how CNPs help plants combat heat and cold stress by modulating physiological processes, enhancing antioxidant defense systems, and inducing gene expression related to stress. The behavior of CNPs to severe temperatures is important for optimizing their use in different agro-climate zones and to ensure the consistency of the effectiveness of stress management practices. Such insights could inspire the development of stable, climate-resilient nanoparticle compositions for sustainable agriculture.

Keywords Chitosan nanoparticles · Temperature extremes · Thermal stability · Cold stress · Heat stress · Stress tolerance

---

N. Ahmad · S. Ahmad
Institute of Biotechnology and Genetic Engineering, The University of Agriculture, Peshawar, Peshawar, Pakistan

Md. S. Islam
Department of Chemical Engineering, Bangladesh University of Engineering and Technology, Dhaka, Bangladesh

M. N. Khan (✉)
Department of Botany, Islamia College Peshawar, Peshawar, Pakistan
e-mail: nomiflora@uop.edu.pk

A. A. H. Abdel Latef
Faculty of Science, Botany and Microbiology Department, Qena University, Qena, Egypt

A. A. H. Abdel Latef (ed.), *Chitosan Nanoparticles for Plant Resilience Under Abiotic and Biotic Stressors*, Smart Nanomaterials Technology,
https://doi.org/10.1007/978-981-95-6422-4_6

1 Introduction

Chitosan is a naturally occurring linear polysaccharide composed of repeating β-D-glucosamine units, which are unsubstituted in the case of chitin and become available through the deacetylation of chitin, a resource abundant in the exoskeletons of marine crustaceans [1]. It is a cationic polymer formed by the protonation of primary amine groups along its backbone [2]. Additionally, its ability to form gels, films, and nanoparticles without the need for toxic organic solvents makes it an ideal material for advanced biomedical system design [3]. Thiolated chitosan and trimethyl chitosan derivatives have enhanced mucosal adhesive properties, improved permeation across the intestinal membrane, and are environmentally responsive [4].

Regarding plant systems, the notable bioactivity of chitosan, including its biocompatibility, antimicrobial properties, and film-forming ability, makes it an appealing candidate for the agricultural sector, especially under abiotic stress conditions. Since then, polymeric nanoparticles such as chitosan nanoparticles (CNPs) have been used as agrochemicals, plant growth regulators, and carriers of genetic material. Their mucoadhesive properties and ability to traverse biological mucosal barriers can be utilized to enhance foliar uptake, root absorption, or targeted delivery into specific plant tissues [5]. Their antimicrobial activity can also protect against pathogens, particularly when plant defenses are weakened by temperature extremes. Additionally, their structural flexibility allows CNPs to be engineered for responsiveness to environmental stimuli in plants, such as oxidative stress or pH changes in the rhizosphere [6].

Importance of Nanoparticle Stability in Diverse Environments

The use of chitosan in nanotechnology has led to many nanoscale chitosan's, making it a highly promising delivery platform [7]. Chitosan-based carriers can facilitate translocation across epithelial layers, entry into intracellular compartments, and provide sustained or controlled release profiles for a wide range of therapeutic agents, from small molecules to proteins and nucleic acids, with particle sizes reduced to the nanometer scale [8]. However, these advantages heavily depend on the physicochemical stability of the nanoparticles throughout their lifecycle, from synthesis and storage to final administration.

As dynamic entities, nanoparticles are inherently sensitive to their environment. Changes in pH, ionic strength, shear forces, and temperature can significantly affect their size, shape, surface charge, and colloidal stability. [9]. Among these, temperature is a key parameter that can cause physicochemical changes in the polymer matrix, hydrate layers, and disrupt hydrogen bonds, promoting hydrolytic or oxidative degradation processes. They may lead to undesirable outcomes such as aggregation, precipitation, or premature release of payloads [10].

Practically, the instability of chitosan nanoparticles under changing temperature conditions can lead to a loss of their colloidal integrity and therapeutic effectiveness during storage, transportation, and clinical use [11]. For example, structural rearrangements of nanoparticles stored improperly can affect drug encapsulation efficiency and release kinetics [12]. Additionally, their behavior was modulated in vivo such that exposure to body temperatures or febrile conditions may alter biodistribution, cellular uptake, and ultimately their therapeutic potential. [13].

Therefore, understanding how the stability landscape of chitosan nanoparticles varies with environmental stimuli is essential for formulation development [14]. Retention of structural and functional attributes, along with the maintenance of regulatory standards for quality, safety, and efficacy of biomedical products, is guaranteed. [15]. Consequently, a comprehensive investigation is needed into the thermal, mechanical, and chemical robustness of chitosan-based nanosystems under various ambient and physiological conditions [16].

In agriculture, the stability of chitosan nanoparticles under fluctuating temperature conditions is crucial for effective field deployment. When applied in open-field or greenhouse settings, CNPs must also withstand large diurnal temperature changes and seasonal variations. Instability under such conditions over time can lead to particle fusion, decreased bioavailability of the encapsulated nutrients or pesticides, and loss of functionality during transportation or storage [17]. Indeed, for example, nanoparticles that degrade too quickly could lead to inefficient delivery of nutrients such as fertilizers or phytohormones needed by plants for growth. Therefore, maintaining NP structure in various agricultural environments is essential for optimizing the effectiveness of NPs in plant-based applications like stress management, crop protection, and nutrition [18].

2 Rationale for Studying Temperature-Induced Effects

There are theoretical benefits and practical evidence in various applications that support considering the influence of temperature on chitosan nanoparticles [19]. Among the most frequent or chronic stress nanomaterials that will endure are temperature changes caused by weather fluctuations during storage or temperature variations during transportation. Khan et al. [20] (i.e., due to accidental exposure to hypothermia or hyperthermia, or intentionally exposing materials to high temperatures for their use as hyperthermic treatments [21]. All these scenarios pose challenges to the stability and performance of the chitosan formulations [22].

Chitosan depolymerizes, undergoes thermal oxidation, and experiences conformational changes that decrease intermolecular interactions and weaken nanoparticle structure at higher temperatures. [23]. These processes may cause [24] a decrease in the mechanical strength, an increase in the polydispersity of particles, or an increase in surface charge density. On the other hand, low temperature, especially freeze–thaw cycles, can lead to the generation and enlargement of ice crystals, destroying the organizational structure of nanoparticles and subsequent irreversible aggregation or drug release [25].

These thermal behaviors are further analyzed at the nanoscale to offer valuable insights into the thermodynamic and kinetic stability of chitosan systems [26]. Furthermore, it finds thresholds where damage is irreversible, which enables the development of mitigation strategies [27]. Such methods of delivering cryoprotectants, stabilizing agents, and crosslinkers can also be utilized. A third and crucial observation is the difference caused by changes in temperature and the related onset, which is important for the rational design of a thermoresponsive drug delivery system that involves temperature gradients or externally applied heat as a release mechanism at the target site [28]. In a translational context, developing robust thermal sonication of chitosan nanoparticles enhances the predictability and reproducibility of nanoparticle performance in applications [29]. Consequently, temperature-triggered impact may not only be an academic exercise but also a vital milestone in the preclinical and clinical development roadmap for chitosan-based nanomedicines [30].

Understanding the temperature-dependent behavior of chitosan nanoparticles is directly relevant to plant stress physiology and precision agriculture. Temperature stress, especially heatwaves and frost, can disrupt plant metabolic balance and immune defenses, making targeted nano-interventions highly valuable. Thermally stable CNPs can function as effective delivery systems for antioxidants, heat shock protein inducers, or protective biostimulants, thereby boosting plant resilience under such conditions [31]. Furthermore, utilizing the thermoresponsive properties of modified chitosan systems enables smart, environmentally triggered delivery, releasing protective agents only when plants reach specific thermal thresholds. This responsive feature can greatly reduce input costs and environmental impact in sustainable agriculture [32].

3 Physicochemical Properties of Chitosan Nanoparticles

3.1 Particle Size, Surface Charge, and Morphology

The use of chitosan nanoparticles in biological systems mainly depends on their physicochemical properties. Key factors include particle size, surface charge, and shape, which influence stability, biodistribution, cellular uptake, and drug release kinetics [33].

Nanoparticle systems exhibit a strong dependence on particle size for both in vitro and in vivo performance, making it a key factor in both traditional and emerging technologies [34]. Chitosan nanoparticles normally range from 50 to 500 nm in diameter and depend on the preparative method, molecular weight of the polymer, and degree of deacetylation [35]. Tighter cellular uptake and deeper tissue penetration are common features of smaller particles that can diffuse through tight intercellular junctions and evade phagocytic clearance [36]. Conversely, larger particles may have longer circulation times or tend to accumulate in specific organs due to the enhanced permeability and retention effect [37]. Colloidal stability is also influenced by size, as smaller nanoparticles with a higher specific surface area are more prone to aggregation unless electrostatic forces (charge) or polymer layers (steric forces) prevent them from clumping [38].

Nanoparticle interactions with biological membranes and macromolecules heavily depend on surface charge, which is typically assessed using zeta potential. Due to its polycationic nature, nanoparticles prepared from chitosan exhibit a positive surface potential under acidic to neutral pH conditions. The positive charge on this antiguanidinium facilitates strong interactions with negatively charged cellular membranes, mucosal surfaces, and extracellular matrix components [39]. While surface charge can promote aggregation, excessive charge also encourages aggregation through electrostatic attraction with serum proteins or anionic polymers in the formulation matrix. Generally, an optimal zeta potential value provides electrostatic stabilization by promoting repulsive forces between particles, helping to keep them in suspension during storage and after administration [40].

Chitosan nanoparticles usually have a spherical or near-spherical shape, but irregular shapes can form depending on the synthesis method, drying method, or cargo loaded within the nanoparticles. A general preference for a uniform spherical shape is given to facilitate predictable in vivo behavior by reducing pharmacokinetic variability and maintaining consistent surface area to volume ratios [41]. The chemical surface composition, surface roughness, and porosity of the nanoparticles further influence their interactions with biological systems, including drug adsorption and desorption, as well as immune recognition and clearance rates [42].

In plant systems, the physicochemical properties of chitosan nanoparticles directly affect their interaction with plant surfaces, tissues, and cellular processes. Particle size influences how well they penetrate leaf cuticles and root epidermal layers, which is essential for nutrient or gene delivery. Nanoparticles in the 100–200 nm range are often optimal for effective stomatal uptake and transport through xylem or phloem (Fig. 1) [43, 44]. Surface charge influences binding to plant cell walls, which are negatively charged,therefore, the cationic nature of chitosan improves adhesion and internalization. Additionally, morphology affects stability in soil, retention on leaf surfaces, and compatibility with spray formulations. Under temperature extremes, these factors become even more important because particle aggregation or loss of surface charge can reduce effectiveness in field applications [45].

3.2 Temperature-Sensitive Behavior of Chitosan

Both the semi-crystalline structure and the hydrogen bonding network make chitosan sensitive to temperature. Because of these features' prominence in structural dynamics, it is vulnerable to thermal transitions (such as glass transition and thermal degradation), which then determine nanoparticle integrity and performance [46].

Without plasticizers or salts, the glass transition temperature of chitosan typically ranges from 100 to 150 °C, but values can vary widely based on the degree of deacetylation, moisture content, and other factors. Chitosan remains in a rigid, glassy state below this temperature, where molecular motion is limited [47]. For temperatures above this, the designer polymer chains can become more mobile and soften, leading to shrinking or rearranging the nanoparticle structure. This transition

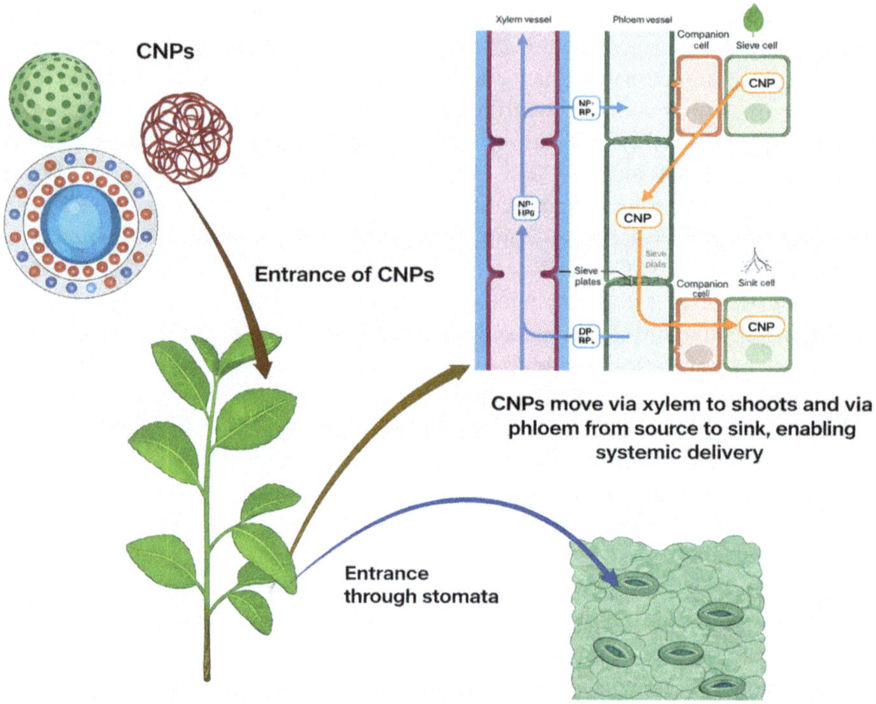

Fig. 1 Schematic representation of the interaction of the CNPs with plants and their movement through plant tissues. When applied to a plant, CNPs first interact with the epidermis and enter through the stomata. Once inside, the CNPs may be transported via xylem and phloem vessels. These vessels facilitate the movement of nanoparticles throughout the plant. Through this pathway, CNPs exert effects on plant growth, defense systems, and stress tolerance by transporting biomolecules or triggering physiological changes. Created with BioRender.com

may also influence release profiles and the diffusion rates of the drug through the matrix [48].

Chitosan generally undergoes thermal degradation at temperatures above 200 °C, leading to depolymerization, deamination, and dehydration reactions that break the polymer backbone. Although these temperatures exceed physiological levels, degradation can occur during extended exposure to less extreme elevated temperatures in processes like manufacturing, sterilization, and storage [49, 50]. The result is decreased molecular weight, changed solubility, and loss of functional groups necessary for bioactivity and mucoadhesion. This leads to thermal stability, an important factor in the formulation and handling of chitosan-based nanoparticles when used in thermally sensitive therapeutic or diagnostic applications [51].

Additionally, in aqueous environments, chitosan exhibits a reversible thermal response, with its solubility and viscosity changing with temperature due to modifications in hydrogen bonding and polymer coil expansion. Temperature-controlled hydrogels or thermoresponsive delivery systems for injection, where shifts in temperature can alter material properties and enable site-specific action, can utilize this thermosensitivity [52, 53].

The thermal sensitivity of chitosan directly affects its application in agricultural settings, where temperature changes are common and unpredictable. In hot regions, nanoparticles may experience phase transitions that influence nutrient release or pesticide effectiveness, while in colder conditions, solubility decreases and systemic absorption in plants slows down [54]. These effects are particularly important when CNPs are used as foliar sprays or soil treatments, since heat waves or frosts can change how they behave. Interestingly, chitosan's thermosensitive properties can also be used to create plant-targeted systems that release encapsulated biostimulants or stress-reducing compounds only at critical temperatures, improving efficiency and resilience in climate-adaptive farming strategies [55].

3.3 Role of Crosslinkers and Formulation Additives on Physical Properties

The structural integrity and functional performance of chitosan nanoparticles can be significantly improved through a commonly used method, namely crosslinking. Crosslinkers can reduce the mobility of chemical polymer chains within the chitosan matrix and increase resistance to environmental stresses such as temperature changes, enzymatic degradation, and mechanical forces by forming covalent or ionic bridges between polymer chains [56, 57].

Chitosan nanoparticles typically form through ionotropic gelation with ionic crosslinkers like tripolyphosphate. Here, electrostatic interactions between negatively charged phosphate groups and protonated amine groups of chitosan create a network that stabilizes the nanoparticle structure [58]. Particle size, porosity, and drug release kinetics can be easily adjusted by changing the ratio and concentration of polymer and crosslinker, allowing control over the extent and density of crosslinking [59].

Along with other agents that create stronger, more permanent bonds, such as covalent crosslinkers like genipin or glutaraldehyde, it enhances and sustains thermal and mechanical stability [60]. Additionally, these crosslinkers can reduce the solubility of chitosan in aqueous solutions, enhancing the durability of the nanoparticles during circulation or over long-term stability [61].

The physicochemical properties of chitosan nanoparticles are also affected by formulation additives. Surface characteristics can be altered, dispersibility increased, and aggregation reduced by using polyethylene glycol, polyvinyl alcohol, and surfactants such as Tween or Span. These same additives may also provide steric stabilization by coating the nanoparticle with a hydrophilic corona, which improves circulation time and decreases nonspecific nanoparticle uptake by the reticuloendothelial system [62, 63].

For freeze-drying or storage under freezing conditions of the chitosan nanoparticles, cryoprotectants such as sucrose, trehalose, and mannitol are required. These agents, however, only freeze the ice shell of the particles in situ during all transition steps, but they leave the ice crystals intact or, to some extent, retain the nanoparticles' structure. In addition, plasticizers, stabilizing salts, and antioxidants might influence the functionality of the nanoparticles under different thermal environments [64].

In plant-based applications, crosslinking methods and formulation additives are necessary to enhance the performance of chitosan nanoparticles due to environmental stresses such as extreme temperatures, UV light, or soil salinity. Ionic crosslinkers like tripolyphosphate enable quick assembly and have low toxicity, making them suitable for agrochemical formulations because of their low cost and non-toxic properties. However, covalent crosslinking with agents like genipin can offer long-term stability for controlled release of nutrients or pesticides [65]. Surfactants and sticking agents like polyethylene glycol improve wetting and retention on leaf foliage when sprayed. Cryoprotectants become especially important for maintaining an effective nanoparticle formulation in moderate climates where sub-zero storage or freezing and thawing are common. These formulation techniques help ensure that such nanoparticles remain stable and active when used in various agricultural settings [66].

A thorough understanding of the relationships between particle size, surface charge, crosslinking, and thermal response in controlling chitosan nanoparticles provides rational guidance for designing plant-targeted nanoformulations. This knowledge will be vital for assessing the stability, efficiency, and safety of nanoparticles across different environments and plant types. For example, such tailored nanoparticle designs for specific crops—such as heat-stable, narrow-diameter CNPs for drought-stressed maize fields or cold-resistant formulations for frost-sensitive horticultural crops—can significantly enhance yields and stress tolerance. Additionally, it enables the development of advanced agricultural nanotechnologies that align with sustainable practices by reducing chemical use while optimizing bioavailability during temperature stresses [67, 68].

4 Thermal Stability Analysis

Thermal stability is an important factor for chitosan nanoparticles used as drug, protein, DNA, flocculating agent, and flocculants. Therefore, a full understanding of the thermal behavior of chitosan nanoparticles is essential to strengthen the robustness of the formulation and predict the product's performance across different thermal environments. Two common methods used to thermally profile chitosan-based NPs are thermogravimetric analysis (TGA) and differential scanning calorimetry (DSC). These methods offer complementary information in the kinetic studies of thermal degradation, polymer matrix transitions, and structural heat effects [69].

In agricultural fields, chitosan nanoparticles should be thermally stable from the time of production to ensure the bioactivity of bioactives, such as fertilizers, pesticides, or plant hormones. Field-use CNPs are often exposed to high temperatures, especially in tropical or semi-arid regions. A composition that degrades or changes behavior upon heating or cooling may be released prematurely or lose effectiveness. TGA and DSC analysis of thermal stability help researchers design CNPs with better field stability, so that when standing crops are treated with nanoparticles through foliar or soil application, or seed coating, the nanoparticles do not decompose. This protects the release kinetics of phytoactives and reduces wastage, supporting sustainable and climate-smart agriculture [70].

4.1 Thermogravimetric Analysis (TGA)

The gravimetric technique that measures changes in a sample's mass over temperature or time in a controlled atmosphere is called TGA. TGA is used to determine the thermal degradation temperatures, moisture content, and residual solvent or crosslinker amount in chitosan nanoparticles [69].

Thermal decomposition of the chitosan backbone is the second major degradation event and occurs between 250 and 350 degrees Celsius. It involves poorly characterized reactions such as the breakdown of glycosidic linkages, deamination, and depolymerization, during which gases like ammonia, acetic acid, and carbon dioxide are released. The onset and peak temperatures of degradation depend on various factors, including the degree of deacetylation, molecular weight, presence of crosslinkers, and the method used to synthesize the nanoparticle [71, 72].

In plant-focused nanotechnology, TGA plays a key role in analyzing CNP formulations for environmental use, especially when used in open-field conditions where temperature shifts are common. Knowing moisture content and degradation points is essential because losing too much water at low temperatures can reduce adhesion to plant leaves, and degradation beyond certain points can weaken the active ingredient. For example, CNPs that carry auxins or micronutrients must stay stable during hot daytime temperatures. Understanding weight loss patterns also helps in choosing stabilizing excipients that can extend shelf life and maintain consistent effectiveness in the field. TGA data thus assist agronomists in creating thermally stable CNPs suitable for various agroecological zones [73, 74].

4.2 Differential Scanning Calorimetry (DSC)

Conversely, DSC provides a free view by measuring heat flow relative to temperature during physical or chemical changes in a material. This method is particularly useful for identifying whether an event is endothermic or exothermic (e.g., glass transition, crystallization, melting, degradation) [75].

DSC also detects one of the most important thermal transitions of chitosan nanoparticles, the glass transition temperature. This is the temperature range where the polymer shifts from a rigid glassy state to a more flexible rubbery phase. When in its glassy state, the chitosan exhibits little chain mobility and high structural integrity; however, in the rubbery state, increased segmental mobility can lead to swelling or relaxation of the polymer matrix [76].

The glass transition temperature of chitosan generally ranges from 100 to 150 °C, depending on the extent of deacetylation, the degree of polymerization, the level of plasticizer, and environmental humidity [77]. This transition is important for understanding the mechanical and diffusional behavior of nanoparticles when subjected to thermal stress. For high ambient temperature shelf life or sterilization processes, this transition temperature must be considered in stability planning. Valuable information about the strength of intermolecular interactions and the degree of structural order in the nanoparticle matrix is provided by the enthalpic effects associated with these transitions.

4.3 Thresholds and Modes of Decomposition

The combination of TGA and DSC data allows for the identification of critical thermal thresholds indicating the effective limits of chitosan nanoparticle systems [76]. Furthermore, thermal analysis decomposition profiles reveal how drug loading, formulation composition, and crosslinking density affect thermal degradation of the drug. Ionically crosslinked particles may decompose at lower temperatures than covalently crosslinked systems because the latter have stronger chemical bonds, resulting in higher thermal stability. Similarly, including thermally labile drugs can decrease the overall thermal stability of the nanoparticles, requiring additional protective mechanisms [78].

Regarding stability requirements, not only for formulation development but also as guideline values for process conditions such as drying, autoclaving, or sterilization, stabilization was proposed [79]. However, stabilization methods should be adapted to types that undergo rapid early degradation, like cryoprotectants, co-polymers, or encapsulation [80].

The decomposition process also needs to be understood to predict shelf life and recommend storage conditions. For example, compositions that start degrading below 100 degrees Celsius are not suitable for tropical environments or can only be safely stored if refrigerated or kept in protective packaging. The stability limits determined through thermal studies are directly useful in developing CNPs for heat-sensitive agro interventions. For example, foliar-applied formulations in warm climates should show dehydration and softening without harming their adhesion or performance. If a formulation breaks down at moderate temperatures, important plant compounds like flavonoids or enzymes may degrade before reaching target tissues. Understanding decomposition patterns also helps in selecting suitable crosslinkers and protective agents to improve stability [17]. This is especially important for high-value horticultural crops, where precise delivery of growth stimulators or pesticides is crucial during temperature fluctuations common in field and post-harvest environments. [18].

5 Effect of High Temperatures

5.1 *Structural Integrity Under Elevated Heat*

Chitosan nanoparticles are highly sensitive to thermal stress. Here, we show that any structural disturbance, from intermolecular hydrogen bonds to macromolecular chain conformation, can occur at high temperatures, leading to a corresponding loss of nanoparticle structural integrity [79]. Several studies have shown the effect of high temperature on chitosan nanoparticles (Table 1).

The chitosan backbone in this polymer is highly charged with hydroxyl and amino groups, which form hydrogen bonds both between and within molecules. The mechanical stability of the polymer and its compact nanoparticle structure depend on these interactions. As the temperature increases, the polymer chain's kinetic energy rises, effectively weakening these non-covalent interactions [79]. When a thermal input exceeds the glass transition temperature of the matrix, a transition from a rigid to a flexible polymer state can occur. This phase change increases polymer chain mobility, which may promote matrix relaxation, swelling, or fusion of particle boundaries [81].

The high surface area to volume ratios of nanoparticle formulations allow heat to penetrate quickly, which eventually damages the particle's structure. However, even moderate heat exposure causes irreversible deformation of the chitosan matrix [79].

In plant applications, it is essential to maintain the structural integrity of CNPs under high temperatures for reliable performance. Agricultural nanoparticles are often exposed to sun-heated leaf surfaces or warm soil conditions exceeding 40 °C. If polymer chain relaxation or matrix swelling occurs, the release of sensitive bioactives may become uncontrolled. For example, CNPs carrying cytokinin or gibberellin might release too quickly during midday heat, disrupting hormonal balance in plants. Additionally, losing particle integrity may impair binding to leaf surfaces, reducing uptake efficiency. Ensuring thermal resistance helps preserve the controlled release function and enhances the consistency of nano-enabled agri-inputs [18, 82].

Table 1 Functional and mechanistic insights into the role of chitosan-based nanoparticles under temperature extremes in plants

S. no	Plant species	Temperature stress	Nanoparticle type	Functional outcome	Mechanism reported	Reference
1	Musa spp. (Safed Velchi Banana)	Low (4 °C)	Chitosan NPs	Increased fresh and dry biomass; improved growth and nutrient concentration	Foliar-applied CNPs reduced oxidative stress (ROS and MDA); enhanced physiological and nutrient status	[83]
2	*Lactuca sativa* (Iceberg lettuce)	Low (4 °C)	Chitosan nanocomposites with Pro, Cys, GB (CHS-Pro NCs, CHS-Cys NCs, CHS-GB NCs)	Improved antioxidant activity, higher phenol and ascorbic acid content; modulation of GSH, sugars	Enhanced SOD isoenzyme activity (FeSOD, Cu/ZnSOD); modulation of phospholipid membrane flexibility; increased antioxidant metabolites (GSH, phenols); altered lipid peroxidation and fresh and dry biomass under chilling stress; nanocomposites more effective than free amino acids	[84]
3	*Fragaria × ananassa*	Low (Chilling stress in open field)	Chitosan oligosaccharide (COS) solution,	Improved vegetative growth and yield; enhanced fruit development	Induction of cold-tolerance-related genes; expression of novel cold-related proteins (e.g., PRPs, specific polypeptides); biomarkers	[85]
4	*Musa acuminata* var. Baxi (Banana)	Low (5 °C for 72 h)	Chitosan nanoparticles	Increased biomass; enhanced growth, nutrient uptake, photosynthetic pigment levels	Reduced oxidative stress via decreased MDA and ROS; increased antioxidant enzyme activity and osmoprotectant accumulation	[68]

(continued)

Table 1 (continued)

S. no	Plant species	Temperature stress	Nanoparticle type	Functional outcome	Mechanism reported	Reference
5	*Gossypium hirsutum*	High (upto 45 °C)	Salicylic acid-functionalized chitosan NPs (SA-NCS),	Improved sucrose metabolism, Rubisco and invertase activity, sucrose phosphate synthase activity	Enhancement of carbon fixation and sucrose translocation; improved source–sink balance under heat stress	[86]
6	*Solanum lycopersicum*	Long-term moderate heat stress (LTMHS) for 10 days during flowering	Chitosan nanoparticles	Improved flowering stage performance; enhanced productivity under heat stress	Increased accumulation of osmoprotectants such as total soluble sugars, phenolics, and flavonoids under heat stress	[31]
7	*Triticum aestivum*	High (37/28 °C)	Nano-sized chitosan-glycinebetaine	Improved growth, grain yield, and grain-filling	Tissue water conservation, osmotic adjustment via glycinebetaine and proline accumulation, ROS scavenging	[87]
8	*Solanum melongena*	High (late-season field heat stress in arid environment)	Chitosan solution	Increased yield, improved growth, and higher recovery under stress	Enhanced proline, glycinebetaine, phenolics, improved linear electron flow and non-photochemical quenching (NPQ)	[88]
9	*Cucumis sativus*	Low (storage at 4 °C)	CEO-loaded chitosan nanoparticles	Reduced fruit rot incidence and severity; improved firmness, water retention	Sustained release of CEO from CSNs; antimicrobial activity against P. drechsleri; enhanced physicochemical stability during cold storage	[89]

(continued)

Table 1 (continued)

S. no	Plant species	Temperature stress	Nanoparticle type	Functional outcome	Mechanism reported	Reference
10	*Capsicum annuum*	Low (wet and cold soil conditions)	Chitosan solution	Improved germination %, faster mean time to germination (MTG), better seedling emergence and growth	Increased chitinase and glucanase activity; potential induction of systemic acquired resistance (SAR) protecting seedlings from fungi	[90]
11	*Diospyros kaki*	Low (4 °C)	Chitosan–Phenylalanine nanoparticles (C-Phe NPs)	Reduced chilling injury, improved firmness, antioxidant capacity and carotenoids	Reduced H_2O_2 and MDA accumulation; enhanced SOD, CAT, APX activities; reduced PPO activity; improved antioxidant and metabolic balance	[91]
12	*Prunus domestica*	Low (1 °C)	Chitosan–Arginine nanoparticles	Reduced chilling injury, weight loss, and firmness degradation; extended shelf life	Increased antioxidant enzyme activity (CAT, POD, APX, SOD); decreased H_2O_2, MDA, electrolyte leakage; increased phenolics, flavonoids, proline	[92]
13	*Phoenix dactylifera*	Low (cold storage postharvest)	Nano-chitosan spray	Reduced weight loss and decay; improved firmness, moisture retention, and shelf life	Maintained moisture content; reduced acidity and chlorophyll degradation; enhanced quality via improved physiological stability	[93]

(continued)

Table 1 (continued)

S. no	Plant species	Temperature stress	Nanoparticle type	Functional outcome	Mechanism reported	Reference
14	*Musa acuminata*	Low (10 °C)	Chitosan nanoparticles	Reduced chilling injury and weight loss; increased firmness, antioxidant activity, and phenolic content	Morphological modification via nanoparticle formation; preservation of tissue integrity and enhancement of antioxidant systems under cold stress	[94]
15	*Fragaria* × *ananassa*	Low (upto 2 °C)	Nano-chitosan coating	Maintained visual quality, firmness, TSS, reduced decay and weight loss; prolonged shelf life	Reduced MDA and PPO activity; preserved antioxidant capacity; better quality retention at 2 °C storage	[95]

5.2 Aggregation, Shrinkage, or Denaturation Behavior

A wide range of physical instabilities in conducting chitosan nanoparticles, caused by thermal stress, such as aggregation, shrinkage, and sometimes protein denaturation of incorporated bioactives like enzymes and antibodies, has also been observed. [96].

The collapse of stabilizing electrostatic or steric forces at high temperatures leads to aggregation. Disruption of hydration layers and a reduction in surface charge weaken the repulsive forces between nanoparticles, allowing Van der Waals attractions to act. This results in a higher polydisperse index and significantly altered biodistribution, especially in intravenous or mucosal delivery routes, due to particle fusion or clumping. Such aggregation in foliar sprays or root applications in plants can reduce uptake efficiency and change bioavailability to plant tissues (Stolte Bezerra Lisboa [97]).

Chitosan nanoparticles can be reduced in size through dehydration or polymer chain retraction caused by the loss of internal water and the softening of the gel-like matrix. This thermal shrinkage during encapsulation may lead to a tighter encapsulation site, but it also reduces drug accessibility, especially for encapsulated hydrophilic compounds. In such cases, the papers discuss how plant science, particularly in soil-based delivery, may be negatively impacted by this shrinkage, potentially decreasing mobility and diffusion rates, and thus making it harder for a drug or nutrient to reach roots [98].

The denaturation of biologically active macromolecules, such as therapeutic proteins, peptides, or plant-based enzymes encapsulated in chitosan nanoparticles, is a significant challenge. These molecules are often sensitive to high temperatures and tend to unfold and aggregate. This results in the loss of their secondary or tertiary structure, which leads to a decrease in bioactivity and therapeutic effectiveness. Monitoring temperature and implementing stabilization strategies are essential in biomedical formulations, including vaccines and protein-based therapeutics, where preserving protein conformation is vital. Additionally, when enzymes or plant extracts are delivered via chitosan nanoparticles for crop enhancement, maintaining functional bioactivity in fluctuating field temperatures is crucial [99].

Aggregation and shrinkage under thermal stress can impede CNPs' dispersal and bioavailability in plant systems. In foliar spray applications, aggregated particles might clog nozzles or fail to stick to leaf surfaces, while shrunken particles could hinder the diffusion of encapsulated compounds into plant tissues. For root-zone delivery, changes in particle size and shape affect soil mobility and uptake by root hairs. Additionally, plant-based bioactives like enzymes or antioxidants encapsulated in CNPs are especially vulnerable to heat-induced denaturation, losing their biological function. Therefore, formulations designed for hightemperature zones should include excipients and cryoprotectants that maintain both structural and functional integrity [100].

5.3 Impacts on Drug Release Kinetics and Encapsulation Efficiency

The release kinetics of chitosan nanoparticles and their encapsulation efficiency are highly influenced by temperature. Heat alters the internal matrix structure, permeability, and diffusion properties of polymers, which are crucial in all controlled release systems. Elevated temperatures increase the segmental mobility of chitosan chains and likely create microvoids within the matrix, facilitating greater diffusion of encapsulated agents. As a result, drug release occurs rapidly, especially when the drug is superficially adsorbed or trapped through weak electrostatic interactions. Such uncontrolled release undermines the goal of nanoparticle-based delivery systems, which are often designed for slow, sustained, or targeted release [101, 102].

Although heat does not directly affect encapsulation efficiency, it influences processes such as spray drying, freeze-drying, or post-synthesis sterilization. During or after nanoparticle formation, partial drug leakage or degradation can occur if elevated temperatures are applied, reducing the loading capacity. This issue is most severe for hydrophilic drugs or plant-derived bioactives like phenolics, flavonoids, or plant growth regulators, which may degrade or volatilize under heat. Additionally, part of the polymer itself can degrade, which not only lowers encapsulation efficiency but also compromises biocompatibility and mechanical stability [103].

Subtherapeutic dosing, alterations in pharmacokinetics, or simply failure to reach the therapeutic window may lead to biomedical issues. Premature release or loss of volatile compounds from the formulation in agricultural applications would make the formulation ineffective; therefore, higher doses or more frequent applications would be necessary, and sustainable factors would decrease while costs increase [17].

To address these challenges, thermal stabilizers are used, along with alternative techniques for encapsulating enzymes (such as coacervation or layer-by-layer deposition) and pre-screening formulations under simulated heat stress conditions. Therefore, the thermal resilience of chitosan nanoparticles can be directly linked to their effectiveness in curing diseases in humans or enhancing plant vitality in agricultural settings [98].

In agricultural nanodelivery systems, temperature-driven changes in release kinetics can cause either premature or delayed action of plant-protective agents. A sudden burst release of herbicides or plant growth regulators during a heatwave may lead to phytotoxicity or ineffective growth regulation (Fig. 2) [104]. Similarly, reduced encapsulation efficiency caused by thermal degradation during processing (e.g., spray drying) can limit the amount of payload available to crops, raising costs, and environmental impact. For biofertilizers or microbial inoculants, maintaining controlled release under field conditions is crucial for aligning availability with plant uptake cycles. Therefore, plant-specific formulations must be tested under simulated thermal stresses to ensure optimal release profiles and encapsulation stability across different agricultural environments [105].

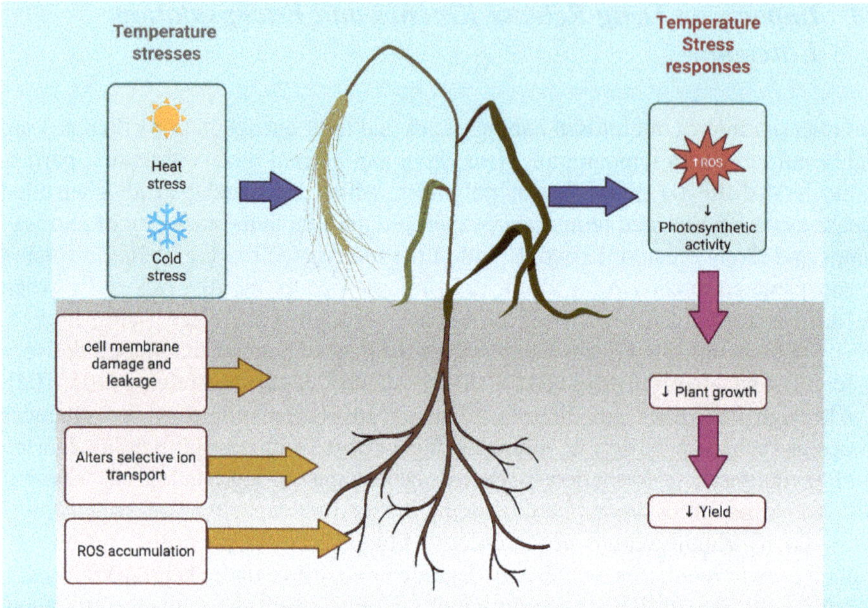

Fig. 2 Adverse effects of high temperature on plant function and overall growth. Heat and cold stress disrupt root and shoot physiology by damaging cell membranes, increasing permeability, reducing water and nutrient uptake, and disturbing metabolic activity. These changes lead to production of reactive oxygen species (ROS), decreased photosynthetic activity, stunted plant growth, and reduced crop yield. Created with BioRender.com

6 Effect of Low Temperatures

Sub-ambient temperatures distort chitosan nanoparticles, just as they do with many other polymeric nanocarriers. Both biomedical and plant product handling processes involve cold chain storage, freeze/thaw cycles, and lyophilization. Although such lowtemperature procedures are often used to preserve bioactivity and extend shelf life, they can also introduce mechanical and chemical stresses that may destabilize nanoparticles [106]. To develop chitosan-based systems suitable for real-world conditions across various climates and storage situations, these cold-induced transitions must be understood.

CNPs are increasingly used in plant systems for targeted delivery of agrochemicals, biostimulants, and phytoactive agents. However, the application and effectiveness of these nanosystems are highly affected by environmental factors, especially temperature changes. Low temperatures, particularly those during freezing events or storage below freezing, cause complex physicochemical stresses that can damage nanoparticle stability, change dispersion behavior, and impact bioavailability. This section systematically examines the different effects of low temperatures on CNPs, with a focus on freeze–thaw cycling, changes in particle size and polydispersity, and preservation through lyophilization with cryoprotectants [107, 108].

6.1 *Freeze–Thaw Stability and Ice Crystallization Effects*

Additionally, biological storage, transportation, or field deployment in agriculture involves high chitosan nanoparticles undergoing repeated freeze–thaw cycles, which impose significant mechanical stress on the nanoparticles. During freezing, water expands and forms crystals that exert force on the surrounding nanoparticle matrix. The resulting mechanical stress fields are uneven and can cause the chitosan shell to burst, leading to leakage of the encapsulated agents or even causing particles to stick together [109].

Ice crystallization is the main cause of damage to CNPs. When water freezes inside or around the nanoparticles, it creates "sharp-edged" crystals that physically shear or compress the polymer matrix. This effect is especially severe in chitosan nanoparticles with high water content or loosely packed polymer chains. Additionally, the formation of extracellular ice can promote osmotic dehydration, pulling water out of the particle core and leading to the shrinking and disruption of the nanoparticle architecture [110].

The extent of mechanical disruption depends on the freezing rate, the duration of the frozen state, and the ionic environment. Nanoparticles stored in agricultural settings experience multiple freeze–thaw cycles at varying temperatures, more common in temperate and subpolar regions. For biomedical uses, which often require freezing during shipping, formulations containing labile biomolecules like enzymes, RNA, and vaccines tend to lose effectiveness after thawing [96, 111]. Additionally, freeze–thaw instability is not always visible on a local level (i.e., on a large scale). Latent instability, which is common in many systems, does not become apparent until reconstitution, leading to aggregation, phase separation, and viscosity changes. These subtle but important changes highlight the importance of rigorous thermal cycling testing during nanoparticle development [109].

In aqueous nanoparticle dispersions, water freezing triggers phase transitions that harm nanoparticulate systems. When temperatures go below freezing, water starts to crystallize, forming ice structures that exclude solutes, including nanoparticles, from the growing crystal matrix. As ice forms, nanoparticles become concentrated in the remaining unfrozen liquid phase, greatly increasing local solute concentrations and ionic strength. This process creates osmotic stress and changes the physicochemical environment around the CNPs [109].

The rising ionic strength may disrupt the electrostatic interaction, which is generally a prerequisite for the agglomeration of particles. In CNPs, these particles are mainly stabilized by protonated amine groups on the chitosan surface, which imparts a positive charge and causes electrostatic repulsion between particles [112]. This stabilization fails during the freezing process, which inevitably leads to collision, coalescence, or irreversible agglomeration of nanoparticles. Additionally, freezing causes the nanoparticle matrix to expand due to ice crystallization as water cools, creating mechanical stress that can cause the polymeric shell to fracture or deform if the matrix is not elastic or sufficiently crosslinked [113].

Freeze-induced nanoparticle damage is critically important in plant systems. If CNPs lose their structural integrity or become excessively aggregated, they may not enter plant tissues or be taken up by cells. For example, to be absorbed through stomata and transported via the apoplast, nanoparticles must retain their nanoscale dimensions and possess suitable surface properties. Damaged particles can be excluded from these pathways, reducing the effectiveness of encapsulated bioactives like phytohormones, micronutrients, or RNA-based bioregulators. Consequently, freeze–thaw stability becomes a key factor in designing CNP formulations for use in temperate and cold climates [114].

6.2 Particle Size and Polydispersity

Cold treatment and multiple cycles of freezing and thawing significantly affect the colloidal properties of CNPs, such as size distribution and polydispersity. These changes occur due to thermodynamically unfavorable interactions during freezing, including ice exclusion, nanoparticle crowding, and surface charge destabilization. When reconstituted, the aggregates formed during freezing can persist, leading to an increase in mean particle size and Polydispersity Index (PDI) [115].

Practically, larger particles will decrease the mobility of CNPs in soil or plant systems. For both root uptake and phloem translocation-targeted nanoparticles, there are hydrodynamic size limits that nanoparticles must stay below to effectively pass through the plant [18]. The average particle size of these agglomerates often exceeds the thresholds, causing particles to become physically entrapped in the extracellular matrix or to fail in effectively adhering to foliar surfaces. Additionally, polydispersity affects the uniformity of bioactive compound release. In monodisperse populations, release profiles are predictable and tunable, whereas size heterogeneity introduces variability in both diffusion rates and degradation rates [116].

6.3 Lyophilization (Freeze-Drying) and Cryoprotectants

Lyophilization allows chitosan nanoparticles to maintain their stability over an extended period, with water removed under vacuum. The process involves three stages: freezing, primary drying (sublimation of ice), and secondary drying (removal of adsorbed water). The system is quickly frozen to sub-zero temperatures to solidify the aqueous solvent and concentrate the solutes. This is followed by sublimation under low pressure, converting ice directly to vapor without melting, which helps prevent water-catalyzed degradation reactions. The final desorption stage eliminates remaining moisture, resulting in a dry, stable powder [11].

Although lyophilization effectively preserves bioactive molecules and prevents microbial degradation, it exposes the nanoparticles to significant mechanical and thermal stresses, which may destabilize their structure [79]. The use of cryoprotectants is essential to prevent the latter. Cryoprotectants, including disaccharides such as trehalose and sucrose, can stabilize CNPs through multiple mechanisms [107]. Furthermore, they hinder the growth of ice crystals by forming super-viscous amorphous solutions and by reducing the mechanical stress during freezing [117]. Secondly, they can form hydrogen bonds with chitosan chains and replace water in the polymer hydration shell, preventing polymer shrinkage or denaturation. Third, they act as a matrix-forming agent during drying, embedding the nanoparticles in the protective glassy matrix, thereby preserving the size and surface properties of the nanoparticles [118].

Reversing the lyophilized state to the hydrated form, lyophilized CNPs cryoprotected reveal slight variations in particle size, zeta potential (ZP), and encapsulation efficiency (EE), justified by suitable reconstitution and application [119]. This preservation is especially advantageous in plant systems where constant reactivity and interaction with plant surfaces are required for the bioactive to be delivered. Furthermore, lyophilized products improve logistical demands by providing a product that can be stored and shipped at ambient temperature while retaining its function [120].

In field agriculture applications, lyophilized CNPs can be quickly reconstituted and dispersed through irrigation systems, applied by foliar spray, or incorporated into seed treatments. The maintained colloidal stability ensures uniform distribution and site-specific delivery, even after long-term storage. Additionally, this strategy allows for formulation scaling and field deployment, thereby expanding the use of CNPs to areas with varying cold exposure [17].

7 Mechanisms of Temperature-Induced Changes

Polymeric nanoparticles based on chitosan are highly sensitive to temperature due to the thermo-response behavior of the polymer: formation of dense and collapsing particles at higher temperatures. They also have a high content of functional groups and rely on hydrogen bond-dependent organization. In this section, we demonstrate that CNPs undergo dynamic internal reorganization at the molecular level, which results in modified morphology, mechanical response, and payload interactions. Since these properties significantly influence the biological behavior of the nanoparticles, it is crucial to understand the molecular mechanisms behind their temperature dependence.

7.1 Effect of Temperature on Proton Transfer and Conformational Rearrangement

Chitosan, a linear polysaccharide, is obtained through the deacetylation of chitin, which has a backbone of D-glucosamine units linked by β-(1 → 4)-bonds and contains varying amounts of N-acetyl-D-glucosamine [121]. This molecular structure is rich in hydroxyl and primary amine groups, with the latter involved in intra- and intermolecular hydrogen bonds [122]. These hydrogen bonds are primarily responsible for the stability of the secondary and tertiary structures of chitosan-based nanoparticles, which is especially important for their function in physiological and environmental conditions ideal for agricultural applications [18].

The stability and reactivity of CNPs within plant environments depend closely on temperature-induced changes in hydrogen bonding networks [123]. The hydrogen bonds become stronger at relatively lower temperatures due to reduced molecular motion and longer donor–acceptor interaction times. This stabilization may be reflected as a higher degree of crystallinity and/or as a stiffening of the chitosan matrix, which could impair the nanoparticle's dynamic properties and slow down the release of encapsulated compounds. For example, a less flexible matrix might reduce the permeability of phytohormones, minor nutrient ions, or antimicrobials in plant tissues, leading to a delay or decrease in their activity [124].

Conversely, high temperatures involve transferring thermal energy to the polymer network, which breaks, either partially or completely, such hydrogen bonds [125]. This structural disruption allows the molecules to change shape and move flexibly within the polymer chains. A less crosslinked chitosan network tends to be more amorphous and may exhibit increased permeability to water and solutes, leading to faster release of entrapped compounds. However, excessive thermal disturbance can also cause the collapse or denaturation of the chitosan framework, especially when there are no reinforcing agents or crosslinkers present [126].

7.2 Temperature-Mediated Modulation of Chitosan Amine Group Protonation

One of the most chemically important features of chitosan is the presence of free primary amine groups. These groups not only give the polymer pH-responsive solubility but also affect the electrostatic behavior of CNPs. These amino groups have a pKa usually between 6.2 and 6.5, which shows that their level of protonation—and thus the polymer's solubility and surface charge—depends heavily on the surrounding pH and temperature [127].

Under lowtemperature conditions, especially in acidic environments, the system's kinetic energy decreases, which favors protonation of the amine groups. Protonated amines increase the cationic nature of the nanoparticle surface, boosting electrostatic interactions with negatively charged plant cell wall components such as pectin, cellulose, and hemicellulose [128]. This condition encourages strong adhesion of CNPs to root or leaf surfaces and helps entry through the apoplastic route or by endocytosis. Increased positive charges under these conditions can also enhance the stability of the dispersion by boosting particle-to-particle (or repulsive) interactions, thus reducing aggregation [50].

Several physicochemical changes, however, occur when the temperature increases. First, deprotonation is favored at higher kinetic energy, especially at near-neutral or alkaline pH levels. This reduction in surface charge impacts the electrostatic stabilizing force of CNPs and, as a result, leads to aggregation [129]. A decrease in charge will also reduce particle affinity to the plant cell wall and may weaken their bioadhesive properties and subsequent cellular internalization. This challenge is even greater in foliage applications since CNPs need to be retained on cuticular surfaces and then transported through either stomatal or trichome pathways into leaf interiors [130].

Additionally, the temperature dependence of protonation equilibria can influence the encapsulation and release of specific payloads in CNPs. For example, ionic crosslinkers (such as tripolyphosphate, TPP) may have weaker interactions with partially deprotonated chitosan at higher temperatures, leading to destabilization of the nanoparticle matrix and uncontrolled cargo release [131]. In controlled release agricultural explosives, the timing of bioactive release is crucial for plant development or stress resistance, and such variations can compromise the effectiveness of the formulation [18].

Cold stress extensively affects plant cellular metabolism, such as cell membrane fluidity, enzymatic activities, and over-accumulation of ROS because it causes oxidative stress. Plants under chilling stress commonly have a higher accumulation of malondialdehyde (MDA), H_2O_2, O_2^-, and OH, which, as a whole, promotes lipid peroxidation, DNA damage, and protein oxidation. The adverse impact has been addressed by the remarkable effectiveness of CNPs in counteracting these adverse effects through the alterations in the antioxidant defense mechanism of plants and stability of intracellular redox homeostasis [68].

After the foliar spraying or root uptake of CNPs, they interact with plant tissue and induce endogenous defense responses. Such interactions could directly modulate signal transduction pathways linked to cold-responsive genes, triggering up-regulation of enzymes, such as superoxide dismutase (SOD), catalases (CAT), and peroxidases (POD). These enzymes are essential in detoxifying ROS, converting superoxide radicals into less reactive molecules, and maintaining redox equilibrium in chloroplasts, mitochondria, and the cytosol. CNPs may also influence the transcriptional regulation of cold-responsive elements (e.g., C-repeat binding factors), thereby enhancing systemic stress resilience (Fig. 3) [132, 133].

Additionally, CNPs have been observed to increase the accumulation of osmolytes, which are low molecular weight organic compounds that stabilize proteins and membranes under osmotic and temperature stress [134]. These include proline, soluble sugars (e.g., trehalose), and free amino acids, which not only protect

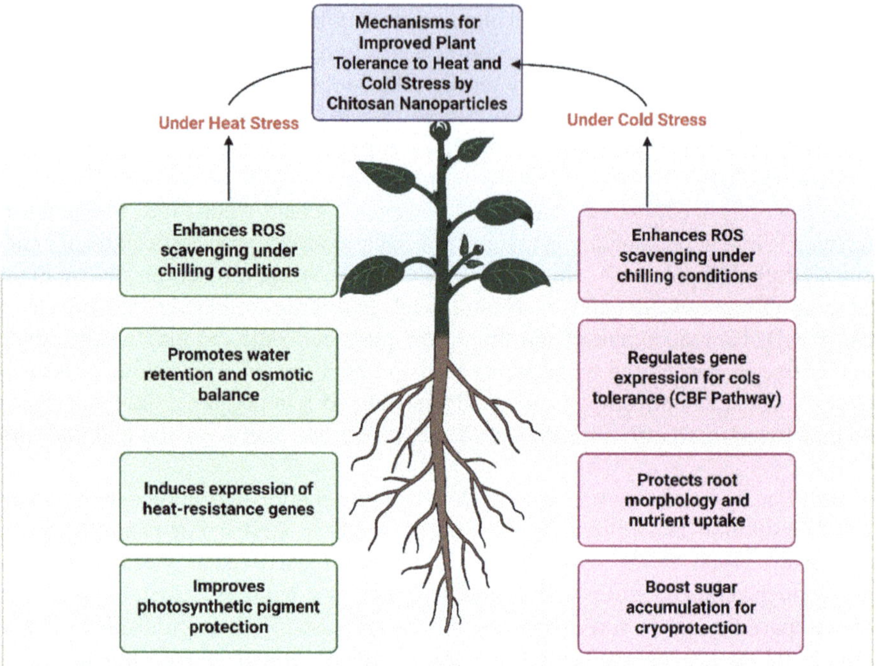

Fig. 3 Chitosan nanoparticles improve plant health by preventing temperature stress by regulating antioxidant activity, gene expression, and osmotic balance and finally enhancing overall physiological stability under heat and cold stress. Created with BioRender.com

macromolecules but also act as signaling molecules in stress adaptation. Additionally, improved retention of photosynthetic pigments, including chlorophylls and carotenoids, under CNP treatment indicates preservation of the photosynthetic apparatus's integrity, allowing for sustained carbon fixation even under lowtemperature stress [68].

Importantly, the nanoscale size of CNPs allows for their efficient absorption and systemic movement within plant tissues, enabling targeted and prolonged action at both cellular and organ levels [135]. Their cationic surface properties improve adhesion to negatively charged plant cell walls and promote internalization through endocytosis or diffusion pathways. These traits are especially beneficial for applications in early spring or high-altitude agriculture, where unexpected cold spells can weaken seedling vigor, flowering, and fruit production [136].

Therefore, the protonation-deprotonation behavior of chitosan's amine groups is a key factor in nanoparticle formulation. Modifying the degree of deacetylation, adding buffer systems, or blending with co-polymers can reduce temperature-driven solubility changes and help maintain electrostatic properties, enhancing performance during field-relevant temperature shifts [137].

7.3 Structural and Functional Implications of Temperature-Induced Changes

CNPs are a multifunctional tool in plant sciences. One of the main reasons why CNPs are important and widely used is because they can encapsulate, protect, and deliver a variety of biologically active agents in a sustained and stimulus-responsive manner. These agents may include, for example, plant growth regulators (such as auxins and gibberellins), micronutrients (like iron and zinc), enzymes, biomolecules such as proteins, antimicrobial peptides, or nucleic acids. The effectiveness of these delivery systems depends on both the structural stability of the nanoparticles and their ability to release their contents in response to environmental stimuli, with temperature playing a key role [138].

Conformational changes of the chitosan matrix at different temperatures affect the physical entrapment and molecular interactions of those agents. However, at low temperatures, the matrix shrinks and hydrogen bonds strengthen, which creates a more compact packaging and retards the release of the enclosed agents. Such behavior is generally desirable to maintain the stability of a compound (e.g., during storage or transportation) [106]. However, if the temperature is low during application, gel-limited delivery might hinder the active agents from being timely available and absorbable, especially under stress conditions that require quick physiological reactions [68].

Upon heating, matrix relaxation occurs as the polymer chains become mobile and hydrogen bonding networks partially break down. This thermal softening can potentially speed up the diffusion of trapped agents from the nanoparticle core, leading to a burst or accelerated release profile. Under certain conditions, this thermal responsiveness could be used to release compounds only when plant metabolism is at its peak (e.g., during peak sunlight hours or in response to thermal stress stimuli such as high temperatures or drought) [14]. However, for this strategy, the release kinetics need to be carefully adjusted to avoid premature depletion of the nanoparticle reservoirs [116].

Additionally, agents encapsulated by this method may experience reduced thermal stability at high temperatures. For instance, some phytohormones are heat-sensitive or prone to isomerization, which can lead to decreased bioactivity or phytotoxic effects. Nutrients stored in ionic forms are also more likely to precipitate or change when exposed to prolonged heat, thereby decreasing their bioavailability [50].

From a materials science perspective, co-encapsulation of stabilizers and thermal buffer materials within the nanoparticle matrix or designing core–shell structures with a thermally protective shell can be used as strategies to enhance thermal robustness. Additionally, preformulation studies should include thermal cycling tests to simulate expected environmental conditions, which will also help optimize the formulation [139].

In plant physiology, the varying temperature sensitivity of CNP systems may be especially important for synchronized release during phenologically responsive stages such as germination, flowering, or fruit development. If the temperature-dependent release kinetics of the coated formulations can be aligned with plant phenology, it could enhance resource use efficiency, reduce susceptibility to biotic and abiotic stresses, and decrease the need for repeated field applications [50].

7.4 The Role of Stabilizing Agents in Preserving the Structural and Functional Integrity of Chitosan Nanoparticles

For the use of chitosan nanoparticles in agricultural nanotechnologies, the thermodynamic stability of the nanoparticles is also a crucial factor that affects their performance in field conditions with varying (and sometimes extreme) temperatures. One of the most common strategies to improve CNPs stability is adding stabilizing agents like polyols, disaccharides, and surfactants to the formulation. These stabilizers work through physicochemical processes to reduce the effects of thermal stress on nanoparticle shape, surface properties, and encapsulation efficiency [140].

The polyols, including glycerol and sorbitol, mainly act as plasticizers, softening the chitosan matrix [141]. Due to their low molecular weight and ability to form multiple hydrogen bonds, they can be placed between chitosan chains, reducing their intermolecular interactions and lowering the glass transition temperature of the polymer. This increased chain mobility under thermal stress gives the NP structure flexibility and likely decreases brittleness, reducing the chances of breakage with irreversible aggregation when cooled and heated. Additionally, the polyol molecules help maintain the hydration shell inside nanovesicles (NAVs), protecting the encapsulated bioactives (e.g., phytohormones or micronutrients) from heat-induced damage [142].

Likewise, disaccharides such as sucrose and trehalose have been shown to be particularly effective in maintaining the stability of CNPs during lyophilization and rehydration. During drying, these sugars vitrify and form an amorphous matrix that surrounds the nanoparticles within a glassy shell, providing structural integrity. The high-glass transition temperature of trehalose during lowtemperature exposure effectively prevents ice crystal growth, thereby reducing mechanical damage from freeze–thaw stress. Additionally, these sugars hydrogen bond with the chitosan matrix of the hydrogel, preserving the native conformation of the polymer chains and maintaining re-dispersibility when the hydrogel is rehydrated in the plant's environment [143].

Type of stabilizer surfactants have different and sometimes opposite stabilization mechanisms. Non-ionic surfactants (such as Tween 80, Pluronic F-68) adsorb onto the surface of nanoparticles to form a steric barrier, which protects against particle aggregation caused by thermal stresses. Thus, surfactants stabilize nanoparticles in an aqueous medium, maintaining them in nanoscale sizes in dispersion. This is a key factor for effective foliar and root absorption of CNPs in plants. Additionally, the surface-active properties of these agents can also be used to control the release kinetics of encapsulated agents by influencing the permeability of the nanoparticle shell. This is especially useful when stimuli-synchronized or controlled release is desired in response to environmental temperature changes [144, 145].

Therefore, due to hydrogen bonding, steric stabilization, and polymer dynamics control, adding polyols, sugars, and surfactants appears to be a reliable method to prevent temperature-induced instability of CNPs for plant application formulations [146].

8 Enhancing the Resilience of Chitosan Nanoparticles to Thermal Degradation

The second method to reduce the thermal sensitivity of chitosan nanoparticles is to modify the encapsulation matrix by blending chitosan with compatible polymers. This co-polymerization approach aims to combine the beneficial properties of chitosan (biocompatibility and mucoadhesiveness) with the enhanced thermal, mechanical, or physicochemical stability provided by other natural or chemically modified polymers [147].

For example, a well-known type of co-polymer includes polysaccharides found in nature, with the most common being alginate, carrageenan, pectin, and guar gum. Ionic or hydrogen-bonded interactions with chitosan can form these materials into polyelectrolyte complexes (PECs) that are much stiffer and better able to withstand thermal stress [148]. For instance, mixtures of chitosan and sodium alginate create a crosslinked matrix where alginate provides negative charges from carboxylate groups, which interact with the cationic amine groups of chitosan. This two-polymer system offers higher strength and lower permeability—two crucial characteristics when the goal is to protect encapsulated agents from leaking or denaturation caused by temperature changes. When developed in plant species, such hybrids can maintain stability during storage as agrochemical formulations and still function effectively as pesticides, even under high soil and atmospheric temperatures [149].

The synthetic polymers, such as poly(lactic-co-glycolic acid) (PLGA), poly(vinyl alcohol) (PVA), and polyacrylic acid (PAA), offer advantages like tunable thermal and physical transitions, biodegradability, and compatibility with industrial fabrication. Chitosan can be combined with these polymers to create nanoparticles with controlled porosity and release profiles. For example, the thermal buffering and slow softening of the PVA coating at elevated temperatures could be useful for a CNP encapsulating a temperature-sensitive auxin analog, enabling a heat-triggered release. This feature is especially beneficial for crops during temperature-sensitive growth stages, such as flowering or fruit set, when hormone signaling needs to be temporally optimized [150, 151]. Additionally, co-polymer systems can be designed to include thermoresponsive polymers like poly (N-isopropylacrylamide) (PNIPAAm), which undergo reversible phase transitions at specific temperatures [152].

In brief, tuning encapsulation with co-polymer blends offers a highly versatile and effective solution to improve the thermostability of CNPs, enabling controlled, responsive, and stable release of agro-bioactives in the field under varying climatic conditions [153].

The surface composition of chitosan nanoparticles also plays a key role in their interactions with environmental and biological targets like plant tissues. Template destabilization caused by temperature usually begins at the surface of the nanoparticle, where thermal stress leads to desorption of the stabilizing layer, surface oxidation, or collapse of the surface-bound functional groups. In response to these challenges, surface functionalization strategies have been developed to enhance the thermal stability of the CNPs and improve their biointeractivity with plant systems [154].

A common method involves covalently attaching functional groups to the chitosan backbone to modify surface features while maintaining the core polymer structure. For example, thiol-functionalized chitosan derivatives with thiol (-SH) groups have been developed, and it is believed they exhibit enhanced mucoadhesion and increased thermal stability because of forming intra- and intermolecular disulfide bonds. These covalent bonds provide structural stability to the nanoparticle and help preserve its integrity during heating and cooling, which could otherwise cause the polymer chains to dissociate [8, 155].

It was also reported that the solubility and surface charge stability of chitosan increased further by quaternizing it through the introduction of quaternary ammonium groups over a wider pH and temperature range. This chemical modification of the particles allows maintaining a high-positive zeta potential so that negatively charged plant cell surfaces are effectively adhered to even with modest temperature fluctuations. Moreover, quaternized chitosan has demonstrated good resistance to enzymatic degradation and a prolonged action time in biological systems [156].

Advanced strategies like layer-by-layer (LbL) assembly enable the formation of multiple coatings on the nanoparticle surface, where layers of anionic and cationic polymers are alternately adsorbed based on electrostatic interactions [157]. This method also allows precise adjustment of surface composition, thickness, and functionality. Mixing in polymers that are naturally thermally insulating or have been observed as such, such as polyethylene glycol (PEG) and polycaprolactone (PCL), can reduce the mean free path of a ULF phonon through its lattice. Incorporating antioxidant-impregnated layers can also effectively enhance resistance to thermal deformation and oxidation of the nanoparticle system. In plant systems, such LbL-modified CNPs might encounter fluctuating temperatures on the phylloplane (leaf interior surface), soil surface, or in the rhizosphere and remain structurally intact until the release of bioactives is needed [158, 159].

Additionally, the CNP's ability to be functionalized from the outer side using ligands that are specifically recognized by plant surface receptors—such as lectins or peptides targeting cuticle waxes or stomatal openings—may enhance the targeting of CNPs. Combined with thermal protection techniques, this surface modification ensures that the nanoparticle remains stable under field conditions and interacts effectively with the plant surface to deliver the payload precisely where needed [160].

Thus, surface engineering offers an additional and synergistic way to improve CNPs performance under temperature stress, strengthening the polymer structure while customizing the nanoparticle interface for the best interaction with plant environments [161].

9 Alleviation of Heat-Induced Cellular Damage in Plants via the Stress-Modulating Functions of Chitosan Nanoparticles

Hightemperature stress, especially when sustained or combined with drought, poses a serious threat to plant survival and growth. It disrupts key physiological processes like stomatal conductance, transpiration, carbon assimilation, and protein folding. One of the most immediate effects of thermal stress is the destabilization of cell membranes, which leads to electrolyte leakage and oxidative stress. Furthermore, heat stress causes denaturation and aggregation of proteins, particularly those involved in photosynthesis, respiration, and stress signaling [162, 163].

Chitosan nanoparticles provide a versatile mitigation strategy by targeting both biochemical and structural components of the plant stress response. Similar to their role under cold stress, CNPs activate the antioxidant defense system, enhancing the enzymatic removal of heat-induced ROS. Increased activities of catalase (which breaks down H_2O_2), ascorbate peroxidase, and SOD after CNPs treatment have been consistently observed, leading to less oxidative damage to lipids, proteins, and nucleic acids [68].

Crucially, CNPs also influence the expression and activity of heat shock proteins (HSPs), a family of molecular chaperones that prevent protein aggregation and help refold denatured proteins. The induction of HSPs is a key part of plant thermotolerance, and CNP-mediated enhancement of HSP pathways suggests a direct role in maintaining protein stability under thermal stress. These effects may occur through chitosan's known ability to act as an elicitor of plant immunity and stress signaling pathways, possibly via calcium-mediated signaling or reactive oxygen species cascades [164, 165].

From a structural perspective, CNPs contribute to stabilizing cell membranes by maintaining lipid bilayer integrity. They may affect the phospholipid components of membranes, lowering phase transition temperatures and slowing oxidative breakdown of unsaturated fatty acids. This helps preserve membrane fluidity and selective and nonselective permeability. Such effects can support essential cellular functions, including ion pumping, enzyme activity, and signal transduction during heat stress [166].

From an agronomic perspective, the benefits of adding CNPs are especially significant during peak summer periods or in regions prone to heatwaves. Foliar sprays or root drenches applied after stress onset could be incorporated into the pre-stress treatment regimen with CNPs formulations, helping the plant prepare and reducing productivity losses. These treatments are particularly relevant for crops with phenophases sensitive to temperature, such as pollen viability in cereals or fruit development in horticultural species [167].

10 Thermal Stabilization and Sustained Delivery of Bioactive Compounds in Climatically Variable Agricultural Environments

Besides their natural physiological functions, natural and synthetic CNPs can be used as protective, long-lasting delivery systems for various bioactive molecules, most of which are thermolabile and easily degrade under different environmental conditions. Phytohormones (like IAA, GA_3, ABA) and sparingly soluble foliar micronutrients (such as zinc, iron, and boron) are sensitive to temperature, which can decrease their stability, availability, and activity when applied through conventional methods [50].

The encapsulation of these chemicals into CNPs offers two levels of protection: entrapment of the active compounds against environmental temperature and chemical stabilization through hydrogen bonding and electrostatic interactions with the chitosan matrix [168]. For example, IAA is vulnerable to oxidative degradation at high temperatures,however, when incorporated into CNPs, its degradation rate can be slowed, extending its presence in the rhizosphere or on the leaves. Additionally, micronutrient ions delivered by CNPs are less prone to precipitation or volatilization under heat stress, leading to their gradual release and continuous supply to plants [169].

The sustained release properties of CNPs are controlled by diffusion, polymer degradation, as well as swelling behavior, and can be adjusted by selecting formulation parameters. Such thermosensitivity characteristics can be incorporated into a matrix with polymers like poly (N-isopropylacrylamide) or crosslinking agents, which can modulate the permeability of the matrix in response to temperature changes [170]. This approach is especially useful in field conditions with extreme daily temperature variations, such as in desert farming or tropical highlands, where the performance of commercial inputs can be highly unpredictable [171].

Furthermore, CNPs can be formulated with complex mixtures of nutrients, biostimulants, and plant protectant products to achieve synergistic effects and reduce application frequency. This not only enhances thermal stability but also supports resource-efficient and eco-friendly farming principles.

11 Conclusion

CNPs possess high thermal stability and can maintain structural integrity as well as bioactivity at high temperatures, which is particularly valuable for enhancing plant heat resistance. With these abilities to modulate antioxidative defense systems, promote heat shock protein expression, and stabilize photosynthesis, CNPs are promising for use in alleviating temperature-induced damage in crops. However, despite the increasing use of these tools, many research gaps remain. Additionally, the long-term ecological consequences of CNPs in thermally variable soils, especially for microorganisms and nutrient cycling, are poorly understood. To address these challenges, interdisciplinary research among plant physiology, nanotechnology, and environmental science is needed, and is considered essential to optimize the formulation, delivery, and strategies for applying CNPs in the field. With their biodegradability, low toxicity, and functional versatility, CNPs stand out as a viable and sustainable solution for next-generation agricultural interventions; however, their real-world deployment will depend on continued research, regulatory development, and validation across diverse agro-climatic zones.

References

1. Attia MS, Osman MS, Mohamed AS, Mahgoub HA, Garada MO, Abdelmouty ES, Abdel Latef AA (2021) Impact of foliar application of chitosan dissolved in different organic acids on isozymes, protein patterns and physio-biochemical characteristics of tomato grown under salinity stress. Plants 10:388
2. Ardean C, Davidescu CM, Nemeş NS, Negrea A, Ciopec M, Duteanu N, Negrea P, Duda-Seiman D, Musta V (2021) Factors influencing the antibacterial activity of chitosan and chitosan modified by functionalization. Int J Mol Sci 22(14):7449
3. Bashir SM, Ahmed Rather G, Patrício A, Haq Z, Sheikh AA, Shah MZuH, Singh H, Khan AA, Imtiyaz S, Ahmad SB (2022) Chitosan nanoparticles: a versatile platform for biomedical applications. Materials 15(19):6521
4. Manna S, Seth A, Gupta P, Nandi G, Dutta R, Jana S, Jana S (2023) Chitosan derivatives as carriers for drug delivery and biomedical applications. ACS Biomater Sci Eng 9(5):2181–2202
5. Arif Y, Siddiqui H, Hayat S (2022) Role of chitosan nanoparticles in regulation of plant physiology under abiotic stress. In: Sustainable agriculture reviews 53: nanoparticles: a new tool to enhance stress tolerance. Springer, pp 399–413
6. Ingle PU, Shende SS, Shingote PR, Mishra SS, Sarda V, Wasule DL, Rajput VD, Minkina T, Rai M, Sushkova S (2022) Chitosan nanoparticles (ChNPs): a versatile growth promoter in modern agricultural production. Heliyon 8(11)
7. Sharifi-Rad J, Quispe C, Butnariu M, Rotariu LS, Sytar O, Sestito S, Rapposelli S, Akram M, Iqbal M, Krishna A (2021) Chitosan nanoparticles as a promising tool in nanomedicine with particular emphasis on oncological treatment. Cancer Cell Int 21(1):318
8. Mikušová V, Mikuš P (2021) Advances in chitosan-based nanoparticles for drug delivery. Int J Mol Sci 22(17):9652
9. Badawy AME, Luxton TP, Silva RG, Scheckel KG, Suidan MT, Tolaymat TM (2010) Impact of environmental conditions (pH, ionic strength, and electrolyte type) on the surface charge and aggregation of silver nanoparticles suspensions. Environ Sci Technol 44(4):1260–1266

10. Feric TG, Hamilton ST, Park A-HA (2021) Insights into the enhanced oxidative thermal stability of nanoparticle organic hybrid materials developed for carbon capture and energy storage. Energy Fuels 35(23):19592–19605
11. Ataide JA, Geraldes DC, Gérios EF, Bissaco FM, Cefali LC, Oliveira-Nascimento L, Mazzola PG (2021) Freeze-dried chitosan nanoparticles to stabilize and deliver bromelain. J Drug Deliv Sci Technol 61:102225
12. Nguyen HD, Ngo HV, Lee BJ (2024) Novel pH-responsive structural rearrangement of myristic acid-conjugated quetiapine nanosuspension for enhanced long-acting delivery performance. Adv Sci 11(40):2405200
13. Nowak-Jary J, Machnicka B (2023) In vivo biodistribution and clearance of magnetic iron oxide nanoparticles for medical applications. Int J Nanomed 4067–4100
14. Hidangmayum A, Dwivedi P (2022) Chitosan based nanoformulation for sustainable agriculture with special reference to abiotic stress: a review. J Polym Environ 30(4):1264–1283
15. Abbasi R, Shineh G, Mobaraki M, Doughty S, Tayebi L (2023) Structural parameters of nanoparticles affecting their toxicity for biomedical applications: a review. J Nanopart Res 25(3):43
16. Cavallaro G, Micciulla S, Chiappisi L, Lazzara G (2021) Chitosan-based smart hybrid materials: a physico-chemical perspective. J Mater Chem B 9(3):594–611
17. Karamchandani BM, Dalvi SG, Bagayatkar M, Banat IM, Satpute SK (2024) Prospective applications of chitosan and chitosan-based nanoparticles formulations in sustainable agricultural practices. Biocatal Agric Biotechnol 103210
18. Kashyap PL, Xiang X, Heiden P (2015) Chitosan nanoparticle based delivery systems for sustainable agriculture. Int J Biol Macromol 77:36–51
19. Shah BR, Dvořák P, Velíšek J, Mráz J (2021) Opening a new gateway towards the applications of chitosan nanoparticles stabilized Pickering emulsion in the realm of aquaculture. Carbohyd Polym 265:118096
20. Khan MN, Wahab S, Wahid N, Shah SN, Ullah B, Kaplan A, Razzaq A, Bibi M, Suleman F, Ali B (2025b) Impact of climate change on yield and quality of legumes. In: Challenges and solutions of climate impact on agriculture. Elsevier, pp 85–111
21. Ribeiro TP, Moreira JA, Monteiro FJ, Laranjeira MS (2022) Nanomaterials in cancer: reviewing the combination of hyperthermia and triggered chemotherapy. J Control Release 347:89–103
22. Wang J, Zhuang S (2022) Chitosan-based materials: Preparation, modification and application. J Clean Prod 355:131825
23. Lima R, Fernandes C, Pinto MM (2022) Molecular modifications, biological activities, and applications of chitosan and derivatives: a recent update. Chirality 34(9):1166–1190
24. Wani TU, Pandith AH, Sheikh FA (2021) Polyelectrolytic nature of chitosan: Influence on physicochemical properties and synthesis of nanoparticles. J Drug Deliv Sci Technol 65:102730
25. Luo W-C, Zhang W, Kim R, Chong H, Patel SM, Bogner RH, Lu X (2023) Impact of controlled ice nucleation and lyoprotectants on nanoparticle stability during Freeze-drying and upon storage. Int J Pharm 641:123084
26. Abomosallam M, Elalfy M, Zheng Z, Nagata K, Suzuki M (2022) Adsorption kinetics and thermodynamics of toxic metal ions onto chitosan nanoparticles extracted from shrimp shells. Nanotechnol Environ Eng 7:1–13
27. Parnsubsakul A, Sapcharoenkun C, Warin C, Ekgasit S, Pienpinijtham P (2022) Selection of cryoprotectants for freezing and freeze-drying of gold nanoparticles towards further uses in various applications. Colloids Surf, B 217:112702
28. Castillo-Henríquez L, Sanabria-Espinoza P, Murillo-Castillo B, Montes de Oca-Vásquez G, Batista-Menezes D, Calvo-Guzmán B, Ramírez-Arguedas N, Vega-Baudrit J (2021) Topical chitosan-based thermo-responsive scaffold provides dexketoprofen trometamol controlled release for 24 h use. Pharmaceutics 13(12):2100
29. Ajayi TA, Okeke CC, Ogunbajo OO, Ayams JN, Oyewole OA, Adetunji CO, Adeyomoye OI, bunmi Adetunji J, Ogundolie FA, Mathew JT (2025) Bioinformatics, instrumentation,

and control for modeling, synthesis, and characterization of chitosan based nanoparticles. In: Chitosan-based nanoparticles for biomedical applications. Elsevier, pp 17–31
30. Yadav S, Singh A, Palei NN, Pathak P, Verma A, Yadav JP (2024) Chitosan-based nanoformulations: preclinical investigations, theranostic advancements, and clinical trial prospects for targeting diverse pathologies. AAPS PharmSciTech 25(8):263
31. Khaled AM, Abdel-Razik A, Tonbol K (2023) Enhancing heat stress resilience of tomato during flowering stage using biostimulants: nano chitosan and spirulina platensis extract. J Agric Chem Biotechnol 14(7):79–87
32. Otari S, Bapat VA, Lakkakula J, Kadam US, Suprasanna P (2024) Advancements in bionanotechnological applications for climate-smart agriculture and food production. Biocatal Agric Biotechnol 57:103117
33. Zhang W, Taheri-Ledari R, Ganjali F, Mirmohammadi SS, Qazi FS, Saeidirad M, KashtiAray A, Zarei-Shokat S, Tian Y, Maleki A (2023) Effects of morphology and size of nanoscale drug carriers on cellular uptake and internalization process: a review. RSC Adv 13(1):80–114
34. Berger S, Berger M, Bantz C, Maskos M, Wagner E (2022) Performance of nanoparticles for biomedical applications: the in vitro/in vivo discrepancy. Biophys Rev 3(1)
35. Wang H, Roman M (2023) Effects of chitosan molecular weight and degree of deacetylation on chitosan− cellulose nanocrystal complexes and their formation. Molecules 28(3):1361
36. Madni A, Rehman S, Sultan H, Khan MM, Ahmad F, Raza MR, Rai N, Parveen F (2021) Mechanistic approaches of internalization, subcellular trafficking, and cytotoxicity of nanoparticles for targeting the small intestine. AAPS PharmSciTech 22:1–17
37. Zi Y, Yang K, He J, Wu Z, Liu J, Zhang W (2022) Strategies to enhance drug delivery to solid tumors by harnessing the EPR effects and alternative targeting mechanisms. Adv Drug Deliv Rev 188:114449
38. Carone A, Emilsson S, Mariani P, Désert A, Parola S (2023) Gold nanoparticle shape dependence of colloidal stability domains. Nanoscale Advances 5(7):2017–2026
39. Arif M, Sharaf M, Samreen KS, Chi Z, Liu C-G (2021) Chitosan-based nanoparticles as delivery-carrier for promising antimicrobial glycolipid biosurfactant to improve the eradication rate of Helicobacter pylori biofilm. J Biomater Sci Polym Ed 32(6):813–832
40. Yan P, Yuan J-l, Kang X, Lv J-y, Liu X-c (2022) Characteristics, formation mechanism and stability of high internal phase emulsions stabilized by porcine plasma protein (PPP)/carrageenan (CG) hybrid particles. Food Biosci 47:101751
41. B Narayanan VH (2022) In-vivo pharmacokinetic studies of Dolutegravir loaded spray dried Chitosan nanoparticles as milk admixture for paediatrics infected with HIV. Sci Rep 12(1):1–12
42. Panico S, Capolla S, Bozzer S, Toffoli G, Dal Bo M, Macor P (2022) Biological features of nanoparticles: protein corona formation and interaction with the immune system. Pharmaceutics 14(12):2605
43. Ahmad N, Ahmad S, Kaplan ABU, Ercisli S, Ahmad MA, Sokan-Adeaga AA, Murtaza G, Rizwana H, Almutairi SM, Iqbal R (2024) Enhancement of rice zinc content using green synthesized ZnO-NPs by foliar and nano-priming applications. Appl Biochem Biotechnol 1–17
44. Avellan A, Yun J, Morais BP, Clement ET, Rodrigues SM, Lowry GV (2021) Critical review: role of inorganic nanoparticle properties on their foliar uptake and in planta translocation. Environ Sci Technol 55(20):13417–13431
45. Attaran Dowom S, Karimian Z, Mostafaei Dehnavi M, Samiei L (2022) Chitosan nanoparticles improve physiological and biochemical responses of Salvia abrotanoides (Kar.) under drought stress. BMC Plant Biol 22(1):364
46. Saeed A, Guizani I, Hanash F, Asnag G, Al-Harthi AM, Alwafi R, Qahtan TF, Morsi M, Assran AS (2024) Enhancing optical, structural, thermal, electrical properties, and antibacterial activity in chitosan/polyvinyl alcohol blend with ZnO nanorods: polymer nanocomposites for optoelectronics and food/medical packaging applications. Polym Bull 81(13):11645–11670
47. Dias MV, de Azevedo VM, Ferreira LF, de Oliveira ACS, Borges SV, de Fatima Ferreira Soares N, Medeiros EAA, de Deus Souza Carneiro J (2021) Chitosan-nanocomposites as a

food active packaging: effect of addition of tocopherol and modified montmorillonite. J Food Process Eng 44(11):e13843
48. Liu G, McEnnis K (2022) Glass transition temperature of PLGA particles and the influence on drug delivery applications. Polymers 14(5):993
49. Deng Z, Wu Z, Tan X, Deng F, Chen Y, Chen Y, Zhang H (2022) Preparation, characterization and antibacterial property analysis of cellulose nanocrystals (CNC) and chitosan nanoparticles fine-tuned starch film. Molecules 27(23):8542
50. Khan MN, Khan Z, Zhu L, Li Y, Wu H (2022) Impact of chitosan and chitosan-based metallic nanoparticles on the regulation of plant hormones. In: Role of chitosan and chitosan-based nanomaterials in plant sciences. Elsevier, pp 343–356
51. Kumar A, Kumar A (2024) Smart pharmaceutical formulations of biopolymeric materials in buccal drug delivery. Biomed Mater Devices 1–9
52. Ullah N, Ullah B, Kaplan A, Dossou-Yovo HO, Wahab S, Iqbal M (2024) Quantitative ethnobotanical study of medicinal plants used by local communities in Chamla Valley, Buner District, Pakistan. Ethnobot Res Appl 28:1–22
53. Yuan Y, Raheja K, Milbrandt NB, Beilharz S, Tene S, Oshabaheebwa S, Gurkan UA, Samia ACS, Karayilan M (2023) Thermoresponsive polymers with LCST transition: synthesis, characterization, and their impact on biomedical frontiers. RSC Applied Polymers 1(2):158–189
54. Rani S, Kumari N, Sharma V (2023) Uptake, translocation, transformation and physiological effects of nanoparticles in plants. Arch Agron Soil Sci 69(9):1579–1599
55. Badiyal A, Mahajan R, Rana RS, Sood R, Walia A, Rana T, Manhas S, Jayswal D (2024) Synergizing biotechnology and natural farming: pioneering agricultural sustainability through innovative interventions. Front Plant Sci 15:1280846
56. Faryal S, Ullah R, Khan MN, Ali B, Hafeez A, Jaremko M, Qureshi KA (2022) Thiourea-capped nanoapatites amplify osmotic stress tolerance in *Zea mays* L. by conserving photosynthetic pigments, osmolytes biosynthesis and antioxidant biosystems. Molecules 27(18):5744
57. Xu W, He K, Lin Z, McClements DJ, Jin Z, Chen L (2025) Progress in using cross-linking technologies to increase the thermal stability of colloidal delivery systems. Crit Rev Food Sci Nutr 65(14):2654–2668
58. Hamidi F, Azadi Aghdam M, Johar F, Mehdinejad MH, Baghani AN (2022) Ionic gelation synthesis, characterization and adsorption studies of cross-linked chitosan-tripolyphosphate (CS-TPP) nanoparticles for removal of As (V) ions from aqueous solution: kinetic and isotherm studies. Toxin Reviews 41(3):795–805
59. Briggs F, Browne D, Asuri P (2022) Role of polymer concentration and crosslinking density on release rates of small molecule drugs. Int J Mol Sci 23(8):4118
60. Wahba MI (2024) A comprehensive review on genipin: an efficient natural cross-linker for biopolymers. Polym Bull 81(16):14251–14305
61. Jonassen H, Kjøniksen A-L, Hiorth M (2012) Stability of chitosan nanoparticles cross-linked with tripolyphosphate. Biomacromol 13(11):3747–3756
62. Abd-Elsalam WH, Ibrahim RR (2021) Span 80/TPGS modified lipid-coated chitosan nanocomplexes of acyclovir as a topical delivery system for viral skin infections. Int J Pharm 609:121214
63. Rizeq BR, Younes NN, Rasool K, Nasrallah GK (2019) Synthesis, bioapplications, and toxicity evaluation of chitosan-based nanoparticles. Int J Mol Sci 20(22):5776
64. Abdelwahed W, Degobert G, Fessi H (2006) Investigation of nanocapsules stabilization by amorphous excipients during freeze-drying and storage. Eur J Pharm Biopharm 63(2):87–94
65. Thapliyal P, Khan M (2024) Biopolymer nanocarriers for improved agrochemical performance. In: Biopolymeric nanoparticles for agricultural applications. Springer, pp 369–391
66. Das A, Ghosh S, Pramanik N (2024) Chitosan biopolymer nanocomposites for agriculture applications. In: Biopolymeric Nanoparticles for agricultural applications. Springer, pp 209–240

67. Hayat F, Khanum F, Li J, Iqbal S, Khan U, Javed HU, Razzaq MK, Altaf MA, Peng Y, Ma X (2023) Nanoparticles and their potential role in plant adaptation to abiotic stress in horticultural crops: A review. Sci Hortic 321:112285
68. Wang A, Li J, Al-Huqail AA, Al-Harbi MS, Ali EF, Wang J, Ding Z, Rekaby SA, Ghoneim AM, Eissa MA (2021) Mechanisms of chitosan nanoparticles in the regulation of cold stress resistance in banana plants. Nanomaterials 11(10):2670
69. Alehosseini E, Shahiri Tabarestani H, Kharazmi MS, Jafari SM (2022) Physicochemical, thermal, and morphological properties of chitosan nanoparticles produced by ionic gelation. Foods 11(23):3841
70. Mondéjar-López M, Rubio-Moraga A, López-Jimenez AJ, Martínez JCG, Ahrazem O, Gómez-Gómez L, Niza E (2022) Chitosan nanoparticles loaded with garlic essential oil: a new alternative to tebuconazole as seed dressing agent. Carbohyd Polym 277:118815
71. Ahmed ME, Mohamed MI, Ahmed HY, Elaasser MM, Kandile NG (2024) Fabrication and characterization of unique sustain modified chitosan nanoparticles for biomedical applications. Sci Rep 14(1):13869
72. Barbosa HF, Francisco DS, Ferreira AP, Cavalheiro ÉT (2019) A new look towards the thermal decomposition of chitins and chitosans with different degrees of deacetylation by coupled TG-FTIR. Carbohyd Polym 225:115232
73. El-Naggar NE-A, Shiha AM, Mahrous H, Mohammed AA (2024) A sustainable green-approach for biofabrication of chitosan nanoparticles, optimization, characterization, its anti-fungal activity against phytopathogenic Fusarium culmorum and antitumor activity. Sci Rep 14(1):11336
74. Othman SH, Ronzi NDA, Shapi'i RA, Dun M, Ariffin SH, Mohammed MAP (2023) Biodegradability of starch nanocomposite films containing different concentrations of chitosan nanoparticles in compost and planting soils. Coatings 13(4):777
75. El-Naggar NE-A, Saber WI, Zweil AM, Bashir SI (2022) An innovative green synthesis approach of chitosan nanoparticles and their inhibitory activity against phytopathogenic Botrytis cinerea on strawberry leaves. Sci Rep 12(1):3515
76. Suhail M, Chiu I-H, Ullah A, Khan A, Ullah H, Al-Sowayan NS, Wu P-C (2024) Formulation and in vitro assessment of polymeric pH-responsive nanogels of chitosan for sustained delivery of madecassoside. ACS Omega 9(17):19345–19352
77. Dong Y, Ruan Y, Wang H, Zhao Y, Bi D (2004) Studies on glass transition temperature of chitosan with four techniques. J Appl Polym Sci 93(4):1553–1558
78. Saeed RM, Dmour I, Taha MO (2020) Stable chitosan-based nanoparticles using polyphosphoric acid or hexametaphosphate for tandem ionotropic/covalent crosslinking and subsequent investigation as novel vehicles for drug delivery. Front Bioeng Biotechnol 8:4
79. Szymańska E, Winnicka K (2015) Stability of chitosan—a challenge for pharmaceutical and biomedical applications. Mar Drugs 13(4):1819–1846
80. Almalik A, Alradwan I, Kalam MA, Alshamsan A (2017) Effect of cryoprotection on particle size stability and preservation of chitosan nanoparticles with and without hyaluronate or alginate coating. Saudi Pharm J 25(6):861–867
81. Whitehead FA, Young SA, Kasapis S (2020) Swelling behaviour and glass transition in genipin-crosslinked chitosan systems. Int J Biol Macromol 164:3075–3083
82. Maluin FN, Hussein MZ (2020) Chitosan-based agronanochemicals as a sustainable alternative in crop protection. Molecules 25(7):1611
83. Nayab N, Alam MA (2023) Chitosan nanoparticles: a promising stimulant for augmenting cold stress resistance in Safed Velchi cultivars of banana plants
84. Kalisz A, Kornaś A, Gil J, Rudolphi-Szydło E, Gawrońska K, Sieprawska A, Jafari H, Mahdavinia GR, Kulak M, Gohari G (2024) Foliar spraying with amino acids and their chitosan nanocomposites as promising way to alleviate abiotic stress in iceberg lettuce grown at different temperatures. Sci Rep 14(1):17208
85. Mahfouze SA, Mahfouze HA, Helmi RY, Fathallah FB, Aboud KA, Ottai ME (2024) The Role of Chitosan in Improving the Cold Stress Tolerance in Strawberry varieties. Jordan Journal of Biological Sciences 17(4)

86. Savani KR, Gajera H, Hirpara DG, Savaliya DD, Kandoliya U (2023) Salicylic acid-functionalised chitosan nanoparticles restore impaired sucrose metabolism in the developing anther of cotton (Gossypium hirsutum) under heat stress. Funct Plant Biol 50(9):736–751
87. Al Masruri MHK, Ullah A, Farooq M (2023) Application of nano chitosan-glycinebetaine for improving bread wheat performance under combined drought and heat stresses. J Soil Sci Plant Nutr 23(3):3482–3499
88. Liaqat A, Ihsan M, Rizwan M, Mehmood A, Ijaz M, Alam M, Abdullah M, Wajid M, Hussain R, Naeem M (2019) inducing effect of chitosan on the physiological and biochemical indices of eggplant (*Solanum Melongena* L.) genotypes under heat and high irradiance. Appl Ecol Environ Res 17(5)
89. Mohammadi A, Hashemi M, Hosseini SM (2015) Chitosan nanoparticles loaded with Cinnamomum zeylanicum essential oil enhance the shelf life of cucumber during cold storage. Postharvest Biol Technol 110:203–213
90. Samarah NH, Al-Quraan NA, Massad RS, Welbaum GE (2020) Treatment of bell pepper (*Capsicum annuum* L.) seeds with chitosan increases chitinase and glucanase activities and enhances emergence in a standard cold test. Sci Hortic 269:109393. https://doi.org/10.1016/j.scienta.2020.109393
91. Nasr F, Pateiro M, Rabiei V, Razavi F, Formaneck S, Gohari G, Lorenzo JM (2021) Chitosan-phenylalanine nanoparticles (Cs-Phe Nps) extend the postharvest life of persimmon (Diospyros kaki) fruits under chilling stress. Coatings 11(7):819
92. Mahmoudi R, Razavi F, Rabiei V, Palou L, Gohari G (2022) Postharvest chitosan-arginine nanoparticles application ameliorates chilling injury in plum fruit during cold storage by enhancing ROS scavenging system activity. BMC Plant Biol 22(1):555
93. El-Gioushy SF, El-Masry AM, Fikry M, El-Kholy MF, Shaban AE, Sami R, Algarni E, Alshehry G, Aljumayi H, Benajiba N (2022) Utilization of active edible films (Chitosan, chitosan nanoparticle, and CACL2) for enhancing the quality properties and the shelf life of Date palm fruits (Barhi cultivar) during Cold storage. Coatings 12(2):255
94. Elbagoury MM, Turoop L, Runo S, Sila DN, Madivoli ES (2022) Postharvest treatments of banana (Musa acuminata cv.'Grand Nain', AAA) during cold and ripening temperatures with chitosan and chitosan nanoparticles to alleviate chilling injury and maintain antioxidant activity. Hortic, Environ, Biotechnol 63(5):677–699
95. Nguyen HHD, Nguyen HV (2021) Effects of storage temperature on postharvest physico-chemical attributes of nano-chitosan coated strawberry (Fragaria x ananassa Duch.)
96. Chouljenko A (2015) Evaluating the effects of vacuum tumbling with chitosan nanoparticles and water-soluble chitosan on the shelf life of cryogenically frozen shrimp. Louisiana State University and Agricultural & Mechanical College
97. Stolte Bezerra Lisboa Oliveira L, Ristroph KD (2024) Critical review: uptake and translocation of organic nanodelivery vehicles in plants. Environ Sci Technol 58(13):5646–5669
98. Mondéjar-López M, García-Simarro MP, Navarro-Simarro P, Gómez-Gómez L, Ahrazem O, Niza E (2024) A review on the encapsulation of "eco-friendly" compounds in natural polymer-based nanoparticles as next generation nano-agrochemicals for sustainable agriculture and crop management. Int J Biol Macromol 136030
99. Lino PR, Leandro J, Amaro M, Gonçalves LM, Leandro P, Almeida AJ (2021) In silico and in vitro tailoring of a chitosan nanoformulation of a human metabolic enzyme. Pharmaceutics 13(3):329
100. Balusamy SR, Rahimi S, Sukweenadhi J, Sunderraj S, Shanmugam R, Thangavelu L, Mijakovic I, Perumalsamy H (2022) Chitosan, chitosan nanoparticles and modified chitosan biomaterials, a potential tool to combat salinity stress in plants. Carbohyd Polym 284:119189
101. Arnaldi P, Pastorino L, Monticelli O (2020) On an effective approach to improve the properties and the drug release of chitosan-based microparticles. Int J Biol Macromol 163:393–401
102. Herdiana Y, Wathoni N, Shamsuddin S, Muchtaridi M (2022) Drug release study of the chitosan-based nanoparticles. Heliyon 8(1)
103. Rezvankhah A, Emam-Djomeh Z, Askari G (2020) Encapsulation and delivery of bioactive compounds using spray and freeze-drying techniques: a review. Drying Technol 38(1–2):235–258

104. Shin H, Oh Y, Kim D (2015) Differences in cold hardiness, carbohydrates, dehydrins and related gene expressions under an experimental deacclimation and reacclimation in Prunus persica. Physiol Plant 154(4):485–499
105. Schoebitz M, López MD, Roldán A (2013) Bioencapsulation of microbial inoculants for better soil–plant fertilization, a review. Agron Sustain Dev 33:751–765
106. Hajji S, Hamdi M, Boufi S, Li S, Nasri M (2019) Suitability of chitosan nanoparticles as cryoprotectant on shelf life of restructured fish surimi during chilled storage. Cellulose 26:6825–6847
107. Butova VV, Bauer TV, Polyakov VA, Minkina TM (2023) Advances in nanoparticle and organic formulations for prolonged controlled release of auxins. Plant Physiol Biochem 201:107808
108. Mishra N, Sharma M, Mishra P, Nisha R, Singh P, Pal RR, Singh N, Singh S, Maurya P, Pant S (2024) Transporter targeted-carnitine modified pectin-chitosan nanoparticles for inositol hexaphosphate delivery to the colon: an in silico and in vitro approach. Int J Biol Macromol 263:130517
109. Bernal-Chávez SA, Romero-Montero A, Hernández-Parra H, Peña-Corona SI, Del Prado-Audelo ML, Alcalá-Alcalá S, Cortés H, Kiyekbayeva L, Sharifi-Rad J, Leyva-Gómez G (2023) Enhancing chemical and physical stability of pharmaceuticals using freeze-thaw method: challenges and opportunities for process optimization through quality by design approach. J Biol Eng 17(1):35
110. Wang Z, Gao D, Shu Z (2024) Mechanisms, applications, and challenges of utilizing nanomaterials in cryopreservation. Adv Eng Mater 26(21):2400800
111. Liu X, Wu J (2024) Application of chitosan preparations as vaccine adjuvants
112. Boroumand H, Badie F, Mazaheri S, Seyedi ZS, Nahand JS, Nejati M, Baghi HB, Abbasi-Kolli M, Badehnoosh B, Ghandali M (2021) Chitosan-based nanoparticles against viral infections. Front Cell Infect Microbiol 11:643953
113. Suresh K, Chowdhury A, Kumar SK, Kumaraswamy G (2019) Critical role of processing on the mechanical properties of cross-linked highly loaded nanocomposites. Macromolecules 52(15):5955–5962
114. Soni K, Sharma Y, Sharma S, Sharma A, Rani A, Rathi A, Kapoor, Naik SB (2024) Nanobiostimulants and plant secondary metabolites. In: Nanobiostimulants: emerging strategies for agricultural sustainability. Springer, pp 283–301
115. Fernández-Quiroz D, Tohidi MM, Paymard B, Lucero-Acuña A (2023) Immobilization of essential oils in biopolymeric matrices: recent approaches for controlled delivery systems. Stud Nat Prod Chem 78:365–401
116. Santo Pereira AE, Silva PM, Oliveira JL, Oliveira HC, Fraceto LF (2017) Chitosan nanoparticles as carrier systems for the plant growth hormone gibberellic acid. Colloids Surf, B 150:141–152
117. Chang T, Zhao G (2021) Ice inhibition for cryopreservation: materials, strategies, and challenges. Advanced Science 8(6):2002425
118. Youm I, West MB, Huang X, Li W, Kopke RD (2022) Enhanced sucrose-mediated cryoprotection of siRNA-loaded poly (lactic-co-glycolic acid) nanoparticles. Colloids Surf, B 220:112880
119. Pignatello R, Matricardi P (2020) Steering the clinical translation of delivery systems for drugs and health products. MDPI
120. Arora S, Dash SK, Dhawan D, Sahoo PK, Jindal A, Gugulothu D (2024) Freeze-drying revolution: unleashing the potential of lyophilization in advancing drug delivery systems. Drug Deliv Transl Res 14(5):1111–1153
121. Qin Z, Zhao L (2019) The history of chito/chitin oligosaccharides and its monomer. Oligosaccharides of Chitin and Chitosan: bio-manufacture and applications 3–14
122. Froimowicz P, Zhang K, Ishida H (2016) Intramolecular hydrogen bonding in benzoxazines: when structural design becomes functional. Chemistry–A Eur J 22(8):2691–2707
123. Khondee S (2011) Polyelectrolyte-based nanoparticles for gene and protein delivery.

124. Detsi A, Kavetsou E, Kostopoulou I, Pitterou I, Pontillo ARN, Tzani A, Christodoulou P, Siliachli A, Zoumpoulakis P (2020) Nanosystems for the encapsulation of natural products: the case of chitosan biopolymer as a matrix. Pharmaceutics 12(7):669
125. Yang Y, Urban MW (2013) Self-healing polymeric materials. Chem Soc Rev 42(17):7446–7467
126. Karimi M, Sahandi Zangabad P, Ghasemi A, Amiri M, Bahrami M, Malekzad H, Ghahramanzadeh Asl H, Mahdieh Z, Bozorgomid M, Ghasemi A (2016) Temperature-responsive smart nanocarriers for delivery of therapeutic agents: applications and recent advances. ACS Appl Mater Interfaces 8(33):21107–21133
127. Huang W, Tsui CP, Tang CY, Yang M, Gu L (2017) Surface charge switchable and pH-responsive chitosan/polymer core-shell composite nanoparticles for drug delivery application. Compos B Eng 121:83–91
128. Shokri Z, Seidi F, Saeb MR, Jin Y, Li C, Xiao H (2022) Elucidating the impact of enzymatic modifications on the structure, properties, and applications of cellulose, chitosan, starch and their derivatives: a review. Mater Today Chem 24:100780
129. Zhao X, Wei Z, Zhao Z, Miao Y, Qiu Y, Yang W, Jia X, Liu Z, Hou H (2018) Design and development of graphene oxide nanoparticle/chitosan hybrids showing pH-sensitive surface charge-reversible ability for efficient intracellular doxorubicin delivery. ACS Appl Mater Interfaces 10(7):6608–6617
130. Hong J, Wang C, Wagner DC, Gardea-Torresdey JL, He F, Rico CM (2021) Foliar application of nanoparticles: mechanisms of absorption, transfer, and multiple impacts. Environ Sci Nano 8(5):1196–1210
131. Hamdi M, Nasri R, Li S, Nasri M (2020) Design of blue crab chitosan responsive nanoparticles as controlled-release nanocarrier: Physicochemical features, thermal stability and in vitro pH-dependent delivery properties. Int J Biol Macromol 145:1140–1154
132. Chaffai R, Ganesan M, Cherif A (2024) Nanoparticles for sustainable agriculture and abiotic stress resistance. In: Plant adaptation to abiotic stress: from signaling pathways and microbiomes to molecular mechanisms. Springer, pp 195–208
133. Somkuwar Subhash R, Meshram JH, Gogle D, Chaudhary R, Mahakhode R, Patil VN, Kamble RB, Sawant R, Gadpayale J, Chaudhary RR (2024) Nanoelicitors mediated abiotic stresses in plant defense response mechanisms: current review and future perspectives. Plant Second Metabltes Abiotic Stress 591–621
134. Kumar D, Singh R, Upadhyay SK, Verma KK, Tripathi RM, Liu H, Dhankher OP, Tripathi RD, Sahi SV, Seth CS (2024) Review on interactions between nanomaterials and phytohormones: Novel perspectives and opportunities for mitigating environmental challenges. Plant Sci 340:111964
135. Sairam AB, Sanmugam A, Pushparaj A, Mahesh Kumar G, Sundarapandian N, Balaji S, Nallal M, Park KH (2023) Toxicity of polymeric nanodrugs as drug carriers. ACS Chem Health Saf 30(5):236–250
136. Jasra AW, Rafi MA (2002) Cash crop farming in the Northern Pakistan: the importance of pollinator diversity and managed pollination in apricots.
137. Dey A, Kamat A, Nayak S, Danino D, Kesselman E, Dandekar P, Jain R (2018) Role of proton balance in formation of self-assembled chitosan nanoparticles. Colloids Surf, B 166:127–134
138. Sangwan S, Sharma P, Wati L, Mehta S (2023) Effect of chitosan nanoparticles on growth and physiology of crop plants. In: Engineered nanomaterials for sustainable agricultural production, soil improvement and stress management. Elsevier, pp 99–123
139. Hefnawy A, Khalil IH, Arafa K, Emara M, El-Sherbiny IM (2020) Dual-ligand functionalized core-shell chitosan-based nanocarrier for hepatocellular carcinoma-targeted drug delivery. Int J Nanomedicine 821–837
140. Lan X, Zhang X, Wang L, Wang H, Hu Z, Ju X, Yuan Y (2023) A review of food preservation based on zein: the perspective from application types of coating and film. Food Chem 424:136403
141. Ma X, Qiao C, Wang X, Yao J, Xu J (2019) Structural characterization and properties of polyols plasticized chitosan films. Int J Biol Macromol 135:240–245

142. Sun H, Cao Y, Kim D, Marelli B (2022) Biomaterials technology for agrofood resilience. Adv Func Mater 32(30):2201930
143. Pinto JT, Faulhammer E, Dieplinger J, Dekner M, Makert C, Nieder M, Paudel A (2021) Progress in spray-drying of protein pharmaceuticals: Literature analysis of trends in formulation and process attributes. Drying Technol 39(11):1415–1446
144. Biondo F, Baldassarre F, Vergaro V, Ciccarella G (2022) Controlled biocide release from smart delivery systems: materials engineering to tune release rate, biointeractions, and responsiveness. In: Nanotechnology-based sustainable alternatives for the management of plant diseases. Elsevier, pp 31–147
145. Cartwright A (2023) Surface-functionalized silica nanocarriers for mitigating water stress in wheat and benefiting the root microbiome
146. Ielo I, Rando G, Giacobello F, Sfameni S, Castellano A, Galletta M, Drommi D, Rosace G, Plutino MR (2021) Synthesis, chemical–physical characterization, and biomedical applications of functional gold nanoparticles: a review. Molecules 26(19):5823
147. Marsili L, Dal Bo M, Berti F, Toffoli G (2021) Chitosan-based biocompatible copolymers for thermoresponsive drug delivery systems: on the development of a standardization system. Pharmaceutics 13(11):1876
148. Luo Y, Wang Q (2014) Recent development of chitosan-based polyelectrolyte complexes with natural polysaccharides for drug delivery. Int J Biol Macromol 64:353–367
149. Kumar R, Najda A, Duhan JS, Kumar B, Chawla P, Klepacka J, Malawski S, Kumar Sadh P, Poonia AK (2021) Assessment of antifungal efficacy and release behavior of fungicide-loaded chitosan-carrageenan nanoparticles against phytopathogenic fungi. Polymers 14(1):41
150. El Amerany F, Aboudamia FZ, Janah I, Taourirte M, Rhazi M (2022) Polymer-based nanoparticles (NPs): a promising approach for crop productivity. In: Plant and nanoparticles. Springer, pp 119–154
151. Ray A, Das P, Chunduri R, Kumar D, Dulta K, Kaushal A, Gupta S, RJ S, Yadav AN, Nagraik R (2025) Nanocomposite-based agricultural delivery systems: a sustainable approach to enhanced crop productivity and soil health. J Nanoparticle Res 27(4):110
152. Raghvendra KM (2018) BSM Solar and Environmental Solution, Unnao, Uttar Pradesh, India. International Inter University Centre for Nanoscience and Nanotechnology, Kottayam, Kerala, India. Indian Institute of Space Science and Technology, Thiruvananthapuram, Kerala, India. Fundam Biomater: Metals 35
153. Lewicka K, Szymanek I, Rogacz D, Wrzalik M, Łagiewka J, Nowik-Zając A, Zawierucha I, Coseri S, Puiu I, Falfushynska H (2024) Current trends of polymer materials' application in agriculture. Sustainability (2071–1050) 16(19)
154. Chen T, Ferris R, Zhang J, Ducker R, Zauscher S (2010) Stimulus-responsive polymer brushes on surfaces: transduction mechanisms and applications. Prog Polym Sci 35(1–2):94–112
155. Žigrayová D, Mikušová V, Mikuš P (2024) Advances in chitosan derivatives: preparation, properties and applications in pharmacy and medicine. Gels 10(11):701
156. Pathak K, Misra SK, Sehgal A, Singh S, Bungau S, Najda A, Gruszecki R, Behl T (2021) Biomedical applications of quaternized chitosan. Polymers 13(15):2514
157. Borges J, Mano JF (2014) Molecular interactions driving the layer-by-layer assembly of multilayers. Chem Rev 114(18):8883–8942
158. Alogla A (2023) Enhancing antioxidant delivery through 3D printing: a pathway to advanced therapeutic strategies. Front Bioeng Biotechnol 11:1256361
159. Pedro AC, Paniz OG, Fernandes IdAA, Bortolini DG, Rubio FTV, Haminiuk CWI, Maciel GM, Magalhães WLE (2022) The importance of antioxidant biomaterials in human health and technological innovation: a review. Antioxidants 11(9):1644
160. Thabit FN, Moursy AR (2023) Sensors efficiency in smart management of the environmental resources. In: Handbook of nanosensors: materials and technological applications. Springer, pp 1–40
161. Yusefi M (2022) Polysaccharide-based magnetic nanocomposites as drug carriers for potential colorectal cancer treatment

162. Yu J, Yang Z, Jespersen D, Huang B (2014) Photosynthesis and protein metabolism associated with elevated CO2-mitigation of heat stress damages in tall fescue. Environ Exp Bot 99:75–85
163. Zahra N, Hafeez MB, Ghaffar A, Kausar A, Al Zeidi M, Siddique KH, Farooq M (2023) Plant photosynthesis under heat stress: effects and management. Environ Exp Bot 206:105178
164. Kang X, Zhao L, Liu X (2023) Calcium signaling and the response to heat shock in crop plants. Int J Mol Sci 25(1):324
165. Pandey V, Singh S (2024) Plant adaptation and tolerance to heat stress: advance approaches and future aspects. Comb Chem High Throughput Screen 27(12):1701–1715
166. Kizaloglu A, Kilicay E, Karahaliloglu Z, Hazer B, Denkbas EB (2020) The preparation of chitosan membrane improved with nanoparticles based on unsaturated fatty acid for using in cancer-related infections. J Bioact Compat Polym 35(4–5):328–350
167. Gupta S, Novák O, Kulkarni MG, Doležalova I, Van Staden J, Doležal K (2024) Unleashing the potential of biostimulants in stimulating pollen germination and tube growth. J Plant Growth Regul 1–32
168. Naghib SM, Ahmadi B, Kangarshahi BM, Mozafari M (2024) Chitosan-based smart stimuli-responsive nanoparticles for gene delivery and gene therapy: recent progresses on cancer therapy. Int J Biol Macromol 134542
169. Gonzalez-Montfort TS, Almaraz-Abarca N, Pérez-y-Terrón R, Ocaranza-Sánchez E, Rojas-López M (2022) Synthesis of chitosan microparticles encapsulating bacterial cell-free supernatants and indole acetic acid, and their effects on germination and seedling growth in tomato (*Solanum lycopersicum*). Int J Anal Chem 2022(1):2182783
170. Tian X, Huang Z, Zhang Q, Wang X, Yang N, Deng N (2024) PNIPAm thermo-responsive nanofibers mats: morphological stability and response behavior under cross-linking. Acta Phys Chim Sin 40(4):2304037
171. Zhang Z, Wang Y, Li Z, Fu H, Huang J, Xu Z, Lai Y, Qian X, Zhang S (2022) Sustainable hierarchical-pored PAAS–PNIPAAm hydrogel with core–shell structure tailored for highly efficient atmospheric water harvesting. ACS Appl Mater Interfaces 14(49):55295–55306

Heavy Metals Stress Mitigation by Chitosan Nanoparticles

Muhammad Adnan, Hansa Gul, Nasir Assad, Muhammad Nauman Khan, Barkat Ullah, Syed Nasar Shah, Fethi Ahmet Ozdemir, and Sarah Abdul Razak

Abstract The pollution of agricultural systems and plant health by heavy metals poses a major threat to human health. Excessive accumulation of cadmium, lead, arsenic, and mercury impairs the physiological function of plants, reduces crop yields, and pollutes the food chain. Chitosan nanoparticles (CNPs), as an environmentally friendly plant, might be an efficient approach to tackle heavy metal pollution. CNPs can bind large amounts of toxic metal ions, reduce bioavailability and mitigate oxidative harm to plants because of their enormous surface area, functional groups, and chelating capacity. The chapter showcases the processes involved in the reduction of heavy metal toxicity by CNPs, such as the increased activity of antioxidant enzymes, improved photosynthetic efficiency, membrane stability and the activity, and level of metal transporters. The use of CNPs is effective in plants exposed to heavy metal stress, resulting in improved plant growth, and biomass creation, as well as physiological stability. Moreover, CNPs may be constructed to provide key nutrients or

M. Adnan (✉)
Department of Chemistry, Bacha Khan University, Charsadda, Khyber Pakhtunkhwa, Pakistan
e-mail: muhammadadnan112255@gmail.com

H. Gul
Department of Zoology, University of Mianwali, Punjab, Pakistan

N. Assad
Institute of Chemistry, University of Sargodha, Sargodha, Pakistan

M. N. Khan · B. Ullah
Department of Botany, Islamia College Peshawar, Peshawar, Pakistan

M. N. Khan
Biology Laboratory, University Public School, University of Peshawar, Peshawar, Pakistan

S. N. Shah
Department of Plant Sciences, Quaid-I-Azam University, Islamabad, Pakistan

F. A. Ozdemir
Faculty of Science and Art, Department of Molecular Biology and Genetics, Bingol University, Bingol, Turkey

S. A. Razak
Faculty of Science, Institute of Biological Sciences, Universiti Malaya, Kuala Lumpur, Malaysia

A. A. H. Abdel Latef (ed.), *Chitosan Nanoparticles for Plant Resilience Under Abiotic and Biotic Stressors*, Smart Nanomaterials Technology,
https://doi.org/10.1007/978-981-95-6422-4_7

signaling molecules, providing a multifunctional aspect of remediation and stimulation of growth. Despite this tremendous potential, issues of standardization of nanoparticle formulation, fate in soil–plant systems, and long-term effects on the soil environment are still unresolved. All in all, the idea of CNPs as a multifunctional, sustainable, and promising method of heavy metal stress in agriculture is gaining ground to serve environmental protection and clean crop production.

Keywords Antioxidant defense · Chitosan nanoparticles · Heavy metal stress · Phytoremediation · Sustainable agriculture

1 Introduction

Heavy metal (HM) contamination in agriculture has become a major environmental and health problem in recent years. Due to the widespread use of chemical fertilizers, pesticides, untreated wastewater, industrial effluents and urban runoff in agriculture, soils are heavily contaminated with hazardous heavy metals such as lead (Pb), cadmium (Cd), arsenic (As), chromium (Cr), mercury (Hg), and nickel (Ni) [1, 2]. As these metals cannot be broken down, they remain in the environment for a long time and end up in plants and the food chain. In contrast to micronutrients, which are only needed in small quantities, an increase in heavy metals in the environment may severely affect plant systems and reduce the sustainability of agriculture.

If heavy metals are present in the soil, seeds can only grow with difficulty, root systems cannot develop normally, and ordinary metabolic functions in plants cannot take place. As a result, the harvest is poorer and of lower quality. To illustrate, cadmium suppresses the process of photosynthesis and disrupts the supply of essential nutrients, and lead breaks the structure and balance of cells and water in plants. He toxic concentrations of HMs in the soil find their way through the vascular system and disrupt the function of biomolecules (such as proteins, DNA, etc.) in plant cells [3]. Metal stress usually leads to the formation of reactive oxygen species (ROS), which in turn damage cell membranes and DNA, and ultimately lead to programmed cell death. Metal stress usually leads to the production of ROS, which in turn damages cell membranes and the DNA, and ultimately leads to programmed cell death. In addition, crops grown on contaminated soils are likely to contain metals in the parts consumed by consumers eat, which can reduce market value and endanger [4]. It is known that prolonged exposure to heavy metals in humans can lead to cancer, poisoning of the nervous system, poisoning of the kidneys, and abnormalities in the development of children (Ehsan et al.). This means that consuming products from contaminated food chains may put people's health at risk, especially in areas where the water used in agriculture is polluted and harmful chemicals are part of the soil conditioners [5].

However, as an adaptation to prolonged HM stress, some plant species have evolved morphological (formation of hairy roots, trichomes, thickened cell walls, and cuticle layers) and physiological (such as secretion of root exudates rich in organic acids, accumulation of proline, and production of phytohormones) strategies that allow them to mitigate metal stress [6]. At the cellular level, tolerant plants utilize a combination of mechanisms, including regulation of heavy metal uptake and release, intracellular transport and compartmentalization, sequestration in vacuoles, and chelation via ligands such as phytochelatins and metallothioneins [7]. Examples of some metal-tolerant plant species are *Brassica juncea* (Indian mustard), *Thlaspi caerulescens,* and *Pteris vittata,* which have been extensively studied for their phytoremediation potential.

The excessive use of agrochemicals or careless handling of them can be dangerous for the environment as well as for a variety of living beings, including humans. Agrochemicals are used excessively and incorrectly, which poses a danger to the viability of the environment. For this reason, more environmentally friendly methods are being used, including biochar, nanoparticles, and biostimulants [8, 9]. As heavy metal pollution in agriculture is so severe and has far-reaching effects, there is an urgent need for sustainable and efficient measures. Removing the soil, using chemicals to extract contaminants, and immobilizing them usually involves considerable costs and a lot of work, and can harm the environment. In phytoremediation, on the other hand, plants clean, stabilize, or reduce the amount of pollutants in the environment at low cost and with the help of the sun [10]. Also, scientists are now focusing on nanomaterials, biochar, microbial inoculants, and integrated soil amendments as possible ways to reduce metals in the soils that farmers use. These practices aim to renew soil health, support crop development, and ensure food safety for humans and the environment [9, 11].

As the world population increases and the demand for healthy and sustainable food increases, the management of heavy metals in agriculture is crucial for both the environment and the safety of all. Continued research and new developments should aim to find resilient crops, find affordable ways to clean up soil, and sensitize people to be aware of how to carefully remediate metal-polluted farmland. Figure 1 shows the mitigation of heavy metal stress in plants by chitosan nanoparticles (CNPs). While heavy metals from the environment lead to oxidative damage (ROS formation) and reduced plant growth, CNPs offer protection by improving photosynthesis and activating the plant's antioxidant defense mechanisms [12, 13].

2 Chitosan and Chitosan Nanoparticles (CNPs)

The biopolymer chitosan is produced by deacetylation of chitin, which is abundant in the exoskeletons of shrimps, lobsters, and crabs as well as in the cell walls of fungi and yeasts [14]. Chitosan is considered an important natural polyaminosaccharide formed by N-deacetylation of chitin, and is the second most common polysaccharide in the world after cellulose [15, 16]. The arrangement of β-(1 $\rightarrow$ 4)-linked D-glucosamine and N-acetyl D-glucosamine, assisted by their occasional disruptions in the polymer chain, sets the degree of deacetylation (DD) percentage and thereby impacts the behavior of chitosan [17]. A pKa of ~ 6.5 in the amine groups of chitosan renders the

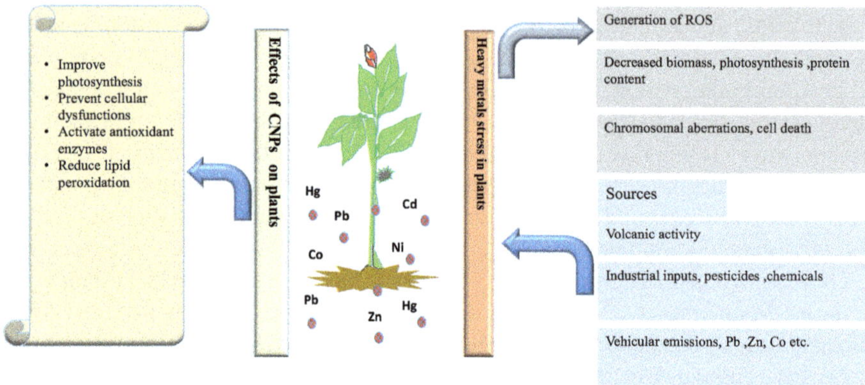

Fig. 1 Mechanistic representation of chitosan nanoparticles (CNPs) mitigating heavy metal-induced stress in plants by improving photosynthesis, preventing cellular dysfunctions, activating antioxidant enzymes, and reducing lipid peroxidation, while counteracting reactive oxygen species (ROS) generation and physiological damage caused by heavy metal exposure from various environmental sources

molecule insoluble at neutral pH. However, if chitosan is protonated with the acid present at a $pH < 6.5$, it dissolves and carries a positive charge. This protonation allows the solution to bind more easily with negatively charged surfaces and contaminants [18]. Records show that compounds without charges can take insoluble forms but can penetrate lipid membranes and attract the attention of biologists [19, 20].

The amount of chitosan that can dissolve in substances depends on its degree of deacetylation and the number of amine groups it can protonate. Conditionally, chitosan dissolves in some organic acids, but not in phosphoric and sulfuric acids. This molecule can be made with different molecular weights and accepts an unknown number of acetyl groups, all of which influence its behavior in solution, such as its particle size, the type of binding, and the willingness to form nanoparticles [21, 22]. The optimal values of these parameters have a strong impact on the degree of demineralization and the degree of purification of the chitosan yield. Ca^{2+} ions are expelled when the pH increases during the process, so it is necessary to re-adjust the pH after demineralization. The proteins that bind to chitin make it difficult to use chitin for diverse biomedical purposes, as it triggers an immune reaction in the human body. Consequently, the removal of proteins is highly essential. Another method to remove proteins is to use proteolytic enzymes during the deproteination stage for demineralized crab or shrimp shells. The sequence of these processes should be demineralization first and then deproteination. CNPs work better when produced at the nanoscale as they are more reactive features and have a better surface area-to-volume ratio. Due to their high surface area, nanoparticles can interact better with contaminants and absorb them better. One important reason for the appreciation of CNPs is that they contain many amino ($-NH_2$) and hydroxyl ($-OH$) groups. They can catch metal ions through chelation or using electrostatic forces, and so CNPs can trap toxic metals [23].

In addition, chitosan becomes positively charged under acidic conditions, which attracts negatively charged materials and makes it even more suitable as an adsorbent. The chemical modification or crosslinking of enzymes allows their performance to be improved by giving them more selectivity and making them more reliable. In general, chitosan and its nanoparticles are suitable for the removal of heavy metals in agriculture and nature due to their ability to change shape, react with various substances and be environmentally friendly. In addition, the use of chitosan-based nanomaterials has been shown to activate plant defenses by inducing systemic acquired resistance (SAR) and triggering increased expression of defense-related genes connected to defense. The effect of phytochemicals on plant cell membranes can facilitate their entry. This allows the toxins to be internalized by chelation and removed by storage in plant vacuoles. This process not only reduces the toxicity of the metals but also increases the antioxidant content of the plant, and helps it to resist the stress caused by free radicals [22, 24].

In addition to soil detoxification, chitin-based nanomaterials supplement plant growth by increasing root system growth, increasing seed germination, and increasing plant chlorophyll content [25]. Reducing plant stress promotes growth, as these microorganisms can moderate the number of hormones and nutrient uptake in a tough situation. Moreover, chitosan has the potential to boost microbial activity in the rhizosphere, resulting in favorable interactions with plant roots. It also improves the soil structure, so that the soil is better aerated and can retain water better. Chitosan nanocomposites can be adapted in various forms for the agricultural industry, as they can take on hydrogel, film, and suspension properties. To achieve this, coatings with chitosan and polymers or ZnO or CuO nanoparticles improve the entity and functions of the nano-carriers, which help in delivering nutrients and safeguarding plants against all kinds of threats [26, 27].

Smart agriculture utilizes knowledge of the use of chitosan nanomaterials to facilitate the resolution of problems caused by the presence of heavy metals in the soil and to facilitate the maintenance of eco-friendly practices by reducing the use of chemical agricultural products. Chitosan also plays a role in activating crop protection genes and inducing systemic resistance to pathogens, which will lead to less dependence on synthetic pesticides. An example of this is the fact that previous research has shown that chitosan is a signaling molecule in plants that stimulates defense responses, such as the presence of lignin and the accumulation of phytoalexin. Similarly, the use of chitosan carbon black particles greatly contributed to the tolerance of wheat fruits to drought through increasing the activities of antioxidant enzymes and maintaining the concentrations of stomatal conductance [28]. The problems caused by heavy metals in the soil are eliminated through smart agriculture with chitosan nanomaterials, which not only eliminate these issues but also enable eco-friendly agricultural practices by reducing the chemicals in agriculture. The application of chitosan nanomaterials in smart agriculture helps to overcome the challenges posed by heavy metals in soil and promotes an environmentally friendly process by reducing the consumption of chemical agricultural products. Future studies should focus on increasing the efficiency of mass production of drones, verifying the environmental effects in the long-term perspective, and developing valid approaches for delivering the materials to crops and fields. It may also be supplemented by integration with remote sensing technologies, artificial intelligence-based applications and monitoring [29].

In addition, treatment with CNPs has the potential to enhance the uptake of nutrients in the plant and the performance of photosynthesis, making plants not only more productive but also more resilient. Future research should look at the efficiency of mass production of drones, inspecting the long-term environmental impact and developing suitable routes for transportation of the materials to the plants and farmland.

3 Mechanisms of Heavy Metal Stress in Plants

Cd, Pb, As, and Hg are some of the most dangerous environmental pollutants that can affect the growth and well-being of plants [30]. They can seriously damage the chemical and physical functions of plants. They disrupt ion control, cause vital ions to be removed from cells, and influence water potential. They also disrupt the function of stomata and membranes, which can lead to problems in plant metabolism [31]. HM-tolerant plants are classified into two types: (i) excluders, which limit HM uptake and prevent accumulation in the shoots and (ii) hyperaccumulators, which take up and tolerate high HM concentrations through internal detoxification [32].

ROS are created in excess, including hydrogen peroxide (H_2O_2), superoxide anions (O_2^-), and hydroxyl radicals (•OH), which leads to damage to lipids, proteins, and nucleic acids in the body. This imbalance between ROS production and the plant's ability to detoxify them leads to oxidative stress, an important side effect of heavy metal exposure [33]. Oxidative stress weakens the outer membrane, which in turn leads to more permeability, excessive ion leakage, and a loss of specific cell compartments. Additionally, damage to cell membranes leads to electrolyte loss, altered osmotic balance, and impaired nutrient uptake. The presence of heavy metals prevents the proper synthesis of chlorophyll and damages the photosystems, seriously blocking photosynthesis. This not only reduces the efficiency of light absorption but also disrupts the electron transport chain in the thylakoid membranes. At the same moment, respiratory activity decreases, the performance of enzymes is reduced, and the process of energy production no longer functions effectively. This combined impairment of the photosynthetic and respiratory pathways leads to a significant reduction in ATP and NADPH production, which are essential for cellular processes and growth. This eventually leads to less productivity and energy in the plant [18].

Nevertheless, various plants, mainly metal hyperaccumulators and halophytes, have found ways to cope with the presence of heavy metals in the soil. Some of these ways include limiting the uptake of metals at the roots, storing metals in vacuoles, and creating peptides that bind to metals, such as phytochelatins (PCs) and metallothioneins. PCs are made by plants in response to metals and bind strongly to Cd^{2+}, As^{3+}, and Pb^{2+}, enabling the compounds to be safely stored in vacuoles. Metallothioneins (MTs) bind heavy metals and help to maintain the metal balance in the body. They trigger signaling through hormones and also stimulate the action

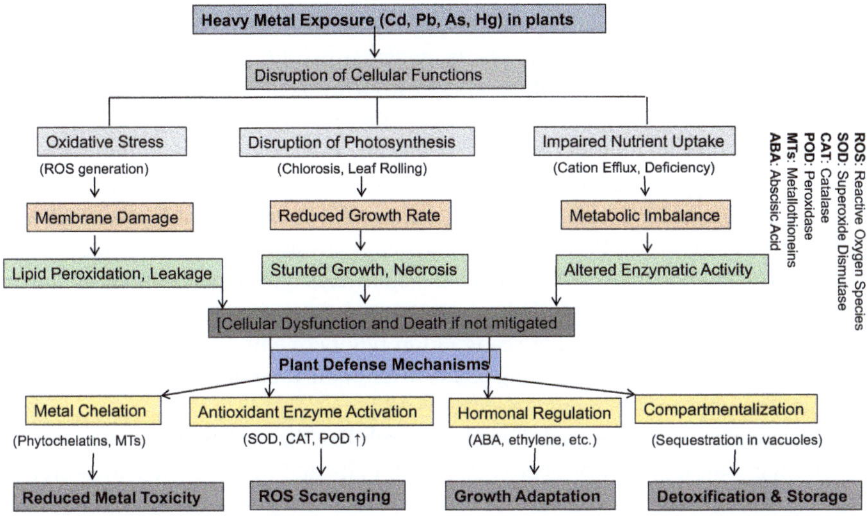

Fig. 2 Illustrative pathway of heavy metal-induced cellular dysfunction in plants and corresponding defense mechanisms, highlighting oxidative stress, disruption of photosynthesis, and impaired nutrient uptake, along with plant adaptive responses including metal chelation, antioxidant enzyme activation, hormonal regulation, and compartmentalization

of superoxide dismutase and other antioxidant enzymes. Salicylic acid, jasmonic acid, and abscisic acid are frequently involved in supportive processes that explain, for example, why stomata close, control oxidation reactions and respond to stress signals. These enzymes help reduce oxidative stress in response to environmental factors [34]. Because of these mechanisms, metals can reduce the harmful effects of stress on cells and allow cells to continue to grow and develop properly [35]. The general mechanism of how plants behave in stress situations is shown in Fig. 2.

4 Role of CNPs in Stress Alleviation

CNPs are well-suited to increase the resistance of plants to both high salt and heavy metal stresses. The reason for their versatility in reducing stress lies in their interesting physicochemical properties and their interactions with various cellular signaling pathways and mechanisms. One of the main ways in which CNPs support protection against heavy metals is through the accumulation and binding of metal ions [36]. As a result of its amino and hydroxyl groups, chitosan binds strongly to metal ions such as cadmium, lead, and copper. Thanks to this binding, the metals are kept closer to the soil particles, so that fewer of them reach the root tissue of the plant. It should be noted that due to its higher metal binding capacity, chitosan interacts better at higher pH and absorbs more metals, which is due to the additional unprotonated amine groups [37, 38].

Apart from metal chelation, the presence of chitosan nanoparticles leads to increased antioxidant activity in plants under stress [27, 39]. When plants are subjected to abiotic stress, they often produce high levels of ROS that damage various cellular components. The presence of chitosan NPs encourages superoxide dismutase (SOD), catalase (CAT), and peroxidase (POD) to work so that the plant can better cope with higher levels of ROS [40, 41]. The activated antioxidant enzymes maintain the right oxidation level in the cells, reduce the stress caused by free radicals, and prevent cell damage. Photosynthesis is a vital physiological process that can be severely affected by stress, and chitosan nanoparticles have been reported to have a protective effect in this regard. The application of chitosan nanoparticles improves photosynthetic efficiency by increasing chlorophyll content and preserving chloroplast structure and function, enabling plants to maintain their growth and productivity under unfavorable environmental conditions [42].

CNPs are also useful because they help to stabilize the membrane. Heavy metals, in particular, can lead to lipid peroxidation in cells and damage their membranes, resulting in the release of cell contents and cell death. They also inhibit the formation of lipid peroxide, which makes the membranes fluid and functional. This enhances the cell culture of living cells and their resistance to stress elements. The cell membranes should be sound to maintain metabolic functions, multi-compartmentalization, and responses to external stressors. Chitosan also plays a role in the release and storage of metals in the plant. They can modify the action of metal transporter genes to transport a smaller quantity of the metal, not to the shoots, but to the roots. This system has resulted in essential micronutrients being taken up into the cells and toxic metals being well managed or stored in a specific area of the cell. The ability of chitosan to mitigate plant stress due to the influence of metals in polluted soils tends to vary and is influenced by numerous factors, including soil and plant type, as well as the nature of the stressor. These compounds promote the healthy transport and excretion of toxins to help manage nutrients and protect against the toxic impacts of certain metals [37]. Timely spraying of leaves with chitosan is useful in mitigating Cd toxicity in wheat by reducing the extent of damage caused by the metal and significantly improving plant health [43]. In sum, the mechanism of chitosan nanoparticles in relieving plant stress involves the application of different techniques, including chemical, physiological, and molecular. They help eliminate and regulate toxic metals and fulfill other fundamental functions in the body, which is why they are used in many cases in organic farming. The effectiveness of chitosan nanoparticles in mitigating the toxic effects of various heavy metals has been well-documented in numerous studies. Table 1 illustrates a comprehensive summary of these results, detailing the specific adverse effects of key metals such as cadmium, lead, and arsenic on plant systems and the corresponding protective outcomes observed following the application of chitosan-based nanoparticles.

Table 1 Summary of heavy metal toxic effects in plants and mitigation by chitosan nanoparticles (CNPs)

Heavy metal	Effect of heavy metals on plants	Effect of chitosan NPs on the plant	Reference
Cadmium	Affects plant morphology, physiology, and biochemistry	The stimulatory effects of chitosan (CTS) could significantly alleviate the toxic effects of cadmium (Cd) on wheat seedlings	[44, 45]
	Numerous studies have shown that this metal may inhibit seed germination, decrease total plant length, inhibit root elongation, and reduce the number of leaves per plant, leading to the death of the plant	Enhances antioxidant enzyme activity (SOD, POD, CAT)	[46–49]
	Cadmium may result in plant physiological disorders, especially those due to increased oxidative stress (overproduction of reactive oxygen species (ROS)	Alters subcellular Cd distribution	[50, 51]
Lead	Decreases enzyme activities (e.g., aryl sulfatase)	Chitosan increased shoot length by 10–11.62% compared to Pb-stressed plants without Chitosan	[52, 53]
	It creates morphological, physiological, and biochemical problems	Reduced the negative impact of Pb on pigments	[52, 54, 55]
	Adsorption of lead in the roots is observed in several species of plants. Absorbed lead is accumulated in the roots, and only a small fraction is transported to aerial plant parts	Decreased the concentration of Pb in plant shoots by 8.46% (at 50 ppm Pb) and 10.43% (at 100 ppm Pb)	[52, 54]
Cupper	Chlorosis, necrosis, root inhibition, and stunted growth	Maintains stomatal conductance	[56, 57]
	Reduces yield and impairs photosynthesis and root development	Promotes better root and shoot development	[56, 57]
Arsenic	Leads to chlorosis, necrosis, growth inhibition, and plant death	Significantly decreases arsenic accumulation in roots and leaves and reduces its translocation from root to shoot	[58, 59]
	Arsenic contamination of soils primarily affects productivity	Increases all growth parameters and significantly mitigates the reductions caused by arsenic	[58, 60]

(continued)

Table 1 (continued)

Heavy metal	Effect of heavy metals on plants	Effect of chitosan NPs on the plant	Reference
	Inorganic arsenic was five times more toxic to corn grown on a sand soil than on a clay soil	Enhances the activity of RuBisCo and carbonic anhydrase (CA), counteracting the inhibition caused by arsenic	[58, 61]
Zinc	The most obvious symptoms of Zn toxicity reported in plants are the inhibition of growth, and chlorosis of young leaves (probably the consequence of lower uptake of Fe^{2+} and Fe^{3+}), and can lead, in some cases, to cell death	Enhanced biomass accumulation, root morphology, chlorophyll content, photosynthetic parameters, the ascorbate–glutathione (AsA-GSH) cycle, antioxidant enzyme activity, and soluble protein content in wheat seedlings	[62, 63]
	Excess Zn may alter plant development	Reduces malondialdehyde (MDA) levels and Zn translocation to the above-ground parts	[62, 64, 65]
	At high concentrations, Zn becomes toxic. Studies demonstrated that in A. thaliana, treatment with 0.075 mM of Zn did not affect plant development, whereas at concentrations above 0.1 mM, Zn became toxic	100 mg L^{-1} CTS increased the tolerance index of shoots and roots by 30.59% and 47.90%, respectively, and significantly reduced oxidative damage caused by Zn stress	[62, 66]
Nickel	The excess of Ni is associated with many adverse effects in plants like reduction in germination	Increased shoot length (39%), aerial biomass (24%), leaf area (24%), and boosted seed yield by 3.1-folds compared to untreated, stressed plants	[67, 68]
	Nickel can create problems in plant growth, cell division, biomass production, nutrient absorption, photosynthesis, transpiration, leaf chlorosis, and necrosis	Significantly lowered the uptake and accumulation of nickel in all parts of the plant	[67–73]
	All the toxic effects due to Ni stress reduce crop yield and quality	Lowered MDA levels, indicating reduced oxidative damage to cell membranes	[74, 75, 68]
Mercury	Negatively impacts both the light-dependent and light-independent (dark) reactions of photosynthesis	Chelates mercury ions, neutralizing their toxicity	[76, 77]
	The transport of water through the plant's cell membranes is adversely affected by the presence of mercury	Triggers plant defense mechanisms for stress tolerance	[36, 77, 78]

5 Crop-Specific Applications and Outcomes

Studies of chitosan and its nanoscale forms in crops have been shown to be very beneficial for rice, maize, and soybeans. There are several reports of corn showing increased plant height, leaves, leaf size, and leaf area in early stages of development. Treatment of plants with chitosan increased the dry weight of shoots and roots. Under cadmium stress conditions, the roots become longer, have a larger surface area, and volume. The addition of chitosan nano-fertilizer in various concentrations, between 0.01 and 0.16% to the foliage resulted in an increase in plant height, root length, root number, and stem diameter, with the highest concentration causing the strongest effects. Chitosan and modified chitosan showed better results in root extension and seedling height, while nano-chitosan improved plant height and leaf area [79]. The application of chitosan to seeds helped to improve seed germination and increased the height of maize and sorghum plants. On the other hand, in some studies, no major changes were observed, and in some cases, the application of polymers hurt seed germination and development. This could be due to various factors like the concentration of polymers, the simultaneous use of fertilizers or salicylic acid, and differences between maize hybrids. Increasing the concentration of Cu-chitosan nanoparticles between 0.01 and 0.12% improved the germination rate, shoot length, root length, root number, seedling length, fresh and dry weight of maize, compared to lower and higher concentrations, whereas too much copper affected the growth of the crop. Both salicylic acid-chitosan nanoparticles and Zn-chitosan nanoparticles resulted in improved germination, shoot–root growth, weight, seedlings vigor, and plant height and stem dimension growth with increased root length and root number. The use of chitosan-silicon nano-fertilizer resulted in higher leaf area, shoot and root growth, and root number than the control group. Although these discoveries are promising, the fact that corn does not always respond in the same way emphasizes the need for experiments to investigate various factors, like the environment, properties of chitosan, and plant species [80–82].

Rice growth has been shown to benefit from the presence of chitosan nanoparticles and related substances. While in some experiments no significant effects on seedling height were observed, in other experiments it was found that soaking the seeds and spraying the leaves with chitosan increased both plant height and dry matter, as well as leaf and root weight. Chitosan, with fewer acetyl groups in the chains, showed the greatest improvement in plant growth. Oligochitosan increased the growth of roots and stems, increased metabolic activities, and promoted efficient photosynthesis in the plant. In addition, chitosan nanoparticles treated with lanthanum increased both the height and fresh weight of rice, demonstrating that modification of chitosan can promote rice growth. Two factors responsible for such outcomes are the product's acetyl group ratio and several other factors that play a role in corn. In soybeans, the use of chitosan has improved seedling growth, seed germination, and all-around vegetative growth. Treating plants with chitosan raises metabolism and causes hormone-like changes that lead to more biomass on shoots and roots, a taller plant, and a larger leaf area. Under abiotic stress such as salinity or drought, chitosan helps improve antioxidant defenses by increasing the action of the enzymes superoxide dismutase, catalase, and peroxidase. It improves the plant's ability to resist damage, prevents membrane breakdown, and helps maintain photosynthesis and water retention at beneficial levels under pressure [24, 83]. Chitosan nanoparticles help plants to better absorb nutrients and fix nitrogen, which leads to a higher protein content in the seed. Chitosan also influences which genes are expressed, making plants better able to resist external pressure. All these benefits contribute to soybeans being healthier on a physiological and biochemical level [84].

All in all, the use of chitosan and its nano versions in rice, maize, and soybeans has been found to increase germination rates, contribute to biomass accumulation, and promote root and shoot system growth. As a result, there is a restoration of physiology and biochemistry, including an increase in antioxidant enzymes, more efficient photosynthesis, strong membranes, and better resilience under abiotic stress [37]. The extent of these effects also depends on the properties of chitosan, the concentration, the way it is used, and the conditions of the plants and the environment. Therefore, chitosan helps farmers to make cropping more sustainable and deal with stress, but further studies should ensure that it is used optimally for the needs of each plant.

6 Future Prospects

To ensure the successful transition of CNPs from laboratory research to real-world agricultural applications, large-scale production must be both economically and environmentally sustainable. A major hurdle is to scale up synthesis from small lab-scale batches to industrial quantities without sacrificing quality or incurring excessive costs. A growing trend is the adoption of green synthesis approaches that avoid toxic solvents and energy-intensive methods. Biological resources such as plant extracts and microbial enzymes are increasingly being used to form nanoparticles, reducing the need for hazardous chemicals. A practical and sustainable source of the starting material for chitosan can be found in the waste streams of the seafood industry, particularly in the shells of crabs, shrimps, and lobsters. Utilizing this bio-waste not only reduces raw material costs but also promotes a circular economy by turning industry by-products into valuable agricultural inputs.

Innovative process technologies are also of central importance. In contrast to traditional batch production, the development of continuous production systems offers the possibility of achieving constant particle sizes and uniform product quality at high volumes. Advanced techniques such as spray drying and high-pressure homogenization are currently being investigated for the production of water-dispersible, stable CNP powders on an industrial scale. These methods allow greater control over the properties of the nanoparticles and ensure that they retain their effectiveness when used in practice. Ultimately, overcoming the challenge of low-cost, high-throughput manufacturing remains a critical step toward the widespread implementation of CNPs in agriculture.

In addition to manufacturing, the real impact of CNPs lies in the smooth integration of their products into sustainable agriculture. In the future, CNPs will be used in crop production, not as stand-alone remedies, but as a multifunctional approach to crop management. Another area of development is multifunctional nano-formulation, in which CNPs will act as carriers for other active ingredients. These can be micronutrients, nano-fertilizers, beneficial microbes, nano-bio fertilizers, or bio-pesticides, so that a single application may provide nutrition, growth, and protection against pathogens. Such integrated solutions minimize the use of chemical inputs, reduce the cost of farming and promote environmentally friendly practices [85, 86].

The synergetic application of other sustainable technologies and CNPs is also quite promising. An example of their potential is the ability to improve soil structure and water retention when used together with biochar, as well as high nutrient uptake and root growth when used together with plant growth-promoting rhizobacteria (PGPR) [87, 88]. Further, precision agriculture tools provide a method to optimize the use of most CNPs. With specialized drones and GPS-guided coordinates ensuring direct foliar inoculation, CNPs on seed layers provide extended protection in the early stages. Intelligent delivery vehicles that apply CNPs only when necessary, for example, in drought or pathogenic conditions, are also part of the future. These precise methods ensure effective distribution, save the environment a lot of waste products, and improve the overall performance of CNPs on crops. The integration of CNPs into innovation systems can help agriculture reach the next level and make it more productive, resilient, and sustainable [86].

7 Conclusion

Contamination with heavy metals poses a significant threat to agriculture on Earth, endangering human health, the fertility of cultivated land, and the nutritional value of crops consumed by humans. As explained in this chapter, CNPs have become one of the most promising and sustainable methods to reduce this pervasive pollution. CNPs offer effective protection to plants by exerting multiple synergistic effects. First, they have a large surface area and numerous amino and hydroxyl groups, which can significantly chelate and immobilize hazardous metal ions such as cadmium, lead, and arsenic directly in the soil, thereby reducing the uptake of these metal ions by plant roots. At the same time, CNPs act as highly effective biostimulants that strengthen the plant's internal defense mechanisms by increasing levels of key antioxidant enzymes, stabilizing cell membranes against oxidative stress, and increasing photosynthetic efficiency, so that plants maintain essential physiological processes even under stress. Aside from the immediate remedies they can provide, the real strength of CNPs lies in their impressive multifunctionality and environmentally friendly status. They have a dual effect, as they not only clean up the plant habitat, but also promote growth by leading to increased biomass, an enhanced root system and an increase in production of important crops such as soybean, maize, and rice. Besides, the fact that they can be used as nano-carriers for beneficial compounds or other protective substances opens up the opportunity of creating sophisticated, high-tech agricultural products that are all in one. Due to their water solubility and biodegradability, resulting from renewable bio-waste, CNPs are an even more environmentally friendly alternative to persistent synthetic chelators and fertilizers, which make even more sense in a world that is moving toward a more sustainable and circular agricultural system. Although this market has enormous potential, the journey from discovery in the laboratory to widespread application practice has not been without its challenges. Along the way, the critical issue of developing standard formulations of nanoparticles, scaling up and economically producing nanoparticles in a cost-effective manner, and a thorough understanding of their long-term ecology and transport in soil–plant systems must be addressed. Thus, there should be a joint and interdisciplinary research initiative. Materials scientists, agricultural scientists, soil ecologists, and toxicologists should work together to perfect these technologies, develop viable safety measures and establish clear guidelines for their effective and responsible use. By bridging this gap, chitosan nanoparticles will mark the beginning of resilient agriculture, ensuring not only our food safety but also environmental sustainability and leading to a healthier planet.

References

1. Kaplan A, Khan MN, Wahab S, Assad N, Adnan M, Hafsa, Iqbal M (2025a) Chapter 13—Thallium in soil environments and its biological availability. In: Saud S, Fahad S, Wang D (eds) Beneficial elements for remediation of heavy metals in polluted soil. Elsevier, pp 373–398. https://doi.org/10.1016/B978-0-443-26522-8.00013-1
2. Khan MN, Kaplan A, Iqbal M, Wahid N, Wahab S, Assad N, Hayat K, Adnan M, Hafsa (2025) Chapter 4—Cadmium dynamics: beneficial elements and chemical reactions in soil. In: Saud S, Fahad S, Wang D (eds) Beneficial elements for remediation of heavy metals in polluted soil. Elsevier, pp 93–139. https://doi.org/10.1016/B978-0-443-26522-8.00004-0
3. Arif N, Yadav V, Singh S, Kushwaha BK, Singh S, Tripathi DK, Vishwakarma K, Sharma S, Dubey N, Chauhan DJPrtx (2016) Assessment of antioxidant potential of plants in response to heavy metals 97–125
4. Malik SY, Arshad M, Batool S, Batool S, Iram F, Batool T, Ali B, Wahab S, Khan MN, Kaplan A, Bibi M, Assad N, Alnemar SK, Farouk Elsadek M, Ercisli S, Iqbal M (2024) Atrazine-induced toxicity and biochemical changes in *Mus musculus*: the ameliorative efficacy of *Nigella sativa* L. (Ranunculaceae) Oil. Pol J Environ Stud 33:5211–5220. https://doi.org/10.15244/pjoes/183076
5. Ashraf A, Dutta J, Farooq A, Rafatullah M, Pal K, Kyzas GZJJoMS (2024) Chitosan-based materials for heavy metal adsorption: recent advancements, challenges and limitations 1309:138225
6. Assad N, Laila MB, Naeem-Ul-Hassan M, Ali B, Kaplan A, Khan MN, Hayat K, Ozdemir FA (2025b) Phytohormones as key players: boosting oilseed crop tolerance to nutrient deficiency stress. In: Abdel Latef AAH (ed) Oilseed crops under abiotic stress: mitigation strategies and future perspectives. Springer Nature Singapore, Singapore, pp 415–440. https://doi.org/10.1007/978-981-96-8346-8_13
7. Viehweger KJBS (2014) How plants cope with heavy metals 55:1–12
8. Assad N, Ansa, Laila MB, Ullah S, Adnan M, Wahab S, Kaplan A, Khan MN, Ullah B (2025a) Nanotechnology in oilseed crops: a new frontier for abiotic stress adaptation. In: Abdel Latef AAH (ed) Oilseed crops under abiotic stress: mitigation strategies and future perspectives.

Springer Nature Singapore, Singapore, pp 507–536. https://doi.org/10.1007/978-981-96-8346-8_16
9. Sebastian A, Nangia A, Prasad MNV (2020) Advances in agrochemical remediation using nanoparticles. In: Agrochemicals detection, treatment and remediation. Elsevier, pp 465–485
10. Haq EU, Afzal E, Aslam M, Zafar F, Assad N (2025) Green approach to phytochemicals: alkaloids purification as used in health therapeutical and wellness activities. Insights-J Health Rehabil 3(1):201–209
11. Khalid F, Asif K, Rasheed Y, Ashraf H, Maqsood MF, Rana S, Zulfiqar U, Naz N, Shahbaz M, Sardar RJB, Biotechnology A (2023) Nano priming for boosting growth and resilience in crops under abiotic stresses 53:102892
12. Ochoa L, Shrivastava M, Srivastava S, Cota-Ruiz K, Zhao L, White JC, Hernandez-Viezcas JA, Gardea-Torresdey JLJESN (2025) Nanomaterials for managing abiotic and biotic stress in the soil-plant system for sustainable agriculture. Environ Sci Nano 12:1037–1058
13. Zhou P, Adeel M, Shakoor N, Guo M, Hao Y, Azeem I, Li M, Liu M, Rui YJN (2020) Application of nanoparticles alleviates heavy metals stress and promotes plant growth: An overview 11(1):26
14. Lárez-Velásquez CJP (2023) Chitosan: an overview of its multiple advantages for creating sustainable development poles 33(1):e20230005
15. Dutta JJMD (2013) Engineering of polysaccharides via nanotechnology 87–134
16. Rinaudo MJPips (2006) Chitin and chitosan: properties and applications 31(7):603–632
17. Cunha RA, Soares TA, Rusu VH, Pontes FJ, Franca EF, Lins RD (2012) The molecular structure and conformational dynamics of chitosan polymers: an integrated perspective from experiments and computational simulations. In: The complex world of polysaccharides. IntechOpen
18. Phurailatpam L, Dalal VK, Singh N, Mishra SJFiSFS (2022) Heavy metal stress alleviation through omics analysis of soil and plant microbiome 5:817932
19. Fu C-C, Tran HN, Chen X-H, Juang R-SJJoI, Chemistry E (2020) Preparation of polyaminated Fe3O4@ chitosan core-shell magnetic nanoparticles for efficient adsorption of phosphate in aqueous solutions 83:235–246
20. Pal P (2018) Methylene blue removal: an approach towards sludge management after adsorption of cadmium onto surfactant modified chitosan beads 95:357–364
21. Kaczmarek MB, Struszczyk-Swita K, Li X, Szczęsna-Antczak M, Daroch MJFib, biotechnology (2019) Enzymatic modifications of chitin, chitosan, and chitooligosaccharides 7:243
22. Vakili M, Rafatullah M, Salamatinia B, Abdullah AZ, Ibrahim MH, Tan KB, Gholami Z (2014) Amouzgar PJCp. Application of chitosan and its derivatives as adsorbents for dye removal from water and wastewater: A review 113:115–130
23. Benettayeb A, Haddou B (2023) New biosorbents based on the seeds, leaves and husks powder of *Moringa oleifera* for the effective removal of various toxic pollutants 103(18):6859–6884
24. Shah W, Ullah S, Ali S, Idrees M, Khan MN, Ali K, Khan A, Ali M, Younas F (2021) Effect of exogenous alpha-tocopherol on physio-biochemical attributes and agronomic performance of lentil (Lens culinaris Medik.) under drought stress. PloS One 16(8):e0248200
25. Ajmal M, Ullah R, Muhammad Z, Khan MN, Kakar HA, Kaplan A, Okla MK, Saleh IA, Kamal A, Abdullah A (2023) Kinetin capped zinc oxide nanoparticles improve plant growth and ameliorate resistivity to polyethylene glycol (PEG)-induced drought stress in *Vigna radiata* (L.) R. Wilczek (Mung Bean). Molecules 28(13):5059
26. Assad N, Abbas A, ur Rehman MF, Naeem-ul-Hassan M (2024) Photo-catalytic and biological applications of phyto-functionalized zinc oxide nanoparticles synthesized using a polar extract of Equisetum diffusum D. RSC Adv 14(31):22344–22358
27. Ghaffar S, Abbas A, Naeem-ul-Hassan M, Assad N, Sher M, Ullah S, Alhazmi HA, Najmi A, Zoghebi K, Al Bratty M (2023) Improved photocatalytic and antioxidant activity of olive fruit extract-mediated ZnO nanoparticles. Antioxidants 12(6):1201
28. Román-Doval R, Torres-Arellanes SP, Tenorio-Barajas AY, Gómez-Sánchez A, Valencia-Lazcano AAJP (2023) Chitosan: properties and its application in agriculture in context of molecular weight 15(13):2867

29. Mukarram M, Ali J, Dadkhah-Aghdash H, Kurjak D, Kačík F, Ďurkovič JJ (2023) Chitosan-induced biotic stress tolerance and crosstalk with phytohormones, antioxidants, and other signalling molecules 14:1217822
30. Angon PB, Islam MS, Kc S, Das A, Anjum N, Poudel A, Suchi SA (2024) Sources, effects and present perspectives of heavy metals contamination: soil, plants and human food chain. Heliyon 10(7):e28357. https://doi.org/10.1016/j.heliyon.2024.e28357
31. Guo Z, Gao Y, Yuan X, Yuan M, Huang L, Wang S, Liu C, Duan C (2023) Effects of heavy metals on stomata in plants: a review. Int J Mol Sci 24(11). https://doi.org/10.3390/ijms24119302
32. Dubey R (2010) Metal toxicity, oxidative stress and antioxidative defense system in plants. In: Reactive oxygen species and antioxidants in higher plants. CRC Press, pp 177–203
33. Ullah S, Khan M, Idrees M, Ali S, Khan MN, Adnan M, Ozdemir FA (2020) 32. Enhancement effect of Hydroxyl Benzoic Acid (HBA) foliar application on growth performance of *Triticum aestivum* L.(Wheat) under induced chromium toxicity. Pure Appl Biol (PAB) 9(4):2529–2542
34. Servin A, Elmer W, Mukherjee A, De la Torre-Roche R, Hamdi H, White JC, Bindraban P (2015) Dimkpa CJJoNR. A review of the use of engineered nanomaterials to suppress plant disease and enhance crop yield 17:1–21
35. Benettayeb A, Morsli A, Elwakeel KZ, Hamza MF, Guibal EJAS (2021) Recovery of heavy metal ions using magnetic glycine-modified chitosan-application to aqueous solutions and tailing leachate 11(18):8377
36. Kanwal M, Sher M, Abbas A, Akhtar S, Siddique AB, ul Hasan MN, Assad N, Alhazmi HA, Amin HM (2025b) Dual colorimetric sensing of Hg (II) and Fe (III) using sulfanilamide-stabilized silver nanoparticles and evaluating their photodegradation and antibacterial properties. J Water Process Eng 75:107981
37. Kashyap PL, Xiang X, Heiden P (2015) Chitosan nanoparticle based delivery systems for sustainable agriculture. Int J Biol Macromol 77:36–51. https://doi.org/10.1016/j.ijbiomac.2015.02.039
38. Zaman MB, Poolla R, Singh P, Gudipati T (2020) Biogenic synthesis of CuO nanoparticles using Tamarindus indica L. and a study of their photocatalytic and antibacterial activity. Environ Nanotechnol, Monit & Manag 14:100346. https://doi.org/10.1016/j.enmm. 2020.100346
39. Khan ZR, Assad N, Naeem-ul-Hassan M, Sher M, Alatawi FS, Alatawi MS, Omran AM, Jame RM, Adnan M, Khan MN (2023) *Aconitum lycoctonum* L.(Ranunculaceae) mediated biogenic synthesis of silver nanoparticles as potential antioxidant, anti-inflammatory, antimicrobial and antidiabetic agents. BMC Chemistry 17(1):128
40. Ali B, Batool R, Arshad A, Azmat A, Hafeez A, Wahid N, Ali S, Khan MN, Singh N, Garhwal V, Fahad S (2025b) Chapter 10—Regulation of enzymatic and nonenzymatic antioxidants in plants by biochar under abiotic stress. In: Abdel Latef AAH (ed) Biochar in mitigating abiotic stress in plants. Academic Press, pp 159–171. https://doi.org/10.1016/B978-0-443-24137-6.00010-0
41. Batool S, Batool S, Batool T, Iram F, Almas T, Faizan M, Assad N, Al-Sadoon MK, Khan MN, Ullah B (2024) Delving the role of the ameliorative effects of *Caralluma tuberculata* NE Br.(Apocynaceae) on diabetes and its effect on the organs weight of alloxan-induced adult male mice. Pol J Environ Stud 33(1)
42. Riseh RS, Vatankhah M, Hassanisaadi M, Kennedy JFJCP (2023) Increasing the efficiency of agricultural fertilizers using cellulose nanofibrils: a review 321:121313
43. Liu J, Gai L, Zong HJPP, Biochemistry (2021) Foliage application of chitosan alleviates the adverse effects of cadmium stress in wheat seedlings (*Triticum aestivum* L.) 164:115–121
44. Hussain I, Iqbal M, Qurat-Ul-Ain S, Rasheed R, Mahmood S, Perveen A, Wahid A (2012) Cadmium dose and exposure-time dependent alterations in growth and physiology of maize (*Zea mays*). Int J Agric Biol 14(6)
45. Zong H, Liu S, Xing R, Chen X, Li P (2017) Protective effect of chitosan on photosynthesis and antioxidative defense system in edible rape (*Brassica rapa* L.) in the presence of cadmium. Ecotoxicol Environ Saf 138:271–278

46. Bae J, Benoit DL, Watson AK (2016) Effect of heavy metals on seed germination and seedling growth of common ragweed and roadside ground cover legumes. Environ Pollut 213:112–118
47. Mani D, Sharma B, Kumar C, Balak S (2012) Cadmium and lead bioaccumulation during growth stages alters sugar and vitamin C content in dietary vegetables. Proc Nat Acad Sci, India Sect B: Biol Sci 82(4):477–488
48. Qu DanYang QD, Gu WanRong GW, Li LiJie LL, Li Jing LJ, Li CaiFeng LC, Wei Shi WS (2018) Regulation of chitosan on the ascorbate-glutathione cycle in *Zea mays* seedling leaves under cadmium stress. Plant Sci J 36(2):291–299
49. Soudek P, Petrová Š, Vaňková R, Song J, Vaněk T (2014) Accumulation of heavy metals using *Sorghum* sp. Chemosphere 104:15–24
50. Moradi R, Pourghasemian N, Naghizadeh M (2019) Effect of beeswax waste biochar on growth, physiology and cadmium uptake in saffron. J Clean Prod 229:1251–1261
51. Zong H, Li K, Liu S, Song L, Xing R, Chen X, Li P (2017) Improvement in cadmium tolerance of edible rape (*Brassica rapa* L.) with exogenous application of chitooligosaccharide. Chemosphere 181:92–100
52. Abdelhameed RE, Abdalla H, Abdel-Haleem M (2024) Offsetting pb induced oxidative stress in *Vicia faba* plants by foliar spray of chitosan through adjustment of morpho-biochemical and molecular indices. BMC Plant Biol 24(1):557. https://doi.org/10.1186/s12870-024-05227-w
53. Brown D, Bates J (1972) Uptake of lead by two populations of *Grimmia doniana.* J Bryol 7(2):187–193
54. Kumar B, Smita K, Flores LC (2017) Plant mediated detoxification of mercury and lead. Arab J Chem 10:S2335–S2342
55. Zeng LS, Liao M, Chen CL, Huang CY (2007) Effects of lead contamination on soil enzymatic activities, microbial biomass, and rice physiological indices in soil–lead–rice (*Oryza sativa* L.) system. Ecotoxicol Environ Saf 67(1):67–74
56. Gomes DG, Debiasi TV, Pelegrino MT, Pereira RM, Ondrasek G, Batista BL, Seabra AB, Oliveira HC (2022) Soil treatment with nitric oxide-releasing chitosan nanoparticles protects the root system and promotes the growth of soybean plants under copper stress. Plants 11(23):3245
57. Paul S, Sasaki C, Matsuyama N, Kato K (2011) Effect of percolation pattern on yields and accumulation of copper and cadmium in the rice plants with soil dressing models. J Environ Sci Eng 5:1464–1473
58. Ahmad B, Dar TA, Khan MMA, Ahmad A, Rinklebe J, Chen Y, Ahmad P (2022) Oligochitosan fortifies antioxidative and photosynthetic metabolism and enhances secondary metabolite accumulation in arsenic-stressed peppermint. Front Plant Sci 13:987746
59. Gulz PA, Gupta S-K, Schulin R (2005) Arsenic accumulation of common plants from contaminated soils. Plant Soil 272(1):337–347
60. Jacobs L, Keeney D, Walsh L (1970) Arsenic residue toxicity to vegetable crops grown on Plainfield Sand 1. Agron J 62(5):588–591
61. Jacobs L, Keeney D (1970) Arsenic-phosphorus interactions on corn. Commun Soil Sci Plant Anal 1(2):85–93
62. Fan L, Zhang J, Xu Z, Jiao Q, Song Z, Yu P, Liu D, Li G, Fahad S, Sahito ZA, Alshehri MA, Abou-Elwafa SF, Yang J, Liu S, Liu H (2024) Exogenous application of chitosan mitigates the inhibitory effects of zinc on the growth of wheat seedlings by modulating Zn transport and ROS scavenging. J Soil Sci Plant Nutr 24(4):7817–7832. https://doi.org/10.1007/s42729-024-02077-1
63. Sturikova H, Krystofova O, Huska D, Adam V (2018) Zinc, zinc nanoparticles and plants. J Hazard Mater 349:101–110
64. Moreira A, Moraes LA, dos Reis AR (2018) The molecular genetics of zinc uptake and utilization efficiency in crop plants. Plant Micronutr Use Effic 87–108
65. Tiecher TL, Ceretta CA, Tiecher T, Ferreira PA, Nicoloso FT, Soriani HH, Rossato LV, Mimmo T, Cesco S, Lourenzi CR (2016) Effects of zinc addition to a copper-contaminated vineyard soil on sorption of Zn by soil and plant physiological responses. Ecotoxicol Environ Saf 129:109–119

66. Jain A, Sinilal B, Dhandapani G, Meagher RB, Sahi SV (2013) Effects of deficiency and excess of zinc on morphophysiological traits and spatiotemporal regulation of zinc-responsive genes reveal incidence of cross talk between micro-and macronutrients. Environ Sci Technol 47(10):5327–5335
67. Rao KM, Sresty T (2000) Antioxidative parameters in the seedlings of pigeonpea (*Cajanus cajan* (L.) Millspaugh) in response to Zn and Ni stresses. Plant Sci 157(1):113–128
68. Sadeghipour O (2021) Chitosan application improves nickel toxicity tolerance in soybean. J Soil Sci Plant Nutr 21(3):2096–2104
69. Aggarwal N, Laura JS, Sheoran IS (1990) Effect of cadmium and nickel on germination, early seedling growth and photosynthesis of wheat and pigeonpea. Int J Trop Agric 8(2):141–147
70. Gajewska E, Skłodowska M, Słaba M, Mazur J (2006) Effect of nickel on antioxidative enzyme activities, proline and chlorophyll contents in wheat shoots. Biol Plant 50(4):653–659
71. Hasinur Rahman HR, Shamima Sabreen SS, Shah Alam SA, Kawai S (2005) Effects of nickel on growth and composition of metal micronutrients in barley plants grown in nutrient solution. J Plant Nutr 28(3):393–404
72. Molas J (2002) Changes of chloroplast ultrastructure and total chlorophyll concentration in cabbage leaves caused by excess of organic Ni (II) complexes. Environ Exp Bot 47(2):115–126
73. Nakazawa R, Kameda Y, Ito T, Ogita Y, Michihata R, Takenaga H (2004) Selection and characterization of nickel-tolerant tobacco cells. Biol Plant 48(4):497–502
74. Ahmad M, Hussain M, Saddiq R, Alvi A (2007) Mungbean: a nickel indicator, accumulator or excluder? Bull Environ Contam Toxicol 78(5):319–324
75. Duarte B, Delgado M, Caçador I (2007) The role of citric acid in cadmium and nickel uptake and translocation, in Halimione portulacoides. Chemosphere 69(5):836–840
76. Chen Z, Zhang C, Tan Y, Zhou T, Ma H, Wan C, Lin Y, Li K (2015) Chitosan-functionalized gold nanoparticles for colorimetric detection of mercury ions based on chelation-induced aggregation. Microchim Acta 182(3):611–616. https://doi.org/10.1007/s00604-014-1365-8
77. Molina JA, Oyarzun R, Esbrí JM, Higueras P (2006) Mercury accumulation in soils and plants in the Almadén mining district, Spain: one of the most contaminated sites on Earth. Environ Geochem Health 28(5):487–498
78. Soostani SB, Ranjbar M, Memarian A, Mohammadi M, Yaghini Z (2025) Investigating the effect of chitosan on the expression of P5CS, PIP, and PAL genes in rapeseed (*Brassica napus* L.) under salt stress. BMC Plant Biol 25(1):215. https://doi.org/10.1186/s12870-025-06187-5
79. Calvo P, Remunan-Lopez C, Vila-Jato JL, Alonso MJJoaps (1997) Novel hydrophilic chitosan-polyethylene oxide nanoparticles as protein carriers 63(1):125–132
80. Faryal S, Ullah R, Khan MN, Ali B, Hafeez A, Jaremko M, Qureshi KA (2022) Thiourea-capped nanoapatites amplify osmotic stress tolerance in *Zea mays* L. by conserving photosynthetic pigments, osmolytes biosynthesis and antioxidant biosystems. Molecules 27(18):5744
81. Khot LR, Sankaran S, Maja JM, Ehsani R, Schuster EWJC (2012) Applications of nanomaterials in agricultural production and crop protection: a review 35:64–70
82. Ullah S, Khan MI, Khan MN, Ali U, Ali B, Iqbal R, Z Gaafar A-R, AlMunqedhi BM, Razak SA, Kaplan A (2023) Efficacy of naphthyl acetic acid foliar spray in moderating drought effects on the morphological and physiological traits of maize plants (*Zea mays* L.). ACS Omega 8(23):20488–20504
83. Tiwari S, Lata CJ (2018) Heavy metal stress, signaling, and tolerance due to plant-associated microbes: an overview 9:452
84. Santos TC, Marques AP, Silva SS, Oliveira JM, Mano JF, Castro AG, van Griensven M, Reis RLJTEPA (2010) Chitosan improves the biological performance of soy-based biomaterials 16(9):2883–2890
85. Kaplan A, Khan MN, Wahab S, Assad N, Adnan M, Iqbal M (2025b) Thallium in soil environments and its biological availability. In: Beneficial elements for remediation of heavy metals in polluted soil. Elsevier, pp 373–398
86. Saberi Riseh R, Vatankhah M, Hassanisaadi M, Varma RS (2024) A review of chitosan nanoparticles: nature's gift for transforming agriculture through smart and effective delivery mechanisms. Int J Biol Macromol 260(Pt 2):129522. https://doi.org/10.1016/j.ijbiomac.2024.129522

87. Ali B, Batool R, Arshad A, Azmat A, Hafeez A, Wahid N, Ali S, Khan MN, Singh N, Garhwal V (2025a) Regulation of enzymatic and nonenzymatic antioxidants in plants by biochar under abiotic stress. Biochar in Mitigating Abiotic Stress in Plants 159–171
88. Gao Z-W, Ding J, Ali B, Nawaz M, Hassan MU, Ali A, Rasheed A, Khan MN, Ozdemir FA, Iqbal R (2024) Putting biochar in action: a black gold for efficient mitigation of salinity stress in plants, review and future directions. ACS Omega 9(29):31237–31253

Suppression of Plant Virus Diseases Through Chitosan Nanoparticles

Aroona Saleem and Muhammad Ammar Javed

Abstract International food security is severely threatened by viral diseases in plants, as they decrease crop productivity and quality. Traditional chemical control methods often fall short because they can harm the environment, are somewhat ineffective, and do not fully eliminate viral pathogens. In this context, chitosan nanoparticles (CNPs) have recently been proposed as a new, practical, and eco-friendly approach to managing plant diseases. Chitosan, derived from chitin, is an antimicrobial, biodegradable, and biocompatible polymer. When engineered at the nano-level, chitosan becomes more soluble, bioavailable, and capable of delivering bioactive agents more effectively. CNPs exhibit a broad spectrum of antiviral actions, both directly by suppressing viral replication or disrupting virion structure, and indirectly by activating the plant's defense system, such as systemic acquired resistance (SAR). They may also serve as carriers to deliver new molecules, like RNA interference (RNAi)-based agents, to specifically silence plant virus genes. Beyond their pathogen control capabilities, CNPs have been found to promote plant growth by enhancing nutrient uptake and stimulating development. Despite these promising features, challenges persist in standardizing nanoparticle production, scaling up manufacturing, ensuring affordability, and obtaining regulatory approval. Their impact on plant tissues, soil ecosystems, and non-target organisms is still not fully understood. Nonetheless, ongoing efforts in designing intelligent nano-formulations, integrating precision agriculture systems, and applying combination therapies suggest that future CNPs could play an even greater role in sustainable crop protection. Overall, chitosan nanoparticles provide a highly versatile, environmentally friendly, and scientifically validated platform for managing viral diseases in plants. Their ability to boost plant immunity and deliver antiviral compounds

A. Saleem
School of Science, Edith Cowan University, Joondalup, WA, Australia
e-mail: aroona.saleem@ecu.edu.au

M. A. Javed (✉)
School of Agriculture and Environment, UWA Institute of Agriculture, The University of Western Australia, Perth, WA, Australia
e-mail: ammarjaved94@gmail.com; muhammadammar.javed@research.uwa.edu.au

A. A. H. Abdel Latef (ed.), *Chitosan Nanoparticles for Plant Resilience Under Abiotic and Biotic Stressors*, Smart Nanomaterials Technology,
https://doi.org/10.1007/978-981-95-6422-4_8

underscores their potential as a cornerstone of environmentally sustainable farming practices in the future.

Keywords Chitosan nanoparticles · Viruses · Transmission · Diagnosis · Pathogens · Disease management

1 Introduction

Viruses cause serious diseases by triggering physiological abnormalities in plants, resulting in significant yield losses of economically important crops worldwide. Such viral diseases result in an annual yield loss of $30 billion globally, threatening agricultural productivity and sustainability [1]. They jeopardize crop plants not only by hindering their growth and vigor, hence reducing gross yields, but also by dishonoring produce quality, hence reducing marketable yields [2]. Viruses cause half of all plant diseases and one-fifth of the global reduction in crop output. About 900 pathogenic virus species induce plant diseases. Food security issues are highly attributed to viral disease outbreaks that result in massive crop destruction in plants [3]. Major outbreaks of viral diseases came to be regularly scheduled events in most herbaceous crops: there are epidemic illustrations of cereal yellow dwarf, wheat streak mosaic, rice tungro, and sugar beet yellows cited by Thresh in 1980 [4]. For field-based plant viral disease, the current management system is weak; for intracellular pathogens like viruses, the chemical-based control strategies are not working. Applications of agrochemicals can prevent the spread of the virus by vectors. However, they are necessities that are not considered essential from an environmental safety perspective. The mechanism of resistance has developed in insect vectors because, in most cases, pesticides are overused. Besides, genetic shifts caused by mutation and genetic recombination helped in the emergence of highly virulent virus strains, which result in gene-mediated plant resistance evasion. Thus, the time required to determine possible alternatives that would have a higher efficacy and would not harm the environment is less [3].

There is a need to develop efficient and novel management measures in order to fight the losses that will be experienced since plant viral infection has affected the economically relevant crop species. Nanotechnology has come in as a potential strategy for plant growing challenges, linked to the diagnosis as well as management of plant viral diseases. Nanophytovirology is also a new field in plant virology by the use of nanotechnology, and this would cover disease diagnostics, drug delivery, genetic transformation, therapeutics, plant defense induction, and bio-stimulation, though in its early stages of development. The nano-sized nature of these particles increased the interaction potential with their surroundings and can even neutralize the infectious virus particles. Thus, it opens a new horizon in plant viral disease control [5]. Viral illnesses cannot be fought in the absence of stopping the spread of the viruses by the insect vectors. The widespread use of chemical pesticides has come with a colossal cost to the environment. Nanotechnology, in this regard, entails a positive twist. Nanoparticles (NPs) are also considered to have a marked effect in improved agricultural output due to their remarkable applicability in the field of highly sensitive, biomolecule-detection, disease diagnosis, antimicrobial and therapeutic moieties[6].

Chitosan is a cationic polymer derived by deacetylation of chitin, and in particular, crustaceans, bugs, arthropod exoskeletons, and mollusk exoskeletons are rich sources of it [7]. Chitosan is a straight polymer composed of two subunits, D-glucosamine and N-acetyl-D-glucosamine united by a 1, 4-glycosidic bond. The fact that the extraction of chitosan can be carried out regardless of the season is very advantageous due to the full independence of this process [8]. The various uses of chitosan in many areas of life have been remarkable, as its application has cut across biomedicine, agriculture and the food sector, among others, due to its characteristics and abundance in biological activities. Chitosan is considered a biosafe, biodegradable, and biocompatible product reported to improve, stimulate plant growth, and promote overall plant defence reaction that makes the plants more resistant to a wide variety of disease agents, including bacteria, fungi, nematodes, as well as viruses. The potential payoff is great: having a way of controlling many plant diseases. To treat plant viral diseases more effectively compared to the use of traditional bulk material, antiviral resistance of the plants by using chitosan nanoparticles (CNPs) will enable them to deal with viruses. The applications of the CNPs were also significantly determined to induce an effect on the pathogenesis/related (PR-1) gene and other defense-related factors. It is claimed that chitosan-based nanomaterials show greater efficacy as a plant virus controller agent, i.e., to help antivirally heal plants, thus indicating that it can be naturally used as an auxiliary in the repertoire of managing viral ailments in the field without or with fewer pernicious consequences to health and without or with fewer environmental expenses and costs [9]. In the given chapter, we aim to comment on the perspectives of CNPs as a recent and eco-friendly management approach to control plant viral diseases. It gives the disadvantages of the conventional control measures and highlights the peculiarities of CNPs as an opportunity to induce antiviral effects and the plant immune system. The chapter provides a general observation of CNPs synthesis, the mechanism of action, the mode of application, and the advantages over conventional ways. It also refers to the matters of biosafety, the existing issues, and the future of CNPs, as long as their integration into plant disease management systems is considered. It is covered with recent scientific findings, case studies, and practical knowledge that will be applicable in the research and field practice.

2 Plant Viral Diseases: Economic and Ecological Relevance

Numerous viruses have been reported to infect plants; what matters mainly to the farmers and the scientists is the group of those viruses that bring about serious yield losses to the crops. Some have been mentioned in Table 1. Quantitatively determining the yearly losses on crop products on account of plant viruses is not easy to estimate as numerous factors may add more difficulties to the estimation of losses inflicted on crop products [1]. Viral risks to the economy are no more, but with programs to produce certified seeds, there has been a dramatic reduction in the prevalence of viruses in potato crops in the majority of the locations. There are budding concerns and objections that should be dealt with [10]. A wide range of agronomic and naturally occurring plant species is infected with viruses. However, the host ranges of the individual viruses are narrow to broad. *Citrus tristesa* virus, as an example, shows that it can infect only a certain number of susceptible species because of host specificity. In contrast, cucumber mosaic virus has a very broad host range and infects over 1000 species in 85 plant families [11]. Diseases caused by bacteria, fungi, and viruses tend to affect the growth and maturity of vegetables and greatly reduce the size and quality of the vegetable products. Among the pathogens, the viral infection is the most difficult to control because infected plants cannot be cured, and it is usually unproductive to control the insect vectors. A long list of viruses infecting tomato and cucurbit crops has been reported: Begomovirus, Tospovirus, Tobamovirus, Cucumovirus, Crinivirus, Polerovirus, and Potyvirus, including Pepper mottle virus (PepMoV). Similarly, it has been reported that viruses infecting beans include begomoviruses, poleroviruses, and potyviruses [12].

Begomoviruses are the plant-infecting viruses transmitted by the whitefly Bemisia tabaci and have been documented to decrease the yield to its barest limit in many vegetable crops of world interest. Several begomoviruses have already been reported in West and Central Africa (WCA) to infect vegetable crops. Begomoviruses have also been found to infect plants with small single-stranded circular molecules, alpha satellites and beta satellites, which are about half the size of the begomovirus helper genome [13, 14]. Recall that begomovirus has been a big challenge to crop production in the tropics and subtropical regions. The Begomovirus genus is also the most diverse to be naturally transferred through whiteflies (*Bemisia tabaci*), with a large host range of crops, which causes major losses [14].

Potyvirides or the members of the potyviridae family of viruses are positive-sense, single-stranded RNA viruses. Potyvirids are transmitted by various vectors through various transmission processes. The number of examples is quite significant; viruses of the Potyvirus and Macluravirus genera have aphid transmission and are non-persistent, viruses of the Rymovirus, Poacevirus and Tritimovirus genera are transmitted by eriophyid mites in a semi-persistent manner and the virus of the genus Bymovirus is transmitted by a group of single-celled eukaryotes known as Cercozoa and the whiteflies to transmit viruses in the genus Ipomovirus [15].

Tobamoviruses, named after the Tobacco mosaic virus (TMV), are a genus of plant-infecting viruses with a natural host range of the members, including Solanaceae, Brassicaceae, Cucurbitaceae, Malvaceae, Cactaceae, Passifloraceae, Fabaceae, Apocynaceae, and Orchidaceae. The symptoms caused by TMV also vary based on the affected host plant; signs and symptoms attributed to the virus have similarly been re-reported as being predominantly mosaic in nature with attendant mottling and yellowing of leaves and plant tissues. The MV is a positive-sense, single-stranded, rod-shaped plant virus, reputedly the most thermostable virus in existence, and which accumulates to high levels in plants. Even the discovery of it as

Table 1 Economically important plant viral diseases

Disease/virus	Host range	Virus family	Transmission	Impact/yield loss	Control measures	Reference
Tobacco mosaic virus (TMV)	Tobacco, Tomato, etc	*Virgaviridae*	Mechanical, seed	Leaf mottling, yield loss	Sanitation, resistant varieties	[17]
Tomato yellow leaf curl virus (TYLCV)	Tomato, Pepper	*Geminiviridae*	Whiteflies	Stunting, fruit loss	Vector control, resistant lines	[18]
Maize dwarf mosaic virus (MDMV)	Maize, grasses	*Potyviridae*	Aphids	20–40% yield reduction	Weed control, resistant varieties	[19, 20]
Potato leaf roll virus (PLRV)	Potato	*Luteoviridae*	Aphids	Leaf rolling, reduced tuber yield	Virus-free seed, insecticides	[21, 22]
Papaya leaf curl virus	Papaya	*Geminiviridae*	Whiteflies	Leaf curling, reduced fruit set	Avoid monoculture, whitefly control	[23]
Okra yellow vein mosaic virus	Okra	*Geminiviridae*	Whiteflies	Yellow veins, growth reduction	Resistant cultivars, vector control	[24]
Cauliflower mosaic virus (CaMV)	Brassicas	*Caulimoviridae*	Aphids	Mosaic, stunted growth	Crop rotation, remove infected plants	[17]
Cassava mosaic disease (CMD)	Cassava	*Geminiviridae*	Whiteflies, cuttings	Major yield loss, poor tubers	Resistant varieties, virus-free cuttings	[25]
Cassava brown streak disease (CBSD)	Cassava	*Potyviridae*	Whiteflies	Root necrosis, low marketability	Clean cuttings, breeding resistance	[26]
Tomato spotted wilt virus (TSWV)	Tomato, Pepper, many crops	*Tospoviridae*	Thrips	Wilting, necrosis, yield loss	Resistant varieties, thrip control	[27]

(continued)

Table 1 (continued)

Disease/virus	Host range	Virus family	Transmission	Impact/yield loss	Control measures	Reference
Citrus tristeza virus (CTV)	Citrus	*Closteroviridae*	Aphids	Tree decline, fruit drop	Clean budwood, resistant rootstock	[28, 29]
Maize streak virus (MSV)	Maize, grasses	*Geminiviridae*	Leafhoppers	Streaks, stunted growth	Vector control, resistant lines	[30]
Rice tungro disease	Rice	*Tungrovirus complex*	Leafhoppers	Stunting, yellowing, yield loss	Resistant cultivars, vector control	[31, 32]
Banana bunchy top virus (BBTV)	Banana	*Nanoviridae*	Aphids	Bunchy appearance, no fruiting	Quarantine, clean planting material	[33, 34]
Barley yellow dwarf virus (BYDV)	Cereals (barley, wheat)	*Luteoviridae*	Aphids	Yellowing, reduced grain yield	Early sowing, aphid control	[35]
Sugarcane mosaic virus (SCMV)	Sugarcane, maize	*Potyviridae*	Aphids, mechanical	Mosaic pattern, poor stalk growth	Resistant clones, rogueing infected plants	[36, 37]
Groundnut rosette disease	Groundnut	*Luteoviridae complex*	Aphids	Ro setting, little/no pod formation	Resistant varieties, vector management	[38]
Wheat streak mosaic virus (WSMV)	Wheat	*Potyviridae*	Mites	Leaf streaking, yield decline	Planting time adjustment, resistant lines	[39]

the early virus was preconditioned by economic interest, as is also the exploitation of it as the subject of virology, molecular and cellular biology [16].

3 Mechanisms of Virus Transmission

Many important plant viruses are transmitted by the sap-sucking insects, such as aphids, whiteflies, leafhoppers, planthoppers, and thrips, in the form of persistent infection as augmented by the vector, as shown in Fig. 1 and Table 2. These viruses

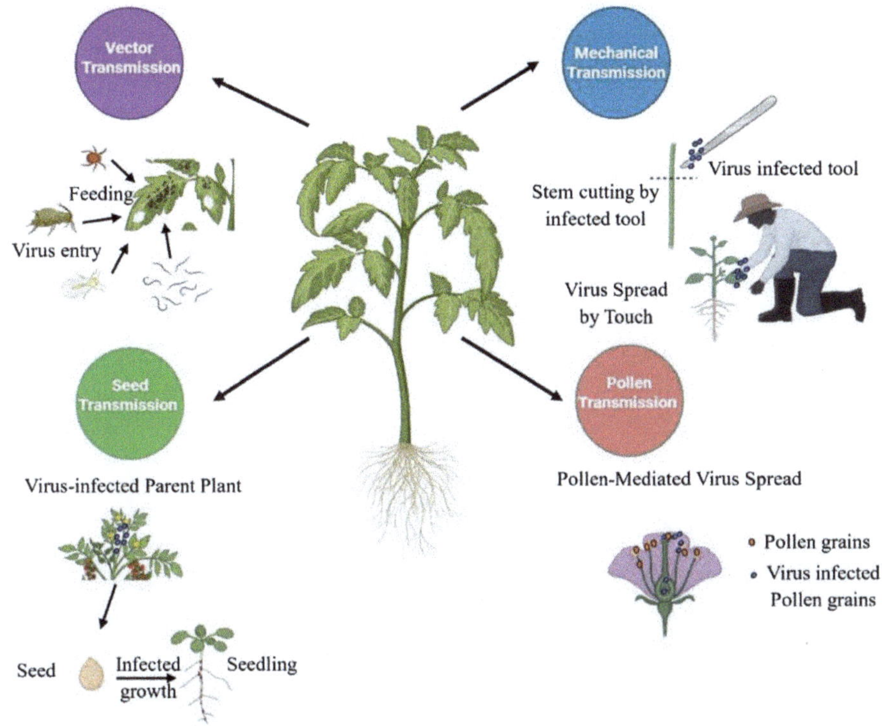

Fig. 1 The transmission mechanisms of plant viruses: vector transmission, seed transmission, mechanical transmission, and pollen transmission. Viruses are transmitted through insects (vectors), seed, pollen, and mechanically by different infected tools or by human touch, from the parent plant to the next generation or from one plant to another

are distinguished as propagative (replicating in vectors) and non-propagative. In propagative transmission, the viruses penetrate the midgut of the insect and are carried through the hemolymph to the salivary glands to be horizontally transmitted or into the ovaries to be vertically transmitted [40]. The intricacy of plant viruses' relationship with their vectors is essential to devising effective control strategies that may aid in lowering the devastating impact of these diseases on crops [41].

3.1 Vector-Transmission

Most plant viral diseases are transmitted by biological vectors. Vector transmission involves specific interactions, and structural and helper proteins are involved in binding to vector transmission sites to facilitate transmission, which allows intervention points to interrupt transmission of the virus [42]. Insect vectors, particularly

Table 2 Mechanism of plant virus transmission

Transmission mechanism	Virus transmission mode	Primary vectors or agents	Transmission specificity	Key examples of viruses	Impact on epidemiology	Control strategy implications	References
Vector-borne (Persistent-Propagative)	Circulative, replicative	Aphids, whiteflies, leafhoppers, planthoppers, thrips	Highly specific—virus replicates in vector	Tospoviruses (e.g., TSWV), Luteoviruses	Long-distance and vertical transmission via vectors	Target vector life cycle, inhibit viral replication in vectors	[47–49]
Vector-borne (Persistent-Non-Propagative)	Circulative, non-replicative	Aphids, whiteflies	Specific—virus circulates but does not replicate in vector	Geminiviruses (e.g., TYLCV)	Efficient vector retention, delayed transmission	Block virus–vector binding, vector population control	[50, 42]
Vector-borne (Semi-persistent)	Non-circulative	Aphids, thrips	Moderately specific—virus retained in stylet for hours	Cauliflower mosaic virus, Citrus tristeza virus	Fast spread, short retention	Timely vector control, disrupt feeding behavior	[42]
Vector-borne (Non-persistent)	Non-circulative, superficial	Aphids, beetles	Low specificity—short retention (seconds to minutes)	Potyviruses (e.g., PVY), Cucumoviruses	Quick epidemic onset, hard to control	Reflective mulching, oil sprays to disrupt transmission	[51, 52]
Seed-borne Transmission	Vertical (transovarial), indirect	Infected parent plant (occasionally beetles or nematodes)	Moderate—species and virus dependent	Nepoviruses, Potyviruses, Fabaviruses	Primary inoculum source in new fields	Certified seed programs, seed disinfection	[44, 52]
Mechanical transmission	Horizontal	Tools, hands, damaged plant surfaces	Low specificity—physical injury required	Tobacco mosaic virus (TMV), CMV	Important in research, greenhouses, local spread	Sanitation, resistant cultivars	[53–55]
Pollen transmission	Vertical and/or horizontal	Pollen (wind, insects, humans)	High—virus must enter embryo or ovule	Ilarviruses, Nepoviruses, Potyviruses, viroids	Cross-pollination can spread disease	Isolate infected plants, control pollinators	[46]

aphids and whiteflies, transmit about 80% of all plant viruses, although nematodes, mites, and fungi also play a definite role in transmitting viruses [43].

3.2 *Seed Transmission*

Nepovirus is another virus that is transmitted through seed and that uses nematodes as a carrier. Beetle-transmitted viruses include those of the Fabavirus group and the Sobemovirus group, and spread between their hosts through the seed transmission process [43]. Seed transmission in the case of the Potyviridae family is not particularly different, but its more specific mechanism is not found, and thus shows how the viruses integrate into the germ line. In addition, the speed of seed transmission can be considered as a combination of the indirect and the direct embryonic invasion [44].

3.3 *Mechanical Transmission*

In mechanical transmission, a virus is inserted artificially into plants by physical means (i.e., by wounding of the plant). The wounds may be done intentionally to introduce the virus in the plant (experimental mechanical transmission), or they may have occurred naturally by accident [45].

3.4 *Pollen Transmission*

Infected plant pollen spreads the virus. Ovules are invaded by the female plant's stigma pollen. Wind, insects, and humans disseminate virus-infected pollen. Pollen spreads the virus horizontally and vertically. Many viroids are in pollen, including Ilarvirus, Nepovirus, Sobemovirus, Idaeovirus, and Potyvirus. Pollen contains 45 viruses from 16 species and five viroids [46].

3.5 *Symptoms*

Viral diseases' symptoms make it possible to distinguish five main types: stunted growth, discoloration, structural deformities, tissue necrosis, and disruption in reproduction [56]. The plant viruses bring about symptoms that negatively affect the production rate and the crop quality. The prolonged effect of viral infection adversely affects the production standard, rate, and growth, which causes financial loss to those individuals who work in the plant industry. A large proportion of the symptoms caused by the virus can also be exhibited as a result of a harsh environment and associated disease, which are caused by other plant pathogens. It is based on this that a lab test is typically involved in the proper diagnosis of viral diseases. [11].

3.6 Diagnosis

Plant viral diagnosis procedures are numerous, and the range is vast. Nonetheless, the efficiency of each of the commonly used techniques is reliant on some aspects, including cost-effectiveness, sensitivity, speed, availability of the instruments, and the level of the illness. Additional quantitative high-throughput imaging-based methods have also been developed. The use of serological and nucleic acid-based diagnosis remains central in the diagnosis of plant diseases due to the advances over the years with respect to molecular biology technologies. That being said, the emergence of Next Generation Sequencing (NGS) technologies provided a platform for viral diagnostics over the past decades [57]. The Enzyme-linked immunosorbent assay (ELISA) and Polymerase Chain Reaction (PCR) have had the most dramatic effects, and very recently developed microchips would have a significant role in the future diagnosis system [58].

3.7 Pathogenicity

Transmission of viruses also occurs through the vertical pathway, i.e., from parents to offspring, and by the horizontal route, i.e., from infected to healthy plant. To infect neighboring cells, viruses use small intercellular channels called plasmodesmata. The viruses often express the proteins that are used in the motility of the virions by modifying the channels through which the infection can be transported to a neighboring cell. That is how a local infection of a plant takes place. A virus must enter the vascular system of a plant to infect the entire plant, and when in the vascular system, a virus is passively moved through the sieve tubes of the phloem as part of the flow of substances. In this manner, a virus can infect cells that are distant from the original site of the infection [56]. Plant viruses constitute a huge economic setback to the crop in the entire globe, both in yield and quality of the agricultural produce. The banana bunchy top disease has an incidence of 5 to > 70% in a banana plantation, and in extreme cases of the disease infection, the eradication of a banana plantation is usually foregone [59]. Cocoa swollen shoot virus in Africa kills 50,000 tonnes of cocoa beans annually, valued at 28 million dollars [11]. Africa's deadliest crop disease is maize streak virus, which affects maize and 80 Poaceae species. According to the authors, Sub-Saharan African maize crops lose 30–100 percent and US$120–480 million yearly [60].

Whiteflies were influenced most by the planting period (42.7%) and the cassava variety (29.5%) based on the Analysis of variance (ANOVA) [61]. Bunyaviridae orthospovirus SWV is widespread in all the countries of the Mediterranean, but not in Morocco, and could cause 40–90% and even 95% loss to the yield and marketable value of tomatoes, respectively [62]. Cassava and sweet potato are good food crops in African countries, as well as rice in Southeast Asian countries [63]. In 2014, the pandemic and epidemics of viral diseases were estimated to cost the global economy more than > US$30 billion/per year. Rising global losses are caused by plant viruses [4].

4 Chitosan and Nano-Enabled Crop Protection

4.1 Chitosan: Source, Structure, and Bioactivity

About 2000 tonnes of chitosan come from prawn and crab shell waste [7]. Chitin creates linear polysaccharide polycationic chitosan, like cellulose. Hypoallergenic, biocompatible, biodegradable, and toxin-free chitosan compounds [64]. Chitin dissolves into chitosan below 50% deacetylation in water. Research reveals 0.05–0.30 chitosan N-acetylation. The US, Japan, India, Malaysia, Canada, China, South Korea, Russia, Norway, Kenya, and EU countries like France and Germany make chitin and chitosan from crab wastes, says Struszczyk. Chitin is first acid-hydrolyzed, followed by alkaline treatment to remove proteins, pigments, and lipids. During this process, the chitin may undergo slight modifications. After two alkaline treatments, part of the chitin becomes deacetylated, converting into chitosan. In some cases, enzymes are used to supplement the acid or alkaline steps [65].

The degree of acetylation (DA) and deacetylation (DDA) are the key characteristics of chitosan. The commercial chitosans range between 50 and 2000 kDa. They can be further classified based on their molecular weights as low (<100 kDa), medium (100–1000 kDa), or high (>1000 kDa) [66]. The DDA is the key property of chitosan; however, its precise determination requires modern equipment such as a nuclear magnetic resonance (NMR) spectrometer. This implies the need for simple, cost-effective, and reliable methods for degree of deacetylation (DD) measurement [67]. There are different analytical methods and they give varying DD values. For instance, Hydrogen bromide (HBr) titrimetry gives lower DD values ($p < 0.005$) compared to infrared (IR) spectroscopy, ultraviolet (UV) spectroscopy, near-infrared, and linear potentiometric titration. Compared to them, NMR spectroscopy gave significantly ($p < 0.005$, $p < 0.05$) higher DD values [68].

Since the primary amino group pKa is tightly connected, DA affects chitosan solubility. Based on natural chitin source and production conditions, size (MW), DA, polydispersity (MW/MN), crystallinity, and acetylation pattern may vary. [69]. Chitosan, a promising natural polymer possessing antimicrobial effects, demonstrates bactericidal action [70]. The business opportunity of this product has enormous promise, bearing in mind the widespread antimicrobial factor of the product known as chitosan. However, the mechanism of antimicrobial action of chitosan presents flexibility, given that it is sensitive to the physical and chemical properties and targets the microorganisms. [71]. Chitosan possesses antibacterial, antiviral, wound-healing, and hydrogel properties and may be a broad-spectrum antiviral. Transfection of NIH-3T3 cells with GFP-adenovirus containing 0.1, 0.5, or 1 percent chitosan will be employed. Fluorescence microscopy and flow cytometry showed lower GFP fluorescence in using 0.1% chitosan on the virus before transfection was 72% to 1.9%, showing a difference of 70% in transfection. Based on these results, it is suggested that chitosan can be a common antiviral medication [72].

4.2 Nanotechnology in Agricultural Disease Management

Nanotechnology is a dynamic technology that has provided new opportunities to enhance agricultural productivity sustainably. The dramatic progress achieved in nano-enabled devices in the past few years has helped to increase the quality of food products, besides helping to minimize the annual economic losses that are incurred because of pests and pathogens. Nanoparticles have caught the interest of researchers and industries worldwide due to their prospects and effective utilization in the management of plant diseases incorporated as nano-based products such as nano fertilizers, nano fungicides, and nanopesticides. Nanotechnology is pretty promising in addressing the challenges linked to the rise in quality, enriched food requirements. Rapid climatic change events might reduce agricultural productivity due to pesticide management and pest and pathogen detection [73]. The key causal organisms of plant diseases include bacteria, fungi, viruses, and nematodes. Phytopathogens cause an estimated 10–40% productivity and quality losses in food crops and horticultural productions. Although plant diseases cannot be eliminated, science has improved their control. The transdisciplinary field of nanotechnology has produced many nanomaterials for agriculture. Agents that reduce pathogen levels, early detection methods, and chemicals that induce host plant immune responses are needed. Evaluation of nanomaterials based on their usefulness in each of these methods of controlling the disease is done [74].

Nanoparticles made of metals have been proven to exhibit non-specific mechanisms of bacterial toxicity, which not only causes bacteria to have a hard time developing resistance to them, but also expands the spectrum of antibacterial activity [75]. It was revealed that SiO_2 NPs can stimulate plant growth and alleviate the adverse impacts of biotic and abiotic stresses, yet a critical parameter that has been omitted in the research on the biological activity of NPs is their bioaccumulation [76]. Zinc oxide nanoparticles (ZnONPs) have become a prospective instrument in plant science that provides good perspectives of plant growth and yield [77].

4.3 *Chitosan Nanoparticles (CNPs): Synthesis and Characterization*

The most promising natural material for use in nanomedicine is chitosan. The modified precipitation method enables the successful obtaining of CNPs when NaOH is used as the precipitant [78]. To develop the versatile functional compounds of Chitosan, it is modified by various physical and chemical means [79].

4.3.1 Fabrication Methods: Ionic Gelation, Emulsion Crosslinking, Nanoprecipitation

Recent advances in CNPs prepared by ionotropic gelation were communicated. Since the reporting of the technique, enormous development has occurred in this sector. Ionotropic gelation using electrospraying and the spinning disc process permits a more homogeneous size dispersion of the nanoparticles. The second approach enables the mass-scale synthesis of the nanoparticles. Using factorial-design approaches, minimal possible experiments can be used to identify optimal conditions to form nanoparticles [80]. To improve the CNPs' emulsion crosslinking, controlled hydrodynamic cavitation (HC) was developed. The experiments were done utilising a circular venturi and several oil types, glutaraldehyde (Glu) solution addition modes, inlet pressure (Pin), and chitosan solution rheology. Glu-added water-in-oil (W/O) emulsion was best for narrowing CNPs distribution [81]. Nanoprecipitation experiments were conducted to synthesize chitosan nanoparticles as well as to establish the influence of process parameters on the characteristics of the nanoparticles. NPs were minimized by using a low polymer concentration or a large volume of nonsolvent, compared to the solvent. Synthesized particles are slightly larger than those prepared by precipitation of N-(methylsulfonic acid) chitosan. Their size notoriously increased when the suspension medium methanol was replaced by water. The ionic strength does not affect the nanoparticle size precipitated in methanol. Good association and loading efficiency of a model material demonstrate the feasibility of these chitosan nanoparticles as drug delivery nanocarriers [82].

4.3.2 Physicochemical Characterization: Particle Size, Morphology, Zeta Potential, Encapsulation Efficiency

Advanced diagnostics methodologies are used to analyse the dimension, surface properties, and the configuration of CNPs [83]. The characterization of chitosan nanoparticles can be performed in different ways: Fourier Transform Infrared (FTIR) Spectroscopy serves the identification and characterization of functional groups on the CNPs' surface, and scanning electron microscopy (SEM) serves the determination of their shape and morphology. Ultraviolet–visible Spectroscopy (UV–Vis)- To confirm the formation of nanoparticles, the solution was measured in the wavelength range of 200 600 nm in a spectrophotometer with a quartz cuvette with water as the reference [84]. The size distribution for the nanoparticle is measured at a 90-degree scattering angle, and the particles are easily spread in the deionized water [85].

Zeta Potential: Zeta potential ± 30 mV demonstrates the charge, nanoparticle toxicity, and the permeability, by affecting the membrane interaction [86, 87]. Chitosan is a positive Z-potential biocompatible polysaccharide that can stabilize negatively charged nanoparticles [88]. Various encapsulation methods are continuously being created to overcome a considerable number of disadvantages associated with the application of pure active compounds, either systemically or topically [89, 90]. There was an attraction of the protein that had a pI greater than the pH of the solution to the positively charged chitosan, which increased encapsulation efficiency (EE%) [91]. The ionic gelation technique can produce high EE values, which can go as high as 90%, depending on the solubility and the interactions of the encapsulated molecule with chitosan [81, 92].

4.3.3 Functionalization Methods as Strategies of Improved Efficacy

Potentiometric titration confirmed a chitosan functionalization layer on different chitosan-functionalized maghemite nanoparticle syntheses, which is significant regarding subsequent bioactive compound immobilization [93]. The particle can be functionalized, especially concerning the imaging modalities from in vitro to in vivo, to enhance the information content that can be extrapolated and facilitate the differentiated imaging due to high contrast. Nevertheless, despite some shortcomings associated with the functionalization of NPs, purposeful functionalization with selected molecules may have an enormous impact on the revolutionization of the nano-medicine sector in the future [94].

5 Mechanisms Underpinning the Antiviral Efficacy of Chitosan Nanoparticles

CNPs have the potential to be used as a way to regulate numerous plant diseases, as shown in Fig. 2. An exemplary plant virus was considered as BYMV that infects faba bean plants and a broad range of other legumes. As a curative mode of application, the antiviral activity of CNPs and chitosan was determined. They discovered that the addition of CNPs curatively at 48 h post-virus inoculation resulted in total prevention of infectivity of the disease and the accumulation of the virus. It was proven that the application of CNP has a significant effect in inducing pathogenesis-related (PR-1) genes and other defense-related factors. The mRNA PR-1 gene was highly accumulated in the treated plants, with the highest accumulation at 400 mg/L (16.22-fold change in relative expression) compared to the untreated control. It exhibits antiviral

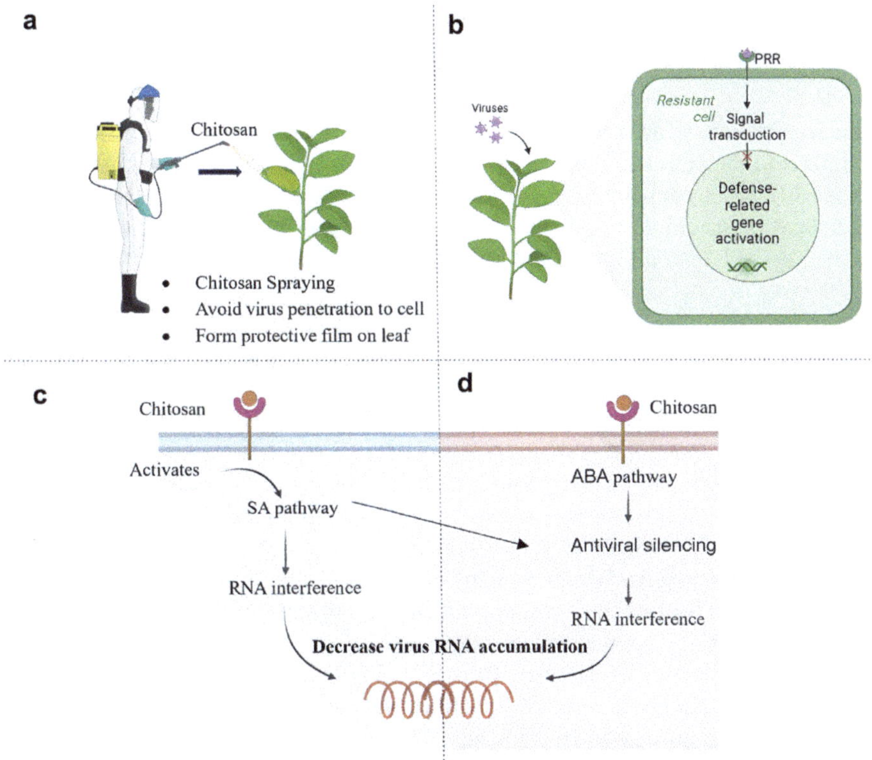

Fig. 2 The role of chitosan in plant defense against plant viruses by **a** preventing the entry of the virus, **b** reproducing and activating the **c** SA (salicylic acid) and **d** ABA (Abscisic acid) pathway. Pattern recognition receptors (PRR)

properties, and this implies that it is likely to be included in the process of controlling plant viral diseases in the field without any serious health or environmental costs [9].

5.1 Inhibition of Viral Replication and Systemic Movement

Plant viruses represent a sustainable crop production concern all over the world. Nanotechnology has been assumed to play an efficient role as part of the existing antiviral methods to avoid plant viruses. The antiviral effect of CNPs, Pepper plant Alfalfa mosaic virus (AMV) infections with CS-NPs and CS-Ag NC at different dosages was determined. AMV systemically infected pepper plants were suppressed by CS-NPs and CS-Ag NC at all doses 24 h later. CNPs and C-Ag NC increased phenol, proline, and capsaicin in infected plants. Height, fresh and dry pod weights, and pod number per plant improved. AMV and plant virus management may benefit from CNPs and C-Ag NC (hitosan–Silver Nanocomposites) [95].

5.2 Disruption of Virus-Host Interactions at Cellular and Molecular Levels

Numerous pathways are induced following the use of chitosan; the majority of them are linked to the activation of plant innate immunity [96]. In addition to this, interspecies differences in plant response to chitosan are significant. Thus, it is difficult to understand the acquired resistance to viruses. Chitosan triggers salicylic acid (SA) building up and Systemic Acquired Resistance (SAR) activation [97, 98]. Nevertheless, the role of SAR in the plant antiviral mechanism differs with the host-virus pair and the infection type [99]. SA was also proven to be a significant component of antiviral defense as a signal molecule that triggers several pathways and causes the beginning of numerous shifts in the physiological status of plant cells and their gene expression [100]. SA was found to have a deleterious impact on replication, viral intercellular transport, and systemic transport. It was noted that the specific stage of viral infection that is inhibited by SA varies depending on the virus-host combination [101, 102]. Accordingly, the hypothesis could be proposed that this characteristic of SA-mediated antiviral defense is one of the factors that explain the different degrees of chitosan activity against different virus-host pairs [103]. Moreover, the viral infection itself causes host defense responses that may or may not share the same pathways as chitosan does or may cause exquisitely balanced interactions between host defense and virus counter-defense. Thus, a vast number of pathways are triggered after chitosan application, most of which are interconnected with the activation of the plant's innate immunity. Besides this, there is significant interspecies variation in the response of plants to chitosan. Accordingly, it is hard to comprehend which of these processes, in particular, is the cause of the acquired resistance to viruses [104]. SA was observed to exert a negative effect on replication, viral intercellular transport, and systemic transport [105]. It was observed that the particular stage of viral infection that is inhibited by SA depends on the virus-host combination [106]. Correspondingly, it might be hypothesized that such a property of the SA-mediated antiviral defense is one of the reasons that can explain the varying extent of chitosan activity against different virus-host pairs. In addition, viral infection per se induces host defense reactions that do or may not involve the same pathways as chitosan does, or can result in the exquisitely balanced host defense-virus counter-defense interactions [104].

5.3 Induction of Plant Immune Responses

A biotrophic oomycete, *Sclerospora graminicola,* causes pearl millet downy mildew, resulting in unimaginable losses in production and cost. The seedling vigour of pearl millet and germination proportion were enhanced significantly following the seed treatment with CNP. In a greenhouse experiment, it was proven that CNP seed treatment generated both systemic and long-lasting downy mildew resistance and fully offered protection against the untreated control. In CNP-treated seeds, genes coding phenylalanine ammonia-lyase, peroxidase, and polyphenol oxidase, catalase accumulation, and superoxide dismutase were strongly up-regulated [107]. The activities of phenylalanine ammonia-lyase (PAL), peroxidase (POX), polyphenol oxidase (PPO), catalase (CAT), and - 1, 3-glucanase were up-regulated considerably in tomato plants subjected to treatment with chitosan and CNPs when compared to the control ones. Unlike pathogen-inoculated plants, the antioxidant enzyme gene expression that is stimulated by chitosan and CNPs is elevated in CNP-treated plants. This case study scenario can be defined as creating a medical emergency due to the at-home use of the artificial heart because of complications that occurred during the operation [108].

Chitosan is a natural biopolymer that has already shown immunomodulatory activity in plants. Discovered the possibility and the mechanism of CNPs in activating and priming the plant immune system. Treatment of leaves resulted in the spectacular induction of the natural immune response of the plant through the induction of defense enzyme activity, up-regulation of defense-related genes, including those of many antioxidant enzymes, as well as elevation in the levels of total phenolics. It is also possible that the defense response upregulation observed in plants is also CNPs' extracellular localization mediated. Accumulation of nitric oxide (NO), a major signaling molecule in plant defense, was also determined to take place upon CNPs treatment. This CNP-mediated immuno-stimulation was, however, significantly negated when NO production inhibitors were present, thus indicating that NO might play a role in this immune stimulation [109].

6 Application of Chitosan Nanoparticles in Plant Virus Suppression

Chitosan was demonstrated to prevent systemic movement of viruses and viroids within the plant and increase the hypersensitive response of the host to infection. The degree of viral infection suppression depended on the chitosan molecular weight. The same was observed with the potato virus X, tobacco mosaic and necrosis viruses, alfalfa mosaic virus, peanut stunt virus, and cucumber mosaic virus [110].

6.1 Experimental Evidence: In Vitro and in Vivo Approaches

The role played by chitosan in the development of the infection by Tobacco mosaic virus (TMV) in the leaves of *Nicotiana tabacum* L. cv. Samsun has been monitored. It was shown that infectivity and the level of viral coat proteins in the leaves infected with a mixture of TMV (12 g/mL) and chitosan (1 mg/mL) have been reduced in the initial stages of infection (3 dpi) by 63% and 66%, respectively, compared to those in leaves infected by TMV alone. It meant that the activation of hydrolases through chitosan treatment induced the intracellular degradation of TMV particles, and, consequently, the activation of hydrolases can be considered [111].

- Developed chitosan-coated lentinan-loaded hydrogel and elucidated their mechanism of action to protect against a variety of virus infections in plants. The synthesized hydrogel procedure was as follows, whereby chitosan was placed over the surface of calcium alginate-lentinan (LNT) hydrogel (SL-gel) to form CSL gel. It is shown that the basal release of lentinan and capacitation can be maintained stably using SL-gels, and that the latter induces Ca^{2+} release. SL gels used to coat the roots of plants provide broad-spectrum protection against plant viruses (TMV, TRV, PVX, and Tum) (TMW/006). *Nicotiana benthamiana* calmodulin-like protein gene 3 (NbCML3) was found to be expressed after an extended stimulus of Ca^{2+} release by the CSL-gel, and silencing and over-expression of NbCML3 altered the response of tobacco to TMV susceptibility and resistance. These findings provide evidence to the notion that this novelty and artificial CSL-gel exerts a tremendous blocking effect on the plant virus infection by the continuous release of LNT and Ca^{2+} [112].
- Report the efficiency of the defense mechanism of chitosan on *Solanum lycopersicum* var. cerasiforme against Cucumber mosaic virus (CMV). The preventative mode of chitosan application was performed by foliar spray prior to inoculating CMV to ensure the effectiveness of application as a preventative treatment modality of CMV infection. They were examining the increase of viruses, photosynthetic capacity, and genes that determine proteins that may influence resistance reactions and biosynthetic pathways. Outstandingly, a drastic reduction in CMV build-up leads to insufflation of chitosan in plants infected with CMV successively after, pluturne, as compared to a single infection with CMV that was observed. A positive effect of chitosan on gas exchange dynamics was also revealed in the same manner. Expression profiling (CEVI-1, NPR1, PSY2, and PAL5) analysis suggested that there was a possibility of chitosan-induced systemic acquired resistance-associated expression patterns by manipulating the oxidation status in the plant [113].
- Abdelkhalek et al. [114] explored the antiviral property (protection, cure, and inactivation) of formulated chitosan/dextran nanoparticles (CDNPs, $100\ g\ L^{-1}$) against *Nicotiana glutinosa* plants. The morphology of synthesized CDNPs, shown by the scanning electron microscope (SEM) and dynamic light scattering images, was regularly shaped and uniform sphere-like nanoparticles with diameters between 20 and 160 nm, with the average diameter of 91.68nm. The treatment based on the inactivation was the most effective one, as the level of the AMV, Acc OK413670, decreased by 100%. Alternately, under greenhouse settings, CDNPs sprayed on the leaves lessened the incidence of the disease and decreased the level of the viral accumulation by 70.43 and 61.65%, respectively, when applied in protective and curative measures. It was also associated with systemic acquired resistance, total phenolic input, and total carbohydrates, as well as it induced the transcriptional activation of peroxidase, phenylalanine ammonia-lyase, and

pathogen-related protein1.El Gamal et al. [9] tested the antiviral efficacy of hNPs and chitosan silver nanocomposites (C-Ag NC) against varying concentrations of AMV (50, 100, 200, 250, 300, and 400 mg/L). It was found that both CNPs and chitosan silver nanoparticles completely suppressed the infectivity of the disease and viral accumulation after 24h of inoculation when used at 300 mg/L and 400 mg/L. The CNPs and C-Ag NC treatment remarkably increased the amount of phenol, proline, and capsaicin compared to the infected plants. In addition to that, there was a significant improvement in terms of agronomic traits (plant height, fresh and dry pod weights, and number of pods per plant as well). The potential applications of the CNPs and the C-Ag NC, as indicated by research findings, may be applicable in the treatment of AMV and other kinds of plant viruses as therapeutic interventions (Table 3).

7 Challenges, Knowledge Gaps, and Future Research Directions

CNPs have become one of the prospective instruments of sustainable agriculture because of their antimicrobial activity, biodegradability, and the ability to be loaded with different bioactive compounds. Although they have advanced significantly in terms of their laboratory-scale applications, there are still a few obstacles that inhibit their development into large-scale farming methods. These issues cut across the standardization, cost-effectiveness, mechanistic insights, and regulatory systems, and also paved the way for future studies in the area of precision farming and intelligent nano-bioformulations. One of the biggest hindrances to the common use of CNPs in agriculture is the lack of standardization of synthesis procedures and characterization techniques. Another important bottleneck is scalability.

The price of high-purity chitosan, as well as the necessity of using sophisticated instrumentation in the characterization and processing of the nanoparticles, further increases the financial toll. Thus, the cost-effectiveness is one of the severe problems that needs to be considered, particularly in the case of low- and middle-income nations, because the agricultural input cannot be too expensive to be considered effective. Despite the demonstrated strong antimicrobial and elicitor effects of CNPs in vitro, the knowledge about their mechanism of action in plants and in the soil ecosystem is lacking substantially.

Also, field-level validations are few. Large-scale field trials in diverse agro-climatic regions are required to gain insight into long-term efficacy, phytotoxicity, persistence, and ecological security. Without such validation, the upscaling of CNPs to the real-world agricultural environment remains a speculation. One appealing future outlook is the application of CNPs as a combination therapy. The CNP encapsulation reduces the RNA molecule speed and enhances target delivery and plant cell uptake, which raises the stability and gene-silencing efficiency. This is the frontier of nanotechnology-enabled gene therapy, which is the next generation of an alternative to conventional agrochemicals. One more significant tendency is integration

Table 3 Antiviral effects of chitosan-based treatments on plant viruses

Study	Plant host	Virus	Virus family	Chitosan form and dose	Formulation type	Application mode	Mechanism proposed	Key outcomes
[111]	*Nicotiana tabacum*	TMV	*Virgaviridae*	Chitosan (1 mg/mL)	Bulk	Mixed or pre-treatment	Hydrolase activation	↓ infectivity/ CP ~ 60%
[112]	*Nicotiana benthamiana*	TMV, TRV, PVX, TuMV	Various RNA virus families	CSL-gel (chitosan-coated LNT hydrogel)	Hydrogel	Root application	$Ca2^{+}$-induced NbCML3 expression	Broad resistance induced
[113]	*Solanum lycopersicum*	CMV	*Bromoviridae*	Chitosan (spray)	Bulk	Preventive (foliar)	SAR pathway, redox balance	100% ↓ virus accumulation
[114]	*Nicotiana glutinosa*	AMV	*Bromoviridae*	CDNPs (100 μg/mL)	Nanoparticles	Pre-, post-, and inactivation	PR gene activation, phenolics	Up to 100% inhibition
[95]	*Capsicum annuum*	AMV	*Bromoviridae*	CNPs (400 ppm), C–Ag NC (200 ppm)	Nanoparticles/ Nanocomposites	24 h post-inoculation	Antioxidant/ defense activation	90–91% inhibition
[9]	*Vicia faba*	BYMV	*Potyviridae*	CNPs (300–400 mg/L)	Nanoparticles	Curative (48 h post)	Viral particle disruption	100% inhibition at high dose

with precision agriculture technologies. Site-specific use: Site-specific CNPs could be applied in combination with remote sensing devices, GPS-controlled sprayers, and Internet of Things (IoT)-connected sensors to apply the correct dosage at the appropriate time. The chitosan nanomaterials are also finding application in the development of bio-nanosensors that detect stress, diseases, or nutrient deficiencies in real time to allow early intervention to be undertaken to better manage the crops. The public–private partnerships, effective regulatory frameworks, and interdisciplinary collaboration will be significant in translating the laboratory innovations into commercially viable products, which are sustainable and safe.

8 Conclusion

CNPs are a group of nanoparticles consisting of the naturally abundant and biodegradable polysaccharide, chitosan, which has evolved to become a new and environmentally friendly tool in the control of plant virus diseases. When compared to the conventional treatment, the following are some of the advantages of CNPs because of their enhanced bioactivity and uptake by plants. They possess natural antimicrobial and elicitor efficiency and can be employed as nanocarriers of agrochemicals and biomolecules, such as double-stranded RNA (dsRNA), to apply in RNA interference (RNAi)-based silencing of viral genes. They can apply their antiviral effect both directly, through inhibition of viral replication, entry, and movement, and indirectly, by inducing plant defense responses including SAR and induced systemic resistance (ISR). Moreover, CNPs unlock plant development by increasing nutrient absorption and root development, and photosynthesis. They can be composed with micronutrients, biological or chemical protection agents, and as they are biocompatible, they can offer two benefits: resist disease and better yield.

CNPs can also be used environmentally as more eco-friendly substitutes to synthetic pesticides since they are biodegradable, less toxic, and do not substantially accumulate in the environment. They can be applied in precision agriculture and controlled-release settings. Although very promising, the wider agricultural implementation of CNPs is hampered by factors like standardization of synthesis, low-cost scale-up, regulations, and large-scale field testing in different climates. Besides, long-term ecological safety, behavior in soil, and off-target effects still need to be evaluated. To summarize, the CNPs are a strategy with high potential to be versatile and environmentally friendly in controlling plant viral diseases. Their combination with RNAi technologies, precision delivery technologies, and sustainable agricultural systems makes them the solutions of the next generation of robust and effective crop protection. With the growing research and development, CNPs will play an important role in changing the management of plant diseases in accordance with global sustainability targets.

References

1. Tatineni S, Hein GL (2023) Plant viruses of agricultural importance: current and future perspectives of virus disease management strategies. Phytopathology® 113:117–141
2. Jones RA, Sharman M, Trębicki P, Maina S, Congdon BS (2021) Virus diseases of cereal and oilseed crops in Australia: current position and future challenges. Viruses 13:2051
3. Mishra P, Kumari S, Saini P, Meena RK (2022) Role of nanotechnology in management of plant viral diseases. Mater Today: Proc 69:1–10
4. Jones RA (2021) Global plant virus disease pandemics and epidemics. Plants 10:233
5. Dutta P, Kumari A, Mahanta M, Biswas K, Dudkiewicz A, Thakuria D, Abdelrhim AS, Singh SB, Muthukrishnan G, Sabarinathan K (2022) Advances in nanotechnology as a potential alternative for plant viral disease management. Front Microbiol 13:935193
6. Warghane A, Saini R, Shri M, Andankar I, Ghosh DK, Chopade BA (2024) Application of nanoparticles for management of plant viral pathogen: current status and future prospects. Virology 592:109998
7. Iber BT, Kasan NA, Torsabo D, Omuwa JW (2022) A review of various sources of chitin and chitosan in nature. J Renew Mater 10:1097
8. Román-Doval R, Torres-Arellanes SP, Tenorio-Barajas AY, Gómez-Sánchez A, Valencia-Lazcano AA (2023) Chitosan: Properties and its application in agriculture in context of molecular weight. Polymers 15:2867
9. El Gamal AY, Atia MM, El Sayed T, Abou-Zaid MI, Tohamy MR (2022) Antiviral activity of chitosan nanoparticles for controlling plant-infecting viruses. S Afr J Sci 118
10. Valkonen JP (2007) Viruses: economical losses and biotechnological potential. In: Potato biology and biotechnology. Elsevier
11. Gergerich RC, Dolja VV (2006) Introduction to plant viruses, the invisible foe. In: The plant health instructor, 478
12. Chan Y-L, Saidov N, Lee L-M, Kuo F-H, Shih S-L, Kenyon L (2022) Survey of viruses infecting tomato, cucumber and mung bean in Tajikistan. Horticulturae 8:505
13. Leke WN, Mignouna DB, Brown JK, Kvarnheden A (2015) Begomovirus disease complex: emerging threat to vegetable production systems of West and Central Africa. Agric Food Secur 4:1–14
14. Pandey V, Srivastava A, Gaur R (2021) Begomovirus: a curse for the agricultural crops. Arch Phytopathol Plant Prot 54:949–978
15. Tsedaley B (2015) A review paper on Potato virus Y (PVY) biology, economic importance and its managements. J Biol, Agric Healthc 5:110–126
16. Dorokhov YL, Sheshukova EV, Komarova TV (2018) Tobamoviruses and their diversity. CRC Press, Plant Viruses
17. Hull R (2013) Plant virology. Academic press
18. Moriones E, Navas-Castillo J (2000) Tomato yellow leaf curl virus, an emerging virus complex causing epidemics worldwide. Virus Res 71:123–134
19. Kannan M, Ismail I, Bunawan H (2018) Maize dwarf mosaic virus: from genome to disease management. Viruses 10:492
20. Sharma K, Misra RS (2011) Molecular approaches towards analyzing the viruses infecting maize (*Zea mays* L.). J Gen Mol Virol 3:1–17
21. SALAZAR, L. F. 1996. *Potato viruses and their control*, International Potato Center.
22. Taliansky M, Mayo MA, Barker H (2003) Potato leafroll virus: a classic pathogen shows some new tricks. Mol Plant Pathol 4:81–89
23. Varma A, Malathi V (2003) Emerging geminivirus problems: a serious threat to crop production. Ann Appl Biol 142:145–164
24. Sanwal S, Venkataravanappa V, Singh B (2016) Resistance to bhendi yellow vein mosaic disease: a review. Indian J Agric Sci 86:835–843
25. Legg JP, Kumar PL, Makeshkumar T, Tripathi L, Ferguson M, Kanju E, Ntawuruhunga P, Cuellar W (2015) Cassava virus diseases: biology, epidemiology, and management. In: Advances in virus research. Elsevier

26. Patil BL, Fauquet CM (2009) Cassava mosaic geminiviruses: actual knowledge and perspectives. Mol Plant Pathol 10:685–701
27. Westmore G (2012) Thrips vectors and resistance to Tomato spotted wilt virus (TSWV) in potato. University of Tasmania
28. Bar-Joseph M, Marcus R, Lee RF (1989) The continuous challenge of citrus tristeza virus control. Annu Rev Phytopathol 27:291–316
29. Moreno P, Ambrós S, Albiach-Martí MR, Guerri J, Pena L (2008) Citrus tristeza virus: a pathogen that changed the course of the citrus industry. Mol Plant Pathol 9:251–268
30. Martin DP, Shepherd DN (2009) The epidemiology, economic impact and control of maize streak disease. Food Security 1:305–315
31. Bunawan H, Dusik L, Bunawan SN, Amin NM (2014) Rice tungro disease: From identification to disease control. World Appl Sci J 31:1221–1226
32. Hibino H (1996) Biology and epidemiology of rice viruses. Annu Rev Phytopathol 34:249–274
33. Dale JL (1987) Banana bunchy top: an economically important tropical plant virus disease. Adv Virus Res 33:301–325
34. Qazi J (2016) Banana bunchy top virus and the bunchy top disease. J Gen Plant Pathol 82:2–11
35. Beoni E, Chrpová J, Jarošová J, Kundu JK (2016) Survey of Barley yellow dwarf virus incidence in winter cereal crops, and assessment of wheat and barley resistance to the virus. Crop Pasture Sci 67:1054–1063
36. Addy HS, Nurmalasari, Wahyudi AHS, Sholeh A, Anugrah C, Iriyanto FES, Darmanto W, Sugiharto B (2017) Detection and response of sugarcane against the infection of Sugarcane mosaic virus (SCMV) in Indonesia. Agronomy 7:50
37. Rishi N (2010) Significant achievements and current status: virology. In: One hundred years of plant pathology in India: an overview. ISMPP, Scientific Publishers, Jodhpur, pp 143–206
38. Waliyar F, Kumar P, Ntare B, Monyo E, Nigam S, Reddy A, Osiru M, Diallo A (2007) A century of research on groundnut rosette disease and its management. Inf Bull 75:40
39. Tatineni S, Hein GL (2018) Genetics and mechanisms underlying transmission of wheat streak mosaic virus by the wheat curl mite. Curr Opin Virol 33:47–54
40. Jia D, Chen Q, Mao Q, Zhang X, Wu W, Chen H, Yu X, Wang Z, Wei T (2018) Vector mediated transmission of persistently transmitted plant viruses. Curr Opin Virol 28:127–132
41. Borah A, Bhausaheb MA, Rana M, Srivastava S (2023) Mechanism of arthropod-mediated transmission of plant viruses-a review. J Adv Zool 44
42. Whitfield AE, Falk BW, Rotenberg D (2015) Insect vector-mediated transmission of plant viruses. Virology 479:278–289
43. Hohn T (2007) Plant virus transmission from the insect point of view. Proc Natl Acad Sci 104:17905–17906
44. Simmons HE, Munkvold GP (2014) Seed transmission in the Potyviridae. In Global perspectives on the health of seeds and plant propagation material, pp 3–15
45. Dijkstra J, Khan JA (2006) Mechanical transmission of plant viruses. CRC Press, Handbook of Plant Virology
46. Bhat AI, Rao GP, Bhat AI, Rao GP (2020) Virus transmission through pollen. Springer
47. Chare ER, Holmes EC (2004) Selection pressures in the capsid genes of plant RNA viruses reflect mode of transmission. J Gen Virol 85:3149–3157
48. Devi BM, Guruprasath S, Balu P, Chattopadhyay A, Thilagar SS, Dhanabalan KV, Choudhary M, Moparthi S, Jailani AAK (2024) Dissecting diagnostic and management strategies for plant viral diseases: what next? Agriculture 14:284
49. Wang P, Liu J, Lyu Y, Huang Z, Zhang X, Sun B, Li P, Jing X, Li H, Zhang C (2022) A review of vector-borne rice viruses. Viruses 14:2258
50. Schönegger D (2023) Analysis of the virome and reciprocal transfers of viruses between cultivated and wild carrot populations. Université de Bordeaux
51. Stevens M, Lacomme C (2017) 15 transmission of plant viruses. Aphids Crop Pests 323–361
52. Yadav S, Chhibbar AK (2018) Plant–virus interactions. Mol Asp Plant-Pathog Interact 43–77
53. Awasthi L (2023) Viral diseases of field and horticultural crops. Elsevier

54. Li R, Baysal-Gurel F, Abdo Z, Miller SA, Ling K-S (2015) Evaluation of disinfectants to prevent mechanical transmission of viruses and a viroid in greenhouse tomato production. Virol J 12:1–11
55. Sacristán S, Díaz M, Fraile A, García-Arenal F (2011) Contact transmission of Tobacco mosaic virus: a quantitative analysis of parameters relevant for virus evolution. J Virol 85:4974–4981
56. Nazarov PA, Baleev DN, Ivanova MI, Sokolova LM, Karakozova MV (2020) Infectious plant diseases: etiology, current status, problems and prospects in plant protection. Acta naturae 12:46
57. Mehetre GT, Leo VV, Singh G, Sorokan A, Maksimov I, Yadav MK, Upadhyaya K, Hashem A, Alsaleh AN, Dawoud TM (2021) Current developments and challenges in plant viral diagnostics: a systematic review. Viruses 13:412
58. Varma A, Singh MK (2020) Diagnosis of plant virus diseases. Elsevier, Applied plant virology
59. Selvarajan R, Balasubramanian V (2014) Plant virus–host interaction molecular approaches and viral evolution
60. Benjamin J, Oyedokun DO, Oziegbe EV, Oni J, Ogundare EB, Ujah GO, Adebayo A (2024) Cereal production in Africa: the threat of current plant pathogens in changing climate-a review. Discov Agric 2:33
61. Mkamilo G, Kimata B, Masinde EA, Masisila FF, Menya RO, Matondo D, Maruthi MN (2024) Impact of viral diseases and whiteflies on the yield and quality of cassava. J Plant Dis Prot 131:959–970
62. Panno S, Davino S, Caruso AG, Bertacca S, Crnogorac A, Mandić A, Noris E, Matić S (2021) A review of the most common and economically important diseases that undermine the cultivation of tomato crop in the mediterranean basin. Agronomy 11:2188
63. Sastry KS (2013) Impact of virus and viroid diseases on crop yields. In: Plant virus and viroid diseases in the tropics: volume 1: introduction of plant viruses and sub-viral agents, classification, assessment of loss, transmission and diagnosis, pp 99–159
64. Paul P, Kolesinska B, Sujka W (2019) Chitosan and its derivatives-biomaterials with diverse biological activity for manifold applications. Mini Rev Med Chem 19:737–750
65. Kardas I, Struszczyk MH, Kucharska M, van den Broek LA, van Dam JE, Ciechańska D (2012) Chitin and chitosan as functional biopolymers for industrial applications. In: The European polysaccharide network of excellence (EPNOE) research initiatives and results, pp 329–373
66. Gonçalves C, Ferreira N, Lourenço L (2021) Production of low molecular weight chitosan and chitooligosaccharides (COS): a review. Polymers 13:2466
67. Czechowska-Biskup R, Jarosińska D, Rokita B, Ulański P, Rosiak JM (2012) Determination of degree of deacetylation of chitosan-comparision of methods. Prog Chem Appl Chitin Deriv 5–20
68. Khan TA, Peh KK, Ch'ng HS (2002) Reporting degree of deacetylation values of chitosan: the influence of analytical methods. J Pharm Sci 5:205–212
69. Kumirska J, Weinhold MX, Thöming J, Stepnowski P (2011) Biomedical activity of chitin/chitosan based materials—influence of physicochemical properties apart from molecular weight and degree of N-acetylation. Polymers 3:1875–1901
70. Yan D, Li Y, Liu Y, Li N, Zhang X, Yan C (2021) Antimicrobial properties of chitosan and chitosan derivatives in the treatment of enteric infections. Molecules 26:7136
71. Ke C-L, Deng F-S, Chuang C-Y, Lin C-H (2021) Antimicrobial actions and applications of chitosan. Polymers 13:904
72. Pauls T (2016) Chitosan as an antiviral
73. Zehra A, Rai A, Singh SK, Aamir M, Ansari WA, Upadhyay RS (2021) An overview of nanotechnology in plant disease management, food safety, and sustainable agriculture. Food Secur Plant Dis Manag 193–219
74. Rajwade JM, Chikte R, Paknikar K (2020) Nanomaterials: new weapons in a crusade against phytopathogens. Appl Microbiol Biotechnol 104:1437–1461
75. Sánchez-López E, Gomes D, Esteruelas G, Bonilla L, Lopez-Machado AL, Galindo R, Cano A, Espina M, Ettcheto M, Camins A (2020) Metal-based nanoparticles as antimicrobial agents: an overview. Nanomaterials 10:292

76. Pan W, Zhang H-J, Zhang Y-F, Wang M, Tsui MT-K, Yang L, Miao A-J (2023) Silica nanoparticle accumulation in plants: current state and future perspectives. Nanoscale 15:15079–15091
77. Islam F, Shohag S, Uddin MJ, Islam MR, Nafady MH, Akter A, Mitra S, Roy A, Emran TB, Cavalu S (2022) Exploring the journey of zinc oxide nanoparticles (ZnO-NPs) toward biomedical applications. Materials 15:2160
78. Bashir SM, Ahmed Rather G, Patrício A, Haq Z, Sheikh AA, Shah MZUH, Singh H, Khan AA, Imtiyaz S, Ahmad SB (2022) Chitosan nanoparticles: a versatile platform for biomedical applications. Materials 15:6521
79. Wang J, Zhuang S (2022) Chitosan-based materials: preparation, modification and application. J Clean Prod 355:131825
80. Desai KG (2016) Chitosan nanoparticles prepared by ionotropic gelation: an overview of recent advances. In: Critical reviews^{TM} in therapeutic drug carrier systems, 33
81. Zhang K, Xu Y, Lu L, Shi C, Huang Y, Mao Z, Duan C, Ren XE, Guo Y, Huang C (2021) Hydrodynamic cavitation: a feasible approach to intensify the emulsion cross-linking process for chitosan nanoparticle synthesis. Ultrason Sonochem 74:105551
82. Luque-Alcaraz AG, Lizardi-Mendoza J, Goycoolea F, Higuera-Ciapara I, Argüelles-Monal W (2016) Preparation of chitosan nanoparticles by nanoprecipitation and their ability as a drug nanocarrier. RSC Adv 6:59250–59256
83. Choudhary RC, Kumari S, Kumaraswamy R, Pal A, Raliya R, Biswas P, Saharan V (2019) Characterization methods for chitosan-based nanomaterials. In: Plant nanobionics: volume 1, advances in the understanding of nanomaterials research and applications, pp 103–116
84. Agarwal M, Agarwal MK, Shrivastav N, Pandey S, Das R, Gaur P (2018) Preparation of chitosan nanoparticles and their in-vitro characterization. Int J Life-Sci Sci Res 4:1713–1720
85. Rampino A, Borgogna M, Blasi P, Bellich B, Cesàro A (2013) Chitosan nanoparticles: Preparation, size evolution and stability. Int J Pharm 455:219–228
86. Benamer Oudih S, Tahtat D, Nacer Khodja A, Mahlous M, Hammache Y, Guittoum AE, Kebbouche Gana S (2023) Chitosan nanoparticles with controlled size and zeta potential. Polym Eng Sci 63:1011–1021
87. Clogston JD, Patri AK (2010) Zeta potential measurement. Springer, Characterization of nanoparticles intended for drug delivery
88. Collado-González M, Montalbán MG, Peña-García J, Pérez-Sánchez H, Víllora G, Baños FGD (2017) Chitosan as stabilizing agent for negatively charged nanoparticles. Carbohyd Polym 161:63–70
89. Kesente M, Kavetsou E, Roussaki M, Blidi S, Loupassaki S, Chanioti S, Siamandoura P, Stamatogianni C, Philippou E, Papaspyrides C (2017) Encapsulation of olive leaves extracts in biodegradable PLA nanoparticles for use in cosmetic formulation. Bioengineering 4:75
90. Yoksan R, Jirawutthiwongchai J, Arpo K (2010) Encapsulation of ascorbyl palmitate in chitosan nanoparticles by oil-in-water emulsion and ionic gelation processes. Colloids Surf, B 76:292–297
91. Jarudilokkul S, Tongthammachat A, Boonamnuayvittaya V (2011) Preparation of chitosan nanoparticles for encapsulation and release of protein. Korean J Chem Eng 28:1247–1251
92. Ismik D, Mansuroglu DS, Buluş E, Sahin YM (2020) The use of chitosan nanoparticles obtained by ionic gelation method as a drug delivery system. J Mater Electron Devices 5:6–11
93. Hojnik Podrepšek G, Knez Ž, Leitgeb M (2020) Development of chitosan functionalized magnetic nanoparticles with bioactive compounds. Nanomaterials 10:1913
94. Thiruppathi R, Mishra S, Ganapathy M, Padmanabhan P, Gulyás B (2017) Nanoparticle functionalization and its potentials for molecular imaging. Adv Sci 4:1600279
95. El-Ganainy SM, Soliman AM, Ismail AM, Sattar MN, Farroh KY, Shafie RM (2023) Antiviral activity of chitosan nanoparticles and chitosan silver nanocomposites against Alfalfa Mosaic Virus. Polymers 15:2961
96. Iriti M, Varoni EM (2015) Chitosan-induced antiviral activity and innate immunity in plants. Environ Sci Pollut Res 22:2935–2944

97. Elsharkawy MM, Omara RI, Mostafa YS, Alamri SA, Hashem M, Alrumman SA, Ahmad AA (2022) Mechanism of wheat leaf rust control using chitosan nanoparticles and salicylic acid. J Fungi 8:304
98. Francesconi S, Steiner B, Buerstmayr H, Lemmens M, Sulyok M, Balestra GM (2020) Chitosan hydrochloride decreases Fusarium graminearum growth and virulence and boosts growth, development and systemic acquired resistance in two durum wheat genotypes. Molecules 25:4752
99. Rahman A, Sinha KV, Sopory SK, Sanan-Mishra N (2021) Influence of virus–host interactions on plant response to abiotic stress. Plant Cell Rep 40:2225–2245
100. Singh DP, Moore CA, Gilliland A, Carr JP (2004) Activation of multiple antiviral defence mechanisms by salicylic acid. Mol Plant Pathol 5:57–63
101. Chivasa S, Murphy AM, Naylor M, Carr JP (1997) Salicylic acid interferes with tobacco mosaic virus replication via a novel salicylhydroxamic acid-sensitive mechanism. Plant Cell 9:547–557
102. Luo Y, Shang J, Zhao PP, Xi DH, Yuan S, Lin HH (2011) Application of jasmonic acid followed by salicylic acid inhibits Cucumber mosaic virus replication. Plant Pathol J 27:53–58
103. Vasyukova N, Ozeretskovskaya O, Chalenko G, Gerasimova N, L'vova A, Il'ina A, Levov A, Varlamov V, Tarchevsky I (2010) Immunomodulating activity of chitosan derivatives with salicylic acid and its fragments. Appl Biochem Microbiol 46:346–351
104. Komarova T, Shipounova I, Kalinina N, Taliansky M (2024) Application of chitosan and its derivatives against plant viruses. Polymers 16:3122
105. Kim T-J, Lim G-H (2023) Salicylic acid and mobile regulators of systemic immunity in plants: transport and metabolism. Plants 12:1013
106. Zhao S, Li Y (2021) Current understanding of the interplays between host hormones and plant viral infections. PLoS Pathog 17:e1009242
107. Siddaiah CN, Prasanth KVH, Satyanarayana NR, Mudili V, Gupta VK, Kalagatur NK, Satyavati T, Dai X-F, Chen J-Y, Mocan A (2018) Chitosan nanoparticles having higher degree of acetylation induce resistance against pearl millet downy mildew through nitric oxide generation. Sci Rep 8:2485
108. Narasimhamurthy K, Udayashankar AC, de Britto S, Lavanya SN, Abdelrahman M, Soumya K, Shetty HS, Srinivas C, Jogaiah S (2022) Chitosan and chitosan-derived nanoparticles modulate enhanced immune response in tomato against bacterial wilt disease. Int J Biol Macromol 220:223–237
109. Chandra S, Chakraborty N, Dasgupta A, Sarkar J, Panda K, Acharya K (2015) Chitosan nanoparticles: a positive modulator of innate immune responses in plants. Sci Rep 5:15195
110. el Hadrami A, Adam LR, el Hadrami I, Daayf F (2010) Chitosan in plant protection. Mar Drugs 8:968–987
111. Nagorskaya V, Reunov A, Lapshina L, Davydova V, Yermak I (2014) Effect of chitosan on tobacco mosaic virus (TMV) accumulation, hydrolase activity, and morphological abnormalities of the viral particles in leaves of N. tabacum L. cv. Samsun. Virol Sin 29:250–256
112. Xiang S, Wang J, Wang X, Ma X, Peng H, Zhu X, Huang J, Ran M, Ma L, Sun X (2023) A chitosan-coated lentinan-loaded calcium alginate hydrogel induces broad-spectrum resistance to plant viruses by activating Nicotiana benthamiana calmodulin-like (CML) protein 3. Plant, Cell Environ 46:3592–3610
113. Rendina N, Nuzzaci M, Scopa A, Cuypers A, Sofo A (2019) Chitosan-elicited defense responses in Cucumber mosaic virus (CMV)-infected tomato plants. J Plant Physiol 234:9–17
114. Abdelkhalek A, Qari SH, Abu-Saied MAA-R, Khalil AM, Younes HA, Nehela Y, Behiry SI (2021) Chitosan nanoparticles inactivate alfalfa mosaic virus replication and boost innate immunity in Nicotiana glutinosa plants. Plants 10:2701

Bacterial Plant Disease Management Using Chitosan Nanoparticles

Nasir Assad, Tooba, Musiab Ullah, Yasir Assad, Muhammad Nauman Khan, Barkat Ullah, Abdul Razzaq, and Hansa Gul

Abstract Bacterial plant diseases are the main cause of high agricultural losses worldwide, requiring the development of effective, sustainable, and environmentally friendly management strategies. Chitosan is a natural chitin-derived biopolymer that has shown promise as an active ingredient due to its biocompatibility, biodegradability, and innate antimicrobial properties. Recent advances in nanotechnology have enabled the production of chitosan nanoparticles (CNPs) that are more effective against bacteria than bulk chitosan due to their larger surface area, improved solubility, and greater affinity for microbial cells. This chapter discusses the synthesis methods, physicochemical properties, and antibacterial activity of CNPs against bacterial phytopathogens. CNPs destabilize bacterial membranes, disrupt biofilm formation, and induce systemic resistance in plants, providing a versatile solution for disease control. Multiple studies have demonstrated the efficacy of CNPs against critical bacterial diseases such as bacterial blight of rice, fire blight of apples and

N. Assad (✉) · Tooba
Institute of Chemistry, University of Sargodha, Sargodha, Pakistan
e-mail: nakhan_98@yahoo.com

M. Ullah
Institute of Biotechnology and Genetic Engineering, The University of Agriculture, Peshawar, Pakistan

Y. Assad
Department of Zoology, Hazara University, Mansehra, Pakistan

M. N. Khan · B. Ullah
Department of Botany, Islamia College Peshawar, Peshawar, Pakistan

M. N. Khan
Biology Laboratory, University Public School, University of Peshawar, Peshawar, Pakistan

A. Razzaq
Qarshi Herb Research Center, Qarshi Industries (Pvt.) Ltd. Hattar, Haripur, Pakistan

Qarshi University, Lahore, Punjab, Pakistan

H. Gul
Department of Zoology, University of Mianwali, Punjab, Pakistan

A. A. H. Abdel Latef (ed.), *Chitosan Nanoparticles for Plant Resilience Under Abiotic and Biotic Stressors*, Smart Nanomaterials Technology,
https://doi.org/10.1007/978-981-95-6422-4_9

pears, and bacterial wilt of tomatoes. Additionally, chitosan nanoparticles act synergistically with traditional bactericides, reducing the use and of chemicals ecological impact. Their potential is further confirmed by field trials and greenhouse experiments in real agricultural environments. Despite these promising results, challenges remain related to large-scale production, standardization, regulatory approval, and understanding the long-term environmental impact. Overall, CNPs are an innovative and sustainable resource that can be integrated into plant disease management strategies, and fit into global efforts to make agriculture greener and safer agriculture. Future research should focus on improving formulations, understanding the interactions between plants, microbes, and nanoparticles, and developing cost-effective delivery systems for practical applications.

Keywords Chitosan nanoparticles · Bacterial plant diseases · Antimicrobial activity · Sustainable agriculture · Plant disease management

1 Introduction

By 2050, the world population will grow to around 9.8 billion people, which will put enormous pressure on the demand for food and water, especially in the developing nations where incomes and food demand will multiply exponentially. The overall demand for food is expected to increase by 35–56% in the period 2010–2050 and the number of people at risk of hunger could fluctuate between −91% and +8% under baseline conditions; under unfavorable climate change scenarios these rates could increase to 30–62% and −91 to +30, respectively [64].

A large part of this threat comes from plant pests and diseases. Yield losses due to plant diseases amounted to around 16% worldwide between 2001 and 2003, and in developing countries, the loss can amount to up to a quarter of the yield. Fungi account for the largest proportion of phytopathogens (42%) and these were followed by bacteria (27%), viruses (18%), and nematodes (13%) [59]. One of the consequences of yield loss is that numerous pathogens can affect food quality by synthesizing mycotoxins (e.g. *Aspergillus*, *Fusarium*), which has had devastating economic impacts in the past and present (e.g. the Irish potato famine caused by *Phytophthora infestans*, and the rice-blast epidemics that spread to the regional and global rice markets). All these possibilities of crop destruction of the crop not only in terms of quantity but also in terms of quality and market disruption, explain how plant diseases can impact food security and livelihoods [57].

According to current estimates, pathogenic bacteria and other phytopathogens cause annual costs of 20–40% in agriculture; therefore, disease control is a viable option for sustainable agriculture [25, 64]. In past decades, the use of synthetic chemical pesticides has been relied upon to achieve control. However, this approach has significant and increasing limitations: Some studies suggest that up to 90% of pesticides applied are not taken up by the environment, whereas in other studies only a fraction of ~0.1% of pesticides sprayed reaches its target, resulting in either environmental contamination or dramatic waste. The extensive application of pesticides has also led to the development of resistant strains of pathogens, and adverse effects on non-target organisms (including useful soil microbes), as well as associated problems for human-health and the environment [47, 54]. The amount of pesticide residue on crops depends on the formulation, quantity and equipment used for application, as well as the quality of application, and most farmers, especially in low-income countries like Laos, Cambodia, and Vietnam, still use the chemicals despite knowing the consequences [55].

These issues have prompted a desire to develop safer, eco-friendly disease control practices. Alternatives include natural bioactive compounds and biodegradable polymers [26, 37, 44]. The reviews have also focused on chitosan as one of the most ubiquitous natural biopolymers; chitosan is a deacetylated product of the biopolymer, chitin, which also has gained special interest due to its intrinsic properties culminating in antimicrobial properties, biodegradability, low toxicity, water solubility, and ability to induce plant defense mechanisms. These characteristics would make chitosan an interesting building block for next-generation crop protection systems, designed to decrease the use of chemicals as pesticides, and increase the resilience of plants [1].

Nanotechnology has expanded these possibilities even further. The engineered nanoparticles (NPs) can be used as nanofertilizers, nanosensors, nanobiotechnology vehicles, and nanopesticides (nanofungicides, nanobactericides, nanoinsecticides), allowing more site-specific nutrient delivery, real time crop testing, and, depending on the engineered characteristics of the nanoparticle; selective pathogen control [11]. Nanoparticles can be divided into two groups: Micronutrient-based nanoparticles such as copper (Cu), zinc oxide (ZnO), and magnesium oxide (MgO) and non-micronutrient nanoparticles such as silver (Ag). Both types have been shown not only to improve plant growth but also to protect plants against harmful pathogens such as *Phoma destructiva*, *Phytophthora infestans*, *Fusarium oxysporum*, *Botrytis cinerea*, and species of *Colletotrichum* [9, 14, 34]. As carrier materials, NPs can enable controlled release of active substances, increase water and nutrient uptake, minimize off-target losses, and reduce the environmental impact of plant protection products [60]. Other purported benefits of nanodelivery systems include longer shelf life, localized uptake, high solubility, decreased soil leaching and lower toxicity [20]. Nanoparticles can be produced by top-down or bottom-up methods using physical, chemical, or biological strategies; the chosen direction has less influence on particle outline, surface activities, and biological capacities [24]. One of the most common chemical methods for the preparation of chitosan NPs described in previous studies is ionic gelation [32]. This method utilizes electrostatic cross-linking between protonated amino groups of chitosan (R-NH $^{3+}$) and polyanionic tripolyphosphate (TPP) to form hydrogel nanoparticles. This process can be performed by using standard laboratory equipment and reagents. It proceeds via solution, clustering, and opalescent suspensions, which are influenced by parameters such as the ratio of chitosan to TPP, stirring conditions, particle aspects (size, polydispersity), and concentration [61].

The potential applications of chitosan nanoparticles (CNPs) have been extensively studied in pharmaceutical fields as well as antimicrobial carriers. Ag-, Cu-, Zn-, Mn-, and Fe- loaded C-NPs produced by ionic gelation exhibited a strong ability to curb the growth of clinically and agriculturally important bacteria. Several studies have highlighted the strong antibacterial and growth-promoting activity of metal-loaded CNPs [18]. For instance, copper-loaded CNPs (Cu-CNPs) have demonstrated significant antifungal activity in vitro by effectively inhibiting spore germination and mycelial growth of crop pathogens such as *Alternaria alternata* (90% inhibition), *Rhizoctonia solani* (60%), and *Macrophomina phaseolina* (63%) [62]. In addition to the direct antimicrobial effect, these nanoformulations also serve as powerful bio-stimulants. Studies on tomatoes have shown that treating seeds with Cu-CNPs significantly enhances growth and germination, while similar treatments in maize enhance seedling development and encourage the activity of important enzymes like α-amylase and protease [35]. This dual capacity to directly combat pathogens while promoting plant defense responses and overall growth underscores its value as a sustainable alternative in modern crop protection [35]. These findings emphasize the dual capability of metal-incorporated nanoparticles, making them promising candidates for the control of pathogenic bacteria. These results emphasize the dual potential of chitosan-based nanocarriers to provide both direct antimicrobial efficacy as well as the ability to reduce the amount of pesticides and environmental pollution when used in crop protection. The overall concept is illustrated in Fig. 1, which summarizes the interconnection between food security threats, plant diseases, pesticide use, and the role of CNPs as eco-friendly antimicrobial nanocarriers.

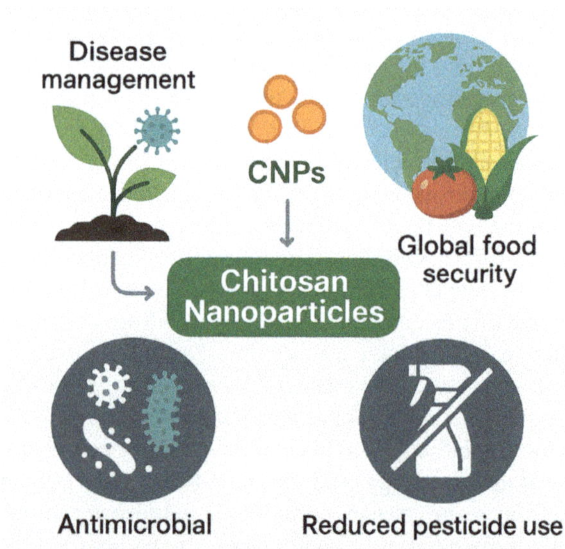

Fig. 1 Role of chitosan nanoparticles in food security through disease management, antimicrobial action, and reduced pesticide use created in (https://BioRender.com)

2 Chitosan and Chitosan Nanoparticles (CNPs)

Chitosan is a natural, linear polysaccharide obtained by deacetylation of chitin, the 2nd most abundant biopolymer on earth after cellulose. Chitin is found everywhere in nature, where it occurs in the form of the exoskeleton of crustaceans, the cell walls of fungi and the cuticle of insects [15, 42]. Chitosan is produced industrially by chemical deacetylation of chitin, whereby the acetyl groups are removed from the polymer chains. The resulting cationic biopolymer has randomly arranged units of 1 to 4 linked D-glucosamines and N-acetyl-D-glucosamines [21]. The importance of this chemical structure and the organizing principle of amino groups containing positively charged groups under acidic conditions for the biological activity of this compound and the determination of the fairly universal application in agriculture.

As a pesticide, chitosan presents several advantages over conventional synthetic pesticides. It is well tolerated by the body, biodegradable, and has no toxic effects on humans, animals or the environment. Chitosan breaks down naturally into harmless substances, unlike synthetic pesticides, which can remain in the soil as harmful substances. It has very low allergenicity, is bioabsorbable [54] and has both a direct and indirect plant-protective effect. Chitosan directly inhibits the growth of other phytopathogens by disrupting the membrane and inhibiting the uptake of nutrients. It then enhances the natural immunity of plants by producing defense-related enzymes, phytotransplantalexins and structural defenses that increase resistance to further pathogen infections [1]. These two features, antimicrobial activity and immunological priming, distinguish chitosan NPs from conventional pesticides, whose activity can only be determined after infection of the individual.

The application of CNPs in nanotechnology has further expanded their potential in the field of plant protection. CNPs are engineered formulations with a nanoscale particle size (1–100 nm) [57], and this change led to an impressive enhancement in the intrinsic properties of chitosan in bulk.

They are so small that high surface to-volume ratio enables more efficient interactions with microbial cell surfaces and better penetration of plant cuticles and microbial cell walls. The increase in the positive surface charge is more pronounced at the nanoscale and leads to electrostatic forces with the negative microbial membranes, resulting in membrane destabilization and loss of intracellular components. CNPs can also be used as effective vehicles for the controlled and sustained release of agrochemicals, including pesticides and fertilizers. This is, due to their intrinsic antimicrobial activity, which is unmatched compared to other available alternatives, resulting in improved and as a result, improve efficacy and a reduction in the total amount of chemicals used [20]. CNPs have higher bioavailability, better stability, and a stronger multimechanistic mode of action than conventional chitosan, which includes direct inhibition of pathogens, strong induction of plant defense mechanisms and potent modulation of microbial gene expression [54]. All these qualities make CNPs an ecologically superior, next-level solution in crop protection that every environmentally conscious person can wish for.

3 Synthesis of Chitosan Nanoparticles

Chitosan nanoparticles can be prepared using various techniques such as micelle formation [50], coacervation [33], spray drying [23], emulsification [52], and solvent evaporation [5]. Of these methods, the ionic gelation is the most commonly used as it has the advantage of being simple, controllable, and does not require organic solvents or toxic reagents [70] (Fig. 2). The proposed method utilizes the electrostatic forces of the protonated amino groups (NH_3^+) of chitosan in an acidic solution (polyelectrolyte) and polyanionic phosphate groups (P 5 O 10 5 1) released by an acid-releasing non-toxic cross-linking agent, usually sodium tripolyphosphate (STPP). When the copolymerization of chitosan and STPP is carried out under mild conditions, usually at ambient temperature and neutral pH, their interaction leads to inter- and intramolecular crosslinking, resulting in the spontaneous formation of nanoparticles [50].

The physicochemical properties of CNPs (size, polydispersity index (PDI), and zeta potential (ZP) are extremely sensitive to the synthesis conditions. The concentration of chitosan and STPP and their mass ratio are crucial for ionic gelation [51]. The migration of chitosan tends to increase with increasing chitosan concentration, leading to nanoparticles with greater sizes as the viscosity of the solution is increased and intermolecular repulsion increases the length of the polymer chains before aggregation [53]. Similarly, a higher concentration of STPP promotes the cross-linking of more chitosan chains into single and bigger polymers [51]. The PDI, which is an indicator of the uniformity of particle size distribution, must be below 0.5 to ensure that the suspension is monodisperse and stable [65]. The surface charge and colloidal stability of nanoparticles, as measured by the zeta potential, are favored at a high chitosan concentration due to the high amount of protonated amino groups, but inhibited at high STPP concentrations as the additional protonation causes blocking of the amino groups [65]. In order to produce nanoparticles with advantageous size, stability, and bioactivity that can meet the expected demand in agriculture, much care must be taken to maximize these parameters.

Besides synthesis, CNPs are characterized through various analytical methods to obtain detailed information. Dynamic light scattering (DLS), performed in a

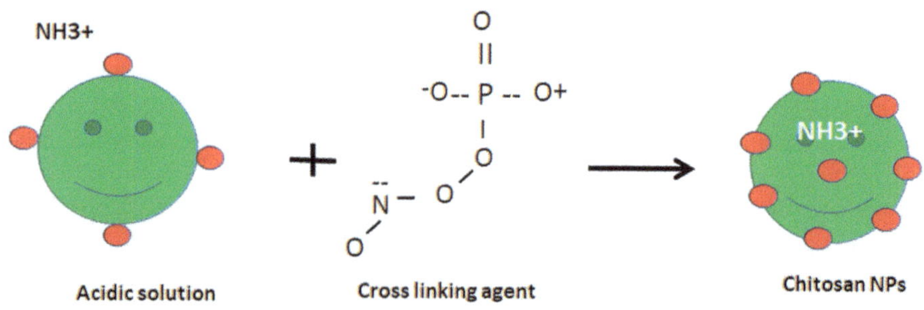

Fig. 2 Ionic gelation (chitosan NPs synthesis)

Zetasizer provides the results for average particle size, PDI, and zeta potential [6]. The morphology, texture, and size of nanoparticles in the dry-state can be directly observed using field emission scanning electron microscopy (FE-SEM) [65]. Attenuated total reflection fourier-transform infrared (ATR-FTIR) is intended to detect the chemical interactions between chitosan and the cross-linker, by spectroscopy. It is employed to verify characteristic functional groups and changes in the absorption peaks [6, 53]. Thermal stability and degradation behavior will be investigated using thermogravimetric analysis (TGA) and differential scanning calorimetry (DSC), and crystallinity and structural arrangement (XRD) will be studied using X-ray diffraction [65], which could indicate the potential change between amorphous and crystalline states after the introduction of the nanoparticles.

Taken together, these synthesis and characterization methods will enable the unlimited generation of CNPs with selected properties to enhance performance and protect plants in the long term, which could indicate a possible shift between amorphous and crystalline states after the presence of nanoparticles.

4 Physicochemical Properties of CNPs and Their Role in Activity

The physicochemical properties of CNPs are decisive for their biological effects and applications. Particle size is considered to be one of the most significant factors determining the work of the nanoparticles and their success. Depending on the synthesis conditions determined by DLS, the size of CNPs, including the polymer swelling in an aqueous medium, can range from about 190 to more than 500 nm [51]. The actual diameter of the liquid dispersion particles in their dried state is often smaller (measured by FE-SEM) [70]. The nanoscale size provides the nanoparticles a significant advantage as the immensely high surface-area-to-volume ratio promotes contact and interaction of the nanoparticles with the microbial cell membranes, thereby enhancing their antimicrobial and biological activity compared to bulk chitosan. Solubility and dispersibility are other important parameters. Although bulk chitosan cannot be dissolved in water under any conditions, the suspension of chitosan in nanoparticles improves its dispersibility in water. The uniformity of this dispersion is proportional to the polydispersity index (PDI), i.e. if this is below 0.5, a homogeneous and uniform distribution of the nanoparticles with low aggregation potential is given [65]. Under suitable synthesis conditions, CNPs with PDI values in the preferred range of 0.18 to 0.48 can be prepared, allowing the formation of a stable and well-dispersed system that can be used in applications, e.g. for the formation of to form stable Pickering emulsions or for microencapsulation of bioactive compounds [6]. Another important characteristic that determines both colloidal stability as well as the antimicrobial activity is the surface charge, expressed in ZP [70]. CNPs therefore have a high positive zeta potential due to the protonated amino groups ($-NH_3^+$)

(e.g. a zeta potential of >30 mV is not uncommon), which leads to a strong electrostatic repulsion between the particles so that the CNPs do not aggregate and thus form a stable suspension [6]. Their antibacterial activity arises from the positive charge interacting with the negatively charged bacterial cell membrane-due to teichoic acids in Gram-positive bacteria and lipopolysaccharides in Gram-negative bacteria. The CNPs carrying positively charged and negatively charged bacterial surfaces electrostatically are attracted, leading to membrane destruction, leakage of intracellular components, and cell death [16]. Therefore, a high positive zeta potential not only influences physical stability, but also confers strong biological efficacy.

5 Mechanisms of Antibacterial Action

The mechanisms of action of chitosan are very extensive: The antibacterial effect of chitosan is aimed at a direct attack on the pathogen as well as at activating the immunity of the host plant [3, 31]. It has three main modes of action, namely destruction of bacterial cell membranes, inhibition of biofilm formation, chelation of nutrients, and stimulation of systemic resistance in plants. The most frequently assumed and investigated mechanism is that of membrane disruption. Chitosan is a polycationic polymer under acidic conditions; it has protonated amino groups (also known as $-NH_3^+$) that bind strongly to negatively charged compounds on the bacterial surface [56]. These targets differ in Gram-negative and Gram-positive bacteria, with Gram-negative bacteria being lipopolysaccharides and Gram-positive bacteria being teichoic and lipoteichoic acids [56]. Electrostatic binding removes the surface charge and alters membrane fluidity and permeability, resulting in the loss of intracellular electrolytes, metabolites and proteinaceous components, which ultimately leads to cell death [27]. Other studies also reveal that chitosan could digest peptidoglycan, further damaging the integrity of bacterial cell walls [22]. This has been observed in plant pathogenic bacteria such as *Acidovorax citrulli* whereby treatment with chitosan resulted in massive membrane damage, cytosolic coagulation, and extracellular release of nucleic material [46].

In addition to destroying bacterial membranes, chitosan is also effective against the formation and survival of bacterial biofilms, i.e. defense colonies surrounded by an extracellular protective shell of self-generated hydrocolloid. The inhibition of the biofilm reduces the capacity of a pathogen to colonize host tissue and be resistant to treatment with antimicrobial agents. Studies on the effect of chitosan on *Xanthomonas oryzae*, the causal agent of bacterial leaf blight in rice, indicated that chitosan causes cell lysis and biofilm disintegration [22]. A significant decrease in biofilm formation was also observed in *Acidovorax avenae* subsp. *avenae* when treated with chitosan [69]. This disruption limits the survival of the pathogen under environmental stress conditions and its ability to sustain infection.

The chelating capacity of CNPs influences their antibacterial effect. They have functional amine groups (NH_2) in their structure that can bind the necessary metal ions required to promote microbial growth and enzyme activity [30] (Fig. 3). Such

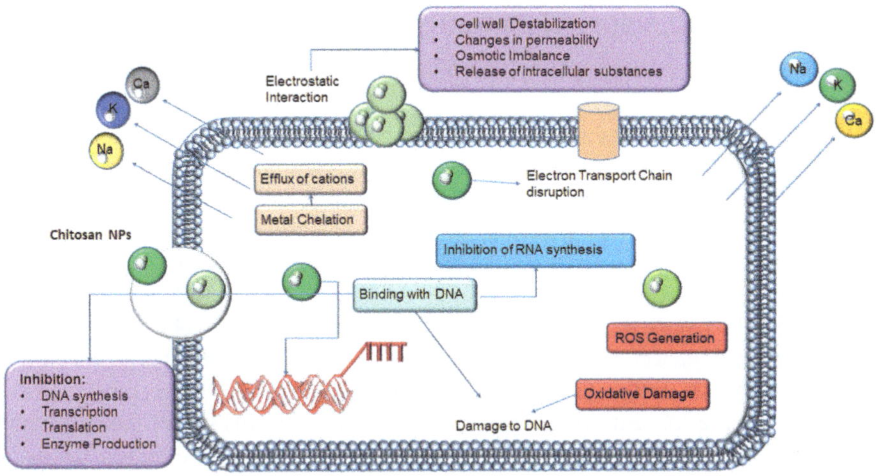

Fig. 3 Overall antibacterial mechanism of chitosan NPs

sequestration robs bacteria of essential nutrients and leads to starvation and limited growth, especially at elevated pH, where the protonation of the amine group is reduced and available to form complexes [22]. Moreover, a high-molecular-weight chitosan can also form a polymeric coating around bacterial cells, providing a further physical barrier to nutrient uptake [68].

In addition to these direct antimicrobial effects, chitosan can also be a highly effective trigger for plant defense responses, activating systemic resistance that provides unspecialized and long-term protection [17]. When the plant cell receptors sense chitosan, they initiate signaling cascades that principally regulate the production of essential phytohormones such as jasmonic acid and salicylic acid, which mediate systemic acquired resistance [38]. This leads to heightened activity of defense-related enzymes like peroxidase and polyphenol oxidase and phenylalanine ammonia-lyase (PAL), production of antimicrobial substances and thickening of the cell wall [22]. These effects have been observed in many host–pathogen interactions, such as the development of rice resistance to *Xanthomonas oryzae* [45] and tomato resistance to *Xanthomonas vesicatoria* [58]. Chitosan interactions: chitosan has been reported to act as an effective antibacterial agent and enhancer of plant immunity through multiple and synergistic interactions, making it a promising sustainable approach for plant disease control.

6 Influencing Factors of Antibacterial Activity of Chitosan NPs

The antibacterial properties of CNPs are the result of a dynamic interaction of intrinsic polymer properties, external environmental factors, and microbial properties that are target-specific. Among the intrinsic factors, the molecular weight (MW) and degree of acetylation (DA) of chitosan are critical determinants of antimicrobial performance. Chitosan with low -MW usually results in smaller nanoparticles with higher surface charge (zeta potential), which increases their capacity to bind to and disrupt bacterial membranes [8]. Thus reduced DA leads to a compaction of positively charged amino groups, which improves electrostatic interactions with negatively charged bacterial surfaces and boosts antibacterial activity. The activity is also influenced by external factors such as the concentration of the nanoparticles and the pH value of the medium. Since they are polycationic and pH-dependent, the antibacterial effect of CNPs is more prominent at low pH values ($pH < 6.5$), where the amino groups remain fully protonated [2]. Moreover, the effect is dose-dependent, and higher concentrations of nanoparticles reduce microbial activity, which also correlates with higher concentrations of nanoparticles. Moreover, microbial vulnerability is influenced by cell the wall structure, and the strong electrostatic interaction of cationic CNPs with the negatively charged lipopolysaccharide can often lead to increased efficacy against Gram-negative bacteria [12, 19, 39].

In order to develop useful application-specific chitosan-based antibacterial systems, it is essential to fully understand and optimize these interrelated areas. The combination of these components proves that the antibacterial performance of CNPs is not determined by a single variable, but by the combination of microbial characteristics, environmental variables and the characteristics of the polymers. Table 1 focuses on the most important factors influencing the antibacterial activity of CNPs to facilitate understanding.

7 The Superiority of Biologically Synthesized Chitosan Nanoparticles

CNPs are still commonly produced on the market using conventional physicochemical techniques such as ionic gelation, emulsion cross-linking, spray drying, and the reverse micellar process [4]. However, these methods have several disadvantages. For example, emulsion cross-linking can be difficult and requires cross-linking agents (like glutaraldehyde) that could be hazardous and incompatible with some agrochemicals. Additionally, the experimental parameters, such as the type of surfactant and stirring speed, have a significant impact on the particle size in this method [48]. Furthermore, the reverse micellar method relies on costly and potentially hazardous surfactants like cetyl trimethyl ammonium bromide (CTAB) and methods such as

Table 1 Factors Affecting Antibacterial Activity of Chitosan Nanoparticles

Factor	Description	Effect on antibacterial activity	References
Molecular Weight (MW)	Low-MW chitosan produces smaller nanoparticles with higher surface charge (zeta potential)	Enhances binding to and disruption of bacterial membranes	[49]
Degree of Acetylation (DA)	Lower DA increases positively charged amino groups	Strengthens electrostatic interactions with negatively charged microbial surfaces, intensifying antibacterial action	[7]
pH (External Factor)	Antibacterial effects are more pronounced under acidic conditions (pH < 6.5)	Protonated amino groups increase cationic activity, boosting microbial inhibition	[66]
Nanoparticle concentration	Antibacterial action is dose-dependent	Higher concentrations correlate with stronger microbial inhibition	[67]
Microbial cell wall architecture	Gram-negative bacteria have a negatively charged lipopolysaccharide (LPS) layer	Stronger electrostatic attraction leads to higher susceptibility compared to Gram-positive bacteria	[66]

sieving and precipitation usually result in nanoparticles with irregular shape, large dispersion and poor stability.

Due to these challenges, new developments have turned to green or biological synthesis routes. In all these green processes, natural biologically produced compounds are used as capping and reducing agents to produce nanoparticles of higher quality in terms of stability, morphology, and biocompatibility without using unfriendly chemicals [10, 28]. It is also not resource- intensive, which is more sustainable in the long term.

Even chitosan itself has unique advantages compared to most other nanomaterials, with intrinsic properties such as biocompatibility, biodegradability and low toxicity that are indistinguishable from metallic or synthetic nanoparticles, which can be considered hazardous elements for the environment and patients' health [40, 41, 43]. These properties enable the safe use of chitosan-based nanomaterials at various levels 3, whether in agricultural crop protection, drug delivery, cosmetics, or waste treatment [48]. Biological synthesis has the advantage of maintaining the amphiphilic and bioadhesive characteristics of chitosan itself and improving its ability to entrap poorly water-soluble agrochemicals in the case of agricultural use. This increases the stability and bioavailability of the active ingredients and minimizes the use of toxic inert additives [35]. The improved antimicrobial potential of synthesized CNPs green has been empirically confirmed. For example, Sathiyabama and Parthasarathy have demonstrated the strong antifungal activity against main phytopathogenic bacteria such as *Alternaria solani, Fusarium oxysporum*, and *Pyricularia grisea* [35, 63]. Moreover, CNPs have been shown to be useful in improving plant resistance to biotic stress, including in tomato plants by activating the plant immune system. Pattern recognition receptors (PRRs) identify these nanoparticles and trigger their defense responses to protect against a variety of bacterial species, fungi such as *Fusarium solani* and plant viruses [13, 36].

The biologically synthesized chitosan nanoformulations represent a decisive sustainable innovation for agriculture. Their versatility as acidifying carriers of controlled release agrochemicals, as effective antimicrobials and as plant growth stimulants make them a promising application, discussed in more detail in the following sections, as one of the solutions to some of the most acute problems in agriculture today.

8 Challenges and Future Perspectives: Bridging the Gap from Laboratory to Field

Despite its well-documented advantages as a biocompatible and biodegradable polymer, and its official GRAS (Generally Recognized as Safe) status as a food additive [29], the translation of chitosan from a promising laboratory material to a widely adopted industrial solution faces significant hurdles. Although chitosan is one of the most researched bio-based polymers, its true potential, remains largely untapped, especially in agricultural applications such as bio-pesticides, nano-fertilizers, and nutrient carriers.

A primary challenge is the successful and economical scale-up of production. While numerous studies have reported the synthesis of chitosan and its nanocomposites on a laboratory scale, a clear and cost-effective pathway to large-scale production is still lacking. This gap is fueling industry hesitation due to perceived risks in product development and uncertainty about farmer acceptance and adoption. To bring chitosan from the laboratory to farmers, research needs to focus on developing standardized, industrially feasible synthesis protocols and exploring novel, more economical sources for its isolation beyond traditional crustacean shells.

Although chitosan has excellent film-forming properties, its use in industrial food packaging and coating is still minimal. More advanced applications, such as the use of chitosan-based nanomaterials for the controlled release of biomolecules or as vectors for genome editing of plants, are still on their infancy. Apart from a few initial studies in gene delivery, the potential for smart, targeted applications in plant genetic engineering has not yet been exhausted.

To overcome these challenges and fully exploit the potential of chitosan, future research must focus on several key areas:

1. **Development of Advanced, Plant-Friendly Formulations**: There is a critical need for novel chitosan blends and derivatives specifically tailored for use in

agriculture. These new materials should be designed to provide higher encapsulation efficiency, improved profiles for controlled-release of active ingredients, and better triggering properties. As the current acidic hydrolysis methods for processing chitosan can occasionally have phytotoxic consequences, a major focus should be on the development of more "plant-friendly" derivatives.
2. **Deepening Mechanistic Understanding**: Although the general antibacterial properties of chitosan are well established, further targeted studies are needed to fully clarify the interaction mechanisms at the molecular level. Especially with regard to the antiviral mode of action of chitosan in plants, there is a considerable knowledge vacuum that needs to be filled.
3. **Growth into High-Value Applications**: Research on chitosan nanocomposites is at a critical crossroads. Future research should explore their potential for cutting-edge agricultural technologies such as genetic engineering tools and intelligent delivery systems that have the potential to completely transform crop protection and enhancement.

9 Conclusion

CNPs, a significant advancement over old-fashioned chemical pesticides, have proven to be an effective and sustainable weapon to combat plant diseases caused by bacteria. Their nanoscale size, high surface charge and solubility allow stronger electrostatic interactions with negatively charged bacterial membranes. This can lead to leakage of internal components, cell integrity and ultimately cell death. The stability, dispensability, and bioavailability of CNPs is better than that of chitosan in bulk, which favors a higher antibacterial activity of CNPs. Besides their direct bacterial action, CNPs also chelate essential metal ions required by microbes for their development, inhibit biofilms, and induce systemic resistance in plants by activating key defense enzymes and phytohormone-mediated signaling pathways. Such a complex mechanism of action makes CNPs a dual-function active ingredient, as it directly inhibits pathogenic bacteria and at the same time strengthens the immune responses of plants. The superiority of biologically generated CNPs due to better properties also increases the potential of CNPs in sustainable agriculture. Green synthesis methods, which use natural reducing and stabilizing agents as opposed to chemically synthesized materials, result in better stability, biocompatibility, and reduced environmental toxicity of nanoparticles. These more environmentally friendly methods not only reduce reliance on hazardous chemicals, but also preserve the natural amphiphilic and bioadhesive properties of chitosan, improving its ability to trap and release agrochemicals in a controlled manner. Despite their potential, large-scale application of CNPs is hampered by production costs, farmer acceptance, and regulations. Their antibacterial effect has been sufficiently proven, but further studies are required to explain the antiviral effect and its environment. The future developments need to target cost-effective synthesis, safer formulations, and new modes of delivery for them to be used in practice. Overall chitosan nanoparticles can be defined as a new, eco-friendly and multifunctional approach for the management of bacterial plant diseases. With their ability to activate plant defense mechanisms and their robust antimicrobial activity, they offer a sustainable alternative to traditional pesticides, and have the potential to minimize the amount of chemicals applied, increase plant resilience and improve food security. As innovations in green synthesis, mechanistic insights, and industrial scalability progress, CNPs will soon become a foundation for combined approaches to plant disease control, as they support the agricultural sector in achieving a global goal of sustainability and environmental health.

References

1. Abdel-Aziz HM, Hasaneen MN, Omer AM (2016) Nano chitosan-NPK fertilizer enhances the growth and productivity of wheat plants grown in sandy soil. Span J Agric Res 14(1):e0902–e0902
2. Abu-Elala NM, Hossam-Elden N, Marzouk MS, El Basuini MFJAoAS (2025) Chitosan for aquaculture: growth promotion, immune modulation, antimicrobial activity, bio-carrier utility, water quality management, and safety considerations–a review 25(2):483–509
3. Afzal M, Ullah S, Assad N, Naeem-ul-Hassan M, Kanwal M, Mubashar B, Shahzad R, Khan MN, Javed MA, Wahab S, Ali B, Adnan M, Kaplan A, Sarfraz MH, Alkahtani J, Mehdaoui I (2024) Photocatalytic degradation of organic dyes and biological potentials of biogenic zinc oxide nanoparticles synthesized using the polar extract of *Cyperus scariosus* R.Br. (Cyperaceae). Green Process Synthesis 13(1). https://doi.org/10.1515/gps-2024-0038
4. Agarwal M, Nagar D, Srivastava N, Agarwal M (2015) Chitosan nanoparticles-based drug delivery: an update. Int J Adv Multidiscip Res 2(4):1–13
5. Al-Nemrawi N, Alsharif S, Dave R (2018) Preparation of chitosan-TPP nanoparticles: the influence of chitosan polymeric properties and formulation variables. Int J Appl Pharm 10(5):60–65
6. Alehosseini E, Shahiri Tabarestani H, Kharazmi MS, Jafari SM (2022) Physicochemical, thermal, and morphological properties of chitosan nanoparticles produced by ionic gelation. Foods 11(23):3841
7. Amor IB, Hemmami H, Laouini SE, Abdelaziz AG, Barhoum A (2024) Influence of chitosan source and degree of deacetylation on antibacterial activity and adsorption of AZO dye from water. Biomass Conversion and Biorefinery 14(14):16245–16255
8. Ardean C, Davidescu CM, Nemeş NS, Negrea A, Ciopec M, Duteanu N, Negrea P, Duda-Seiman D, Musta VJIJoMS (2021) Factors influencing the antibacterial activity of chitosan and chitosan modified by functionalization 22(14):7449
9. Assad N, Abbas A, Fayyaz Ur Rehman M, Naeem-Ul-Hassan M (2024) Photo-catalytic and biological applications of phyto-functionalized zinc oxide nanoparticles synthesized using a polar extract of *Equisetum diffusum* D. RSC Adv 14(31):22344–22358. https://doi.org/10.1039/d4ra03573a
10. Assad N, Ahmad H, Abbas A, Rehman MFu, Naeem-ul-Hassan M, Sher M, Riaz T, Laila MB, Zahra I, Ullah F, Assad Y (2025a) Green synthesis of selenium nanoparticles using *Equisetum diffusum*: characterization, antibacterial potential, effects on plant growth. IEEE Transa NanoBioscience, 1. https://doi.org/10.1109/TNB.2025.3567327
11. Assad N, Ansa, Laila MB, Ullah S, Adnan M, Wahab S, Kaplan A, Khan MN, Ullah B (2025b) Nanotechnology in oilseed crops: a new frontier for abiotic stress adaptation. In: Abdel Latef AAH (ed) Oilseed crops under abiotic stress: mitigation strategies and future perspectives. Springer Nature Singapore, Singapore, pp 507–536. https://doi.org/10.1007/978-981-96-8346-8_16

12. Assad N, Laila MB, Hassan MNU, Rehman MFu, Ali L, Mustaqeem M, Ullah B, Khan MN, Iqbal M, Ercişli S, Alarfaj AA, Ansari MJ, Malik T (2025c) Eco-friendly synthesis of gold nanoparticles using *Equisetum diffusum* D. Don. with broad-spectrum antibacterial, anticancer, antidiabetic, and antioxidant potentials. Scientific Reports 15(1):19246. https://doi.org/10.1038/s41598-025-02450-9
13. Assad N, Laila MB, Naeem-Ul-Hassan M, Ali B, Kaplan A, Khan MN, Hayat K, Ozdemir FA (2025d) Phytohormones as key players: boosting oilseed crop tolerance to nutrient deficiency stress. In: Abdel Latef AAH (ed) Oilseed crops under abiotic stress: mitigation strategies and future perspectives. Springer Nature Singapore, Singapore, pp 415–440. https://doi.org/10.1007/978-981-96-8346-8_13
14. Assad N, Naeem-ul-Hassan M, Ajaz Hussain M, Abbas A, Sher M, Muhammad G, Assad Y, Farid-ul-Haq M (2025) Diffused sunlight assisted green synthesis of silver nanoparticles using *Cotoneaster nummularia* polar extract for antimicrobial and wound healing applications. Nat Prod Res 39(8):2203–2217. https://doi.org/10.1080/14786419.2023.2295936
15. Attia MS, Osman MS, Mohamed AS, Mahgoub HA, Garada MO, Abdelmouty ES, Abdel Latef AAH (2021) Impact of Foliar Application of Chitosan Dissolved in Different Organic Acids on Isozymes, Protein Patterns and Physio-Biochemical Characteristics of Tomato Grown under Salinity Stress. Plants 10(2):388
16. Baruch L, Machluf M (2006) Alginate–chitosan complex coacervation for cell encapsulation: effect on mechanical properties and on long-term viability. Biopolymers: Original Res Biomolecules 82(6):570–579
17. Beauséjour J, Clermont N, Beaulieu C (2003) Effect of Streptomyces melanosporofaciens strain EF-76 and of chitosan on common scab of potato. Plant Soil 256(2):463–468
18. Bhushan B (2017) Springer handbook of nanotechnology. Springer
19. Chandrasekaran M, Kim KD, Chun SCJP (2020) Antibacterial activity of chitosan nanoparticles: a review 8(9):1173
20. Corradini E, De Moura M, Mattoso L (2010) A preliminary study of the incorparation of NPK fertilizer into chitosan nanoparticles. Express Polymer Lett 4(8)
21. El-Naggar NE-A, Shiha AM, Mahrous H, Mohammed AA (2022) Green synthesis of chitosan nanoparticles, optimization, characterization and antibacterial efficacy against multi drug resistant biofilm-forming *Acinetobacter baumannii.* Sci Rep 12(1):19869
22. Elagamey E, Abdellatef MAEAE, Kamel SM (2022) Chitosan is the ideal resource for plant disease management under sustainable agriculture. In: Kumar B (ed) Chitin and chitosan-isolation, properties, and applications. IntechOpen, Rijeka
23. Fan W, Yan W, Xu Z, Ni H (2012) Formation mechanism of monodisperse, low molecular weight chitosan nanoparticles by ionic gelation technique. Colloids Surf B 90:21–27
24. Faqir Y, Ma J, Chai Y (2021) Chitosan in modern agriculture production. Plant Soil Environ 67(12):679–699
25. Food OA (2021) Pathogens, precipitation and produce prices. Nat Clim Change 11:635–635
26. Ghaffar S, Abbas A, Naeem-ul-Hassan M, Assad N, Sher M, Ullah S, Alhazmi HA, Najmi A, Zoghebi K, Al Bratty M, Hanbashi A, Makeen HA, Amin HMA (2023) Improved photocatalytic and antioxidant activity of olive fruit extract-mediated ZnO nanoparticles. Antioxidants 12(6):1201
27. Goy RC, Morais ST, Assis OB (2016) Evaluation of the antimicrobial activity of chitosan and its quaternized derivative on *E. coli* and *S. aureus* growth. Revista Brasileira de Farmacognosia 26(1):122–127
28. Gul H, Khan HA, Nasreen Z, Assad N, Turab SA, Hanif M (2025) Exploring the antibacterial, anticoagulant, and hemolytic potential of green-synthesized Fe(2)O(3) nanoparticles by cucurbita pepo pulp. IEEE Trans Nanobioscience Pp. https://doi.org/10.1109/tnb.2025.3563307
29. Gutiérrez T (2017) Chitosan applications for the food industry. Chapter in a book: Chitosan: derivatives, composites and applications. Scrivener Publishing LLC,
30. Halim V, Vess A, Scheel D, Rosahl S (2006) The role of salicylic acid and jasmonic acid in pathogen defence. Plant Biol 8(03):307–313

31. Hashmi SS, Ibrahim M, Adnan M, Ullah A, Khan MN, Kamal A, Iqbal M, Kaplan A, Assad N, Elshikh MS, Almutairi SM, Zaman W (2024) Green synthesis of silver nanoparticles from *Olea europaea* L. extracted polysaccharides, characterization, and its assessment as an antimicrobial agent against multiple pathogenic microbes. Open Chem 22(1). https://doi.org/10.1515/chem-2024-0016
32. Hoang NH, Le Thanh T, Sangpueak R, Treekoon J, Saengchan C, Thepbandit W, Papathoti NK, Kamkaew A, Buensanteai N (2022) Chitosan nanoparticles-based ionic gelation method: a promising candidate for plant disease management. Polymers 14(4):662
33. Hosseini SF, Zandi M, Rezaei M, Farahmandghavi F (2013) Two-step method for encapsulation of oregano essential oil in chitosan nanoparticles: Preparation, characterization and in vitro release study. Carbohyd Polym 95(1):50–56
34. Hussain SB, Zahra S, Hussain S, Mahmood H, Haq Au, Shafique M, Gill AR, Ali A, Javed MA, Khan MN, Kaplan A, Assad N, Aljowaie RM, Elshikh MS (2025) Characterization and comparative antibacterial activities of gold nanoparticles synthesized by *Tagetes patula* L. (Asteraceae) flower extract. Microscopy Res Tech 88(6):1869–1880. https://doi.org/10.1002/jemt.24829
35. Ingle PU, Shende SS, Shingote PR, Mishra SS, Sarda V, Wasule DL, Rajput VD, Minkina T, Rai M, Sushkova S, Mandzhieva S, Gade A (2022) Chitosan nanoparticles (ChNPs): a versatile growth promoter in modern agricultural production. Heliyon 8(11):e11893. https://doi.org/10.1016/j.heliyon.2022.e11893
36. Iriti M, Varoni EM (2015) Chitosan-induced antiviral activity and innate immunity in plants. Environ Sci Pollut Res 22(4):2935–2944
37. Jabbar A, Abbas A, Assad N, Naeem-Ul-Hassan M, Alhazmi HA, Najmi A, Zoghebi K, Al Bratty M, Hanbashi A, Amin HMA (2023) A highly selective Hg($^{2+}$) colorimetric sensor and antimicrobial agent based on green synthesized silver nanoparticles using *Equisetum diffusum* extract. RSC Adv 13(41):28666–28675. https://doi.org/10.1039/d3ra05070j
38. Jia X, Meng Q, Zeng H, Wang W, Yin H (2016) Chitosan oligosaccharide induces resistance to Tobacco mosaic virus in Arabidopsis via the salicylic acid-mediated signalling pathway. Sci Rep 6(1):26144
39. Kanwal M, Sher M, Abbas A, Akhtar S, Siddique AB, MNu, Hasan, Assad N, Alhazmi HA, Amin HMA (2025) Dual colorimetric sensing of Hg(II) and Fe(III) using sulfanilamide-stabilized silver nanoparticles and evaluating their photodegradation and antibacterial properties. J Water Process Eng 75:107981. https://doi.org/10.1016/j.jwpe.2025.107981
40. Kaplan A, Khan MN, Wahab S, Assad N, Adnan M, Hafsa, Iqbal M (2025) Chapter 13 - Thallium in soil environments and its biological availability. In: Saud S, Fahad S, Wang D (eds) Beneficial elements for remediation of heavy metals in polluted soil. Elsevier, pp 373–398. https://doi.org/10.1016/B978-0-443-26522-8.00013-1
41. Karlsson HL, Toprak MS, Fadeel B (2022) Toxicity of metal and metal oxide nanoparticles. In: Handbook on the toxicology of metals. Elsevier, pp 87–126
42. Kaur S, Dhillon GS (2014) The versatile biopolymer chitosan: potential sources, evaluation of extraction methods and applications. Crit Rev Microbiol 40(2):155–175
43. Khan MN, Kaplan A, Iqbal M, Wahid N, Wahab S, Assad N, Hayat K, Adnan M, Hafsa (2025) Chapter 4 - Cadmium dynamics: beneficial elements and chemical reactions in soil. In: Saud S, Fahad S, Wang D (eds) Beneficial elements for remediation of heavy metals in polluted soil. Elsevier, pp 93–139. https://doi.org/10.1016/B978-0-443-26522-8.00004-0
44. Khan ZuR, Assad N, Naeem-ul-Hassan M, Sher M, Alatawi FS, Alatawi MS, Omran AME, Jame RMA, Adnan M, Khan MN, Ali B, Wahab S, Razak SA, Javed MA, Kaplan A, Rahimi M (2023) *Aconitum lycoctonum* L. (Ranunculaceae) mediated biogenic synthesis of silver nanoparticles as potential antioxidant, anti-inflammatory, antimicrobial and antidiabetic agents. BMC Chemistry 17(1):128. https://doi.org/10.1186/s13065-023-01047-5
45. Li B, Liu B, Shan C, Ibrahim M, Lou Y, Wang Y, Xie G, Li, Hy, Sun G (2013) Antibacterial activity of two chitosan solutions and their effect on rice bacterial leaf blight and leaf streak. Pest Manag Sci 69(2):312–320

46. Li B, Shi Y, Shan C, Zhou Q, Ibrahim M, Wang Y, Wu G, Li H, Xie G, Sun G (2013) Effect of chitosan solution on the inhibition of *Acidovorax citrulli* causing bacterial fruit blotch of watermelon. J Sci Food Agric 93(5):1010–1015
47. Malik SY, Arshad M, Batool S, Batool S, Iram F, Batool T, Ali B, Wahab S, Khan MN, Kaplan A, Bibi M, Assad N, Alnemar SK, Farouk Elsadek M, Ercisli S, Iqbal M (2024) Atrazine-Induced Toxicity and Biochemical Changes in Mus musculus: the Ameliorative Efficacy of *Nigella sativa* L. (Ranunculaceae) Oil. Pol J Environ Stud 33:5211–5220. https://doi.org/10.15244/pjoes/183076
48. Maluin FN, Hussein MZ (2020) Chitosan-based agronanochemicals as a sustainable alternative in crop protection. Molecules 25(7):1611
49. Másson M (2024) The quantitative molecular weight-antimicrobial activity relationship for chitosan polymers, oligomers, and derivatives. Carbohyd Polym 337:122159
50. Mudhakir D, Wibisono C, Rachmawati H (2014) Encapsulation of risperidone into chitosan-based nanocarrier via ionic binding interaction. Procedia Chem 13:92–100
51. Mwangi WW, Ho K-W, Ooi C-W, Tey B-T, Chan E-S (2016) Facile method for forming ionically cross-linked chitosan microcapsules from Pickering emulsion templates. Food Hydrocolloids 55:26–33
52. Nair RS, Morris A, Billa N, Leong C-O (2019) An evaluation of curcumin-encapsulated chitosan nanoparticles for transdermal delivery. AAPS PharmSciTech 20(2):69
53. Nallamuthu I, Devi A, Khanum F (2015) Chlorogenic acid loaded chitosan nanoparticles with sustained release property, retained antioxidant activity and enhanced bioavailability. Asian J Pharm Sci 10(3):203–211
54. Nandhini R, Rajeswari E, Harish S, Sivakumar V, Gangai Selvi R (2025) Role of chitosan nanoparticles in sustainable plant disease management. J Nanopart Res 27:13
55. Noor A, Bin MM, Mujahid A (2022) Characterization and performance evaluation of lignin-based bioflocculants for organic micropollutants in adsorptive coagulation-flocculation process. Universiti Teknologi Malaysia
56. Pasquina-Lemonche L, Burns J, Turner R, Kumar S, Tank R, Mullin N, Wilson J, Chakrabarti B, Bullough P, Foster S (2020) The architecture of the Gram-positive bacterial cell wall. Nature 582(7811):294–297
57. Prasad R, Bhattacharyya A, Nguyen QD (2017) Nanotechnology in sustainable agriculture: recent developments, challenges, and perspectives. Front Microbiol 8:1014
58. Ramkissoon A, Francis J, Bowrin V, Ramjegathesh R, Ramsubhag A, Jayaraman J (2016) Bio-efficacy of a chitosan based elicitor on Alternaria solani and *Xanthomonas vesicatoria* infections in tomato under tropical conditions. Ann Appl Biol 169(2):274–283
59. Riaz N, Zubair F, Ali A, Fiaz S, Batool S, Batool T, MNu, Hassan, Jafarli BS, Khan MN, Wahab S, Azmat A, Kaplan A, Alarjani KM, Elsheikh MS, Iqbal M (2025) Plant Based antimicrobials against multi drug resistant bacterial strains (MDRS). Pol J Environ Stud 34:4799–4806. https://doi.org/10.15244/pjoes/190564
60. Roheen T, Bibi M, Rehman MFu, Assad N, Gondal HY, Nadeem M, Atif FA, Batool F, Perveen N, Qadir R, Rehman SU, Maqbool M, Shahzad (2025) Enhanced antibacterial and anti-inflammatory efficiency of serratiopeptidase immobilized on CMC-silver nanoparticles. J Basic Microbiol n/a (n/a):e70060. https://doi.org/10.1002/jobm.70060
61. Román-Doval R, Torres-Arellanes SP, Tenorio-Barajas AY, Gómez-Sánchez A, Valencia-Lazcano AA (2023) Chitosan: Properties and its application in agriculture in context of molecular weight. Polymers 15(13):2867
62. Saharan V, Mehrotra A, Khatik R, Rawal P, Sharma S (2013) Pal AJIjobm. Synthesis of chitosan based nanoparticles and their in vitro evaluation against phytopathogenic fungi 62:677–683
63. Sathiyabama M, Parthasarathy R (2016) Biological preparation of chitosan nanoparticles and its in vitro antifungal efficacy against some phytopathogenic fungi. Carbohyd Polym 151:321–325
64. Savary S, Willocquet L, Pethybridge SJ, Esker P, McRoberts N, Nelson A (2019) The global burden of pathogens and pests on major food crops. Nature Ecol Evol 3(3):430–439. https://doi.org/10.1038/s41559-018-0793-y

65. Sharkawy A, Barreiro MF, Rodrigues AE (2020) Chitosan-based Pickering emulsions and their applications: a review. Carbohyd Polym 250:116885
66. Wu K, Yan Z, Wu Z, Li J, Zhong W, Ding L, Zhong T, Jiang T (2024) Recent advances in the preparation, antibacterial mechanisms, and applications of chitosan. J Funct Biomaterials 15(11):318
67. Yadav H, Malviya R, Kaushik N (2024) Chitosan in biomedicine: a comprehensive review of recent developments. Carbohydrate Polymer Technol Appl 8:100551
68. Yan D, Li Y, Liu Y, Li N, Zhang X, Yan C (2021) Antimicrobial properties of chitosan and chitosan derivatives in the treatment of enteric infections. Molecules 26(23):7136
69. Yang C, Li B, Ge M, Zhou K, Wang Y, Luo J, Ibrahim M, Xie G, Sun G (2014) Inhibitory effect and mode of action of chitosan solution against rice bacterial brown stripe pathogen Acidovorax avenae subsp. avenae RS-1. Carbohyd Res 391:48–54
70. Zhao L-M, Shi L-E, Zhang Z-L, Chen J-M, Shi D-D, Yang J, Tang Z-X (2011) Preparation and application of chitosan nanoparticles and nanofibers. Braz J Chem Eng 28:353–362

Chitosan Nanoparticles in Fungal Plant Disease Control

Rabeya Islam Tamanna, Raihan Ferdous, Sumaya Jesmin, Hasiba Kabir Mim, Sadia Akter Tonny, Abu Noman Faruq Ahmmed, and Md. Belal Hossain

Abstract Fungal pathogens cause significant yield losses in global agriculture, with species such as *Magnaporthe oryzae*, *Fusarium* spp., and *Botrytis cinerea* posing major threats. Although effective, conventional chemical fungicides can lead to the development of resistance, environmental contamination, and health risks. Chitosan nanoparticles (CNPs), derived from the deacetylation of chitin, offer an eco-friendly and biodegradable alternative for managing fungal diseases. Their nanoscale size, positive surface charge, and functional versatility enable them to directly combat fungi through cell wall disruption, changes in membrane permeability, induction of reactive oxygen species, inhibition of spore germination, and activation of the host plant immune system. CNPs also act as carriers for antifungal agents, facilitating controlled release and targeted action. They are applied as foliar sprays, soil and seed treatments, and postharvest coatings, with proven effectiveness in greenhouse and field trials. Studies comparing CNPs with traditional fungicides and other nanomaterials highlight their advantages, especially in resistance management and reducing chemical use. This chapter aims to explore the potential of chitosan nanoparticles in controlling fungal plant diseases, highlighting their mechanisms of action, applications, benefits, and limitations in sustainable crop protection. However, challenges such as scale-up, cost, stability, phytotoxicity, and regulatory issues remain. Combining CNPs technology with precision agriculture, smart delivery systems, and multifunctional nanocomposites can further enhance their role in sustainable crop protection.

R. I. Tamanna · H. K. Mim · S. A. Tonny · A. N. F. Ahmmed · Md. B. Hossain
Department of Plant Pathology, Sher-e-Bangla Agricultural University, Dhaka, Bangladesh

R. Ferdous (✉)
Mushroom Development Institute, Dhaka, Bangladesh
e-mail: raihanf.agri-2011106@sau.edu.bd

S. Jesmin
Department of Agronomy and Agricultural Extension, University of Rajshahi, Rajshahi, Bangladesh

A. A. H. Abdel Latef (ed.), *Chitosan Nanoparticles for Plant Resilience Under Abiotic and Biotic Stressors*, Smart Nanomaterials Technology,
https://doi.org/10.1007/978-981-95-6422-4_10

Keywords Antifungal activity · Biodegradable nanomaterials · Eco-friendly fungicide alternatives · Nanobiotechnology · Nanocomposites · Sustainable crop protection

1 Introduction

Worldwide fungal pathogens are causing significant damage to agricultural productivity. There are several species causing diseases in plants, such as *Magnaporthe oryzae*, *Fusarium* spp., *Colletotrichum* sp., *Rhizoctonia* sp., *Ustilago* sp., *Botrytis* sp., *Bipolaris* sp., *Alternaria* sp., which are responsible for significant economic loss in important crops [3]. *Ascomycetes*, *Basidiomycetes*, and *Deuteromycetes* fungi are the most important in causing diseases in plants and produce symptoms such as necrosis, wilting, reduced grain filling, rust, blast, smut, shrivelled grain, yellowing leaves, chlorosis, fruit rot, anthracnose, defoliation, etc. [21].

Chemical fungicides are widely used to control fungal diseases; however, the extreme use has been influencing resistance against fungicides in pathogens through mechanisms like target site alteration, efflux pump activation, and genetic mutations. Most importantly, resistance observed in major fungal pathogens, for instance, *Aspergillus fumigatus*, *Botrytis cinerea*, and *Fusarium* spp. [14, 43]. Fungicides can be detrimental to beneficial microorganisms, freshwater, and soil, and they also have an impact on antifungal resistance in non-target fungi [105]. Fungicides' residues like Flutolanil and Azoxystrobin inhibit arbuscular mycorrhizal fungi (AMF), root colonization, and spore germination [16] and can cause human health risk and harm aquatic life, such as endocrine abnormalities, reduced larval growth and development, and reproductive difficulties [10, 50, 104]. To get rid of the above problems, eco-friendly, biodegradable alternatives are now becoming the most inevitable demand due to rising regulatory and public health concerns, and governments and international organizations are also imposing strict regulations on chemical fungicides due to their environmental and human health risks [45]. So, global agriculture is now shifting to the use of biological control agents (BCAs), which are highly specialized, environmentally safe, sustainable, and profitable, and can be antagonistic to pathogens by apoptosis, parasitism, cell wall degradation, competition, or by triggering the immune system of the plant [46] and biodegradable nanotechnologies where nanoparticles (NPs) can directly reduce fungal pathogen infections as well as enhance crop growth [31, 99].

Nanotechnology provides an innovative solution to plant disease management by delivering direct antimicrobial actions, accurate plant pathogen detection, inducing plant defense responses, and targeted delivery at the nanoscale with minimum phytotoxic effects. Nanoparticles have excellent physico-chemical properties such as nanoscale dimension, large surface area coverage, high stability and durability, broad spectrum antimicrobial action, which make them eco-friendly and biocompatible [98]. Nanoparticles counter fungal pathogens by interacting with the physiological activity of the pathogen for a long time and leading to environmental safety [37].

Chitosan is a naturally abundant polysaccharide prepared by deacetylation of chitin, where chitin is a biopolymer found mostly in crustaceans and fungal cell walls [48, 71]. Chitosan has a protonated amine group, which makes it positively charged and soluble in acidic pH [35]. There are various techniques to synthesize chitosan nanoparticles (CNPs) from chitosan [22]. The most prevalent methods used to synthesize CNPs include ionic gelation, microemulsion, emulsion-based solvent evaporation, and emulsification solvent diffusion [66]. CNPs show effective delivery, high permeability through membranes, and effective antimicrobial activity [78, 88].

The purpose of this chapter is to investigate the potential of CNPs in the management of plant fungal infections. This chapter will discuss the synthesis of chitosan nanoparticles, their characteristics, and the mechanism of antifungal activity against fungal diseases. The chapter will provide a review of recent studies that how efficiently CNPs can manage fungal diseases in plants and their contribution to sustainable agricultural development. Lastly, this chapter will discuss the limitations and prospects of CNPs on environment-friendly approaches [30, 73].

2 Chitosan Nanoparticles: Synthesis and Characterization

2.1 Methods of Synthesis

The synthesis of chitosan nanoparticles can be achieved through various methods, each offering specific advantages in terms of particle size, stability, and functionalization potential. The most commonly employed method is Ionic Gelation, which involves the electrostatic interaction between the positively charged amino groups of chitosan and negatively charged cross-linkers like sodium tripolyphosphate (TPP) [17, 33]. This technique is preferred due to its simplicity, mild reaction conditions, and absence of organic solvents.

Another widely used method is Emulsion-based Synthesis, where chitosan is dispersed in an oil phase to form nano-emulsions, followed by solvent evaporation or crosslinking to stabilize the nanoparticles [28, 102]. Reverse micelle and spray drying techniques are also employed for producing CNPs with controlled particle size and surface properties [77]. Microfluidics-based synthesis is gaining attention for its high reproducibility and scalability [5].

2.2 Physicochemical Properties Relevant to Antifungal Action

The antifungal efficacy of CNPs is largely influenced by their physicochemical attributes, including particle size, surface charge (zeta potential), molecular weight, and degree of deacetylation (DD). Smaller nanoparticles (<100 nm) exhibit higher surface area-to-volume ratios, enhancing their interaction with fungal cell membranes [82]. A positive zeta potential facilitates binding to the negatively charged microbial surfaces, leading to membrane disruption and leakage of cellular contents. For instance, particle size 174 nm, zeta potential 39 mV, and concentration of CNPs is 0.25 mg/mL can inhibit the 90% growth of *Fusarium solani* [44]

The molecular weight and DD of chitosan influence its solubility, charge density, and bioactivity. Higher DD results in more available amino groups, thereby increasing the nanoparticle's cationic nature and antifungal performance [56].

2.3 Encapsulation and Loading of Antifungal Agents

CNPs can serve as nanocarriers for loading conventional or natural antifungal agents, improving their stability, solubility, and targeted delivery. Both hydrophilic and hydrophobic compounds can be encapsulated depending on the synthesis method. Encapsulation efficiency depends on the drug-polymer interaction, molecular weight of chitosan, and nanoparticle size [72].

For example, to determine the antifungal activity of the encapsulation of *Cymbopogon martinii* essential oil on post-harvest plant pathogens such as *Fusarium graminearum,* which revealed the presence of 46 compounds, and geraniol was abundant (19.06%) that enabling to control of the disease [51]. The controlled release from the nanocarrier ensures prolonged protection of the plant tissue and minimizes frequent applications.

2.4 Surface Modification for Targeted Delivery

Functionalization of CNPs enhances their target specificity, stability, and compatibility with plant tissues. Surface modification can be achieved by conjugating ligands such as lectins, peptides, antibodies, or polymers that recognize pathogen-specific markers or plant receptors [29]. PEGylation (coating with polyethylene glycol) is a common strategy to improve stability in complex environments and reduce premature degradation.

Incorporating magnetic or pH-responsive moieties into CNPs also enables smart delivery, releasing the antifungal agent in response to environmental stimuli such as pH changes in infected tissues [97].

2.5 Characterization Techniques

Comprehensive characterization is essential to ensure the efficacy and reproducibility of CNPs. The primary techniques used include:

- Scanning Electron Microscopy (SEM) and Transmission Electron Microscopy (TEM): Provide detailed morphological and size information [36]
- Fourier-Transform Infrared Spectroscopy (FTIR): Confirms functional group interactions and encapsulation [36].
- Dynamic Light Scattering (DLS): Measures particle size distribution and zeta potential [47].
- X-ray Diffraction (XRD) and Thermogravimetric Analysis (TGA): Assess crystallinity and thermal stability.
- UV–Visible and Fluorescence Spectroscopy: Used for determining encapsulation efficiency and drug release kinetics.

These characterization tools are essential for optimizing the design of CNPs for maximum antifungal activity and environmental compatibility.

3 Mechanisms of Antifungal Action

3.1 Disruption of Fungal Cell Walls and Plasma Membranes

Generally, the fungal cell wall is composed of α-glucans, β-glucans, chitin- a structural polysaccharide, mannoprotein, and glycoprotein which give mechanical strength in response to the stressed environment [85] and the compartmentalized fungal membrane composed of lipid (ergosterol, sphingolipids and phosphoinositides) which maintain cell integrity [11]. CNPs are positively charged due to their protonated $-NH_3^+$ group, and the fungal cell/membrane is negatively charged. So, an electrostatic interaction occurs between the fungal cell/membrane and CNPs, causing membrane leakage and disruption of the cell and interfering with metabolic pathways [106]. Antifungal mechanisms of CNPs are attributed to their polycationic nature, which allows them to interact electrostatically with the negatively charged phospholipid components of fungal cell membranes. This interaction compromises membrane integrity, leading to increased permeability and leakage of intracellular contents, ultimately resulting in fungal cell death [44]. CNPs inhibit spore germination, cause cytoplasmic condensation, disappearance of membranous organelles, massive vacuolation, separation between plasma lemma and cell wall, decrease protein concentration, release nucleic acids, and thus cause damage to the cell membrane integrity, disrupt hyphal and spore morphology [106]. CNPs produce antifungal proteins (chitinase and glucanase), which degrade the fungal cell wall and inhibit fungal establishment [31]. Microscopic studies provide direct evidence of fungal structural damage by CNPs- SEM analysis of *Verticillium dahliea* shows wrinkled, collapsed and folded hyphae due to lack of cytoplasm and TEM analysis shows plasmalemma separation from inner cell wall, cytoplasm disordered, tabular materials from cytoplasm, plasmalemma ruptured and form debris due to extensive vacuolation in cytoplasm which cause cytolysis of intracellular membranous organelles [106]. There are other factors such as particle size and zeta potential, which are the characteristics of CNPs that enable the nanoparticles to spread over the fungal surface and enter into the cell [42, 61]. CNPs from high molecular weight (HMW) chitosan have a bigger zeta potential and produce bigger size nanoparticles that can degrade the cell membrane of fungus (*Fusarium solani* and *Aspergillus niger*) and cause cell death [82]. On the other hand, low molecular weight chitosan (LMW) gives small-sized nanoparticles and low zeta potential [44]. High molecular weight fractions disrupt the membrane structure, enabling the low molecular weight fractions to enter the fungal cell and interfere with its internal functions [82]. HMW and LMW chitosan nanoparticles can damage the cell wall of *F. solani* and *C. albicans* [44].

3.2 Induction of Reactive Oxygen Species (ROS)

ROS are highly reactive molecules that possess superoxide radicle ($O_2^{\cdot}$), hydrogen peroxide (H_2O_2), hydroxyl radicle (OH^-), single oxygen (O_2), peroxy radicla (ROO) and alkoxy radicles (RO), which can be produced in stressed condition, leading to oxidative stress that damages cells by destroying lipids, protein and nucleic acid and eventually causes death [86]. Generally, CNPs produce superoxide and hydrogen peroxide CSNPs, induce ROS accumulation in fungal cells, and cause mitochondrial dysfunction, where CNPs disturb the electron transport chain and cause electron leakage, and the consequence of ROS formation in fungal cells causes cellular damage by increasing protein levels, which suppresses fungal growth [113]. CNPs lead to fungal cell death by inducing oxidative stress through lipid oxidation, protein oxidation, and DNA fragmentation [6].

Application of CNPs along with wilting pathogen (*Fusarium andiyazi*) in tomato plant inhibits ROS by activating SOD (superoxide dismutase) and CAT (catalase) enzymes in tomato leaf by 4.54-fold and 1.12-fold after 48 hpi, which increases the amount of SOD and CAT, potentially protecting plants against pathogen attack could be induced by CNPs [20].

3.3 Inhibition of Fungal Spore Germination and Growth

Generally, conidia, sclerotia, or clamydospores of fungi cause infectious diseases in plants. Under favorable conditions, successful spore germination occurs, leading to disease development. So, to effectively manage fungal plant diseases, it is essential to inhibit spore germination at an early stage [3]. Positively charged CNPs have an electrostatic interaction with the negatively charged spore wall, thus affecting the spore wall integrity and causing leakage of the intracellular organelles and loss of spore viability (Fig. 1) [83]. Small particle size ~174 nm CNPs from LMW chitosan and high zeta potential (+54 mV) from HMW chitosan can easily penetrate fungal spores or germ tubes and disrupt metabolic processes and inhibit germination [82]. CNPs inhibit 100% sclerotia formation and germination of *Rhizoctonia solani* in rice

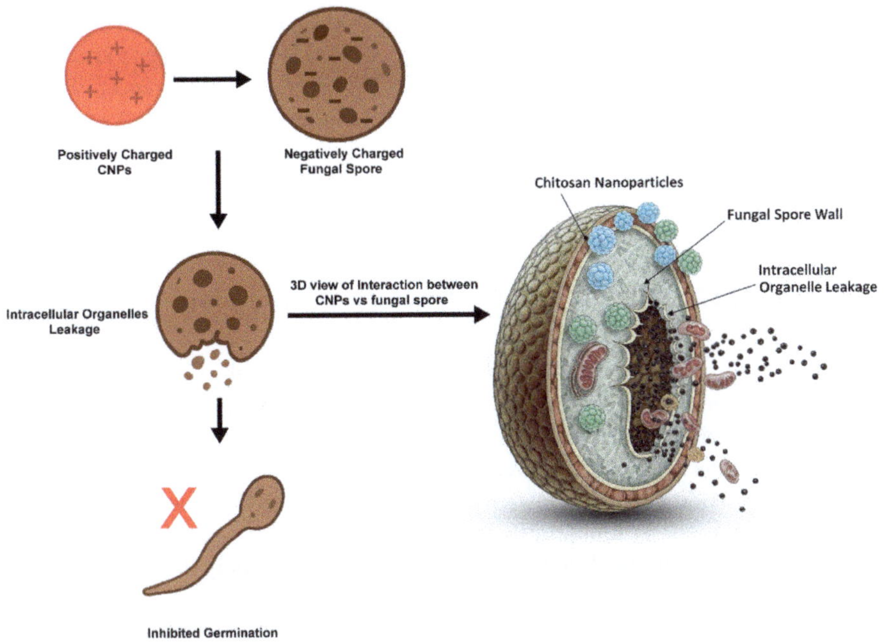

Fig. 1 Inhibition of fungal spore germination by the interaction of chitosan nanoparticles (CNPs)

field [113]. CNPs inhibit spore germination and mycelial growth, interfering with energy metabolism, sugar biosynthesis, and protein synthesis in several species, including *Candida albicans*, *Fusarium solani*, *Rhizoctonia solani*, and *Aspergillus niger* under *in-vitro* conditions [6]. Chitosan guar nanoparticles (CGNPs) inhibit the mycelial growth of the rice blast fungus*(Pyricularia grisea)* [92]. *Fusarium* spp. shows dose-dependent inhibition, while *Aspergillus niger* exhibited resistance to most CNPs except HMW CNPs because *A. niger* has a stronger cell wall or different membrane compositions [6].

3.4 Immunomodulation and Elicitor Activity in Host Plants

Chitosan is recognized by the plant as an elicitor molecule which initiates different signaling events leading to induced systemic resistance (ISR) or localized defense responses, activation of MAPK (Mitogen-Activated Protein Kinase) pathways, and helps to produce antimicrobial compounds such as chitinase/glucanase, which have direct antifungal activity and restrict pathogen growth [100].

Chitosan nanoparticles can boost the immune system in plants when the plant faces pathogen attack. In a study, wilt of tomato caused by *Fusarium andiyazi* had been controlled using CNPs, where some PR-gene (PR-1, PR-2 (β 1,3-glucanase),

PR-8 (chitinase) and PR- 10) activated with application of 5.0 mg/ml over 1.48 ×, 1.15 ×, 1.15 × and 1.14 × after 24 h of applying CNPs [20]. CNPs help to upregulate pathogenesis-related (PR) genes PR1–PR5 and PR10; among these genes, PR3 and PR4 encode fungal cell wall-degrading proteins (chitinase and glucanase) in wheat plants against wheat rust pathogen (*Puccinia triticina*) and degrade cell wall and inhibit disease establishment [31].

3.5 Synergistic Effects with Conventional Fungicides

CNPs can control *Rhizoctonia solani* in rice leaf and sheath with an efficacy of (85.9–81.1%) against the conventional fungicide azoxystrobin (51.5–66.5%) [113]. Garlic essential oil-CNPs can control *Aspergillus versicolor*, *A. niger,* and *Fusarium oxysporum* fungus against the conventional fungicide tebuconazole [67]. Chitosan-loaded copper oxide nanoparticle at 1 mg/L gives better performance in controlling 91.5% Fusarium wilt in tomato plants caused by *F. oxysporum* f. sp. *lycopersici,* whereas classical chemical fungicide Kocide 2000 achieves only 90% [68].

4 Application of Chitosan Nanoparticles in Controlling Fungal Diseases

4.1 Foliar Application of Chitosan Nanoparticles

Chitosan nanoparticles can be used for various applications because of their unique properties, biodegradability, high permeability, non-toxicity to humans, and cost-effectiveness [57]. From various types of applications, foliar application of CNPs has emerged as a promising and sustainable strategy for managing fungal diseases in various crops (Table 1), such as rice [24], wheat [19], tomato [109], soybean [79], watermelon [15], chilli [89], and cassava [41]. Due to their nanoscale size, CNPs can readily penetrate through the leaf cuticles or stomatal openings, facilitating deeper tissue penetration and systemic distribution within the plant [53]. CNPs in combination with salicylic acid foliar application effectively controlled wheat (*Triticum aestivum*) leaf rust caused by *Puccinia triticina* by enhancing the activity of peroxidase, chitinase, and polyphenol oxidase, as well as modulating the expression of defense-related genes [32]. These findings indicate that foliar application of CNPs, either alone or in combination with other bioactive agents, can serve as an effective strategy for managing a variety of plant-pathogenic fungal diseases. Collectively, these studies underline the potential of foliar-applied CNPs as a promising nanocomposite product for crop protection against fungal diseases [68] and effectively combat harmful fungi, making them a valuable resource for minimizing environmental losses and negative effects of chemicals [58].

Table 1 Application of Chitosan Nanoparticles (CNPs) Against Fungal Diseases and Growth Enhancement (Foliar Application, Soil and Seed Treatments, and Postharvest Disease Management)

	Crops	Target Pathogen • Disease	Key outcomes	CNPs application details	References
Foliar application	Rice (*Oryza sativa*)	*Rhizoctonia solani* • Sheath blight	CNPs suppressed 75% under greenhouse conditions; increased activity of defense enzymes (POD, PAL, chitinase); enhanced plant immunity	Foliar spray (greenhouse)	[24]
	Wheat (*Triticum aestivum*)	*Fusarium solani* • Fusarium rot	Ni-chitosan nanoconjugate (NiCNC) at 0.04 mg/mL reduced disease incidence by 83.33% in seedlings; enhanced stress resilience	Foliar and seedling treatment	[19]
	Tomato (*Solanum lycopersicum*)	*Alternaria alternata* • Early blight	Lesion area reduced by 29.52%; induced defense enzymes and secondary metabolites	Em-CNPs (foliar, synthesized from E. modestus)	[109]
	Watermelon (*Citrullus lanatus*)	*Fusarium oxysporum* f. sp. *niveum*	~27% disease reduction; enhanced stress gene modulation; 70% increase in fruit yield in uninfected plants	CTS-MSNs applied once at the seedling stage (500 mg/L)	[15]
	Chilli (*Capsicum annuum*)	*Alternaria alternata* • Leaf spot	Enhanced defense enzymes, phenolics, flavonoids, NO signaling; reduced cell death and disease severity	Foliar spray of CNPs (0.001%) in tolerant and susceptible cultivars	[89]
Soil and seed treatments	Tomato (*Solanum lycopersicum*)	*Rhizoctonia solani*, *F. oxysporum*, *Alternaria solani* • Early blight	87.7% early blight control; 61.1% wilt control; improved nutrient uptake	Soil drench and Cu–CNPs seed treatment	[88]
	Broad Bean (*Vicia faba*)	*Rhizoctonia solani* • Root rot	89% disease suppression with Arbuscular mycorrhizal fungi (AMF) + CNPs; enhanced immunity	Soil drench with AMF + CNPs	[27]

(continued)

Table 1 (continued)

	Crops	Target Pathogen • Disease	Key outcomes	CNPs application details	References
	Wheat (*Triticum aestivum*) Oat (*Avena sativa*), Barley (*Hordeum vulgare*)	*Aspergillus versicolor*, *A. niger*, *Fusarium oxysporum*	Suppressed fungal growth; improved emergence, shoot, and root biomass	Seed treatment with garlic essential oil-loaded chitosan nanoparticles (200–400 nm)	[67]
	Groundnut (*Arachis hypogaea*)	*Fusarium graminearum*, *F. oxysporum*, *Alternaria tenuis*, *Aspergillus niger*, *A. flavus*, *Penicillium* sp., *Beauveria bassiana*, *Sclerotium rolfsii*	COS-NPs (~180 nm, nanorod-shaped) exhibited strong antifungal activity in a dose-dependent manner; inhibited *F. graminearum* mycelial growth and zearalenone (ZEN) production at 800 ppm	Seed treatment with COS-NPs	[1]
Postharvest disease management	Mango (*Mangifera indica*)	*C. gloeosporioides*, *B. theobromae*, *A. niger*	Delayed ripening; reduced lesion size and incidence	Chitosan coating (0.5–2.0%) during storage	[38, 114]
	Apple (*Malus* spp.)	*P. expansum*, *B. cinerea*, *Mucor* spp.	Reduced fungal rot; improved moisture and firmness	Chitosan/CNPs coatings and composite packaging	[111]
	Papaya (*Carica papaya*)	*Colletotrichum* spp.	Stimulated defense enzymes; enhanced antifungal resistance	Chitosan dispersion	[7]
	Dragon fruit (*Selenicereus undatus*)	*C. gloeosporioides*	Induced PR-proteins; delayed rot	Submicron chitosan dispersion (1.0%)	[8]

(continued)

Table 1 (continued)

	Crops	Target Pathogen • Disease	Key outcomes	CNPs application details	References
	Citrus fruits	*Lasiodiplodia pseudotheobromae, Alternaria alternata, Penicillium digitatum*	SCNs (200–250 ppm) suppressed infection symptoms in vivo after 12 days of storage at high humidity	Postharvest dipping/ coating	[40]
	Strawberry (*Fragaria ananassa*)	*Botrytis cinerea, Rhizopus stolonifer, Aspergillus niger*	Edible fungal CNPs coating suppressed fungal growth; altered fungal morphology; outperformed CNP gel	Postharvest coating with fungal chitosan nanoparticles	[65]
	Tomato, Chilli, Brinjal	*Rhizoctonia solani, Fusarium oxysporum, Colletotrichum acutatum, Phytophthora infestans*	CNPs showed strong antifungal and antioxidant activity; reduced weight loss and extended shelf life when used as a vegetable coating	Postharvest edible coating	[23]

4.2 Soil and Seed Treatments with Chitosan Nanoparticles

Soil and seed-borne fungal diseases are one of the major threats to world crop productivity that resulting in the loss of seed viability, early seedling mortality, and reduced yield. Soil and seed-borne fungal diseases are one of the major threats to world crop productivity that result in the loss of seed viability, early seedling mortality, and reduced yield [34]. To deal with this challenge, CNPs have emerged as promising antifungal agents in seed treatment and soil application (Table 1) against *Alternaria alternata*, *Macrophomina phaseolina*, *Rhizoctonia solani* [87], *Fusarium graminearum* [55], *Fusarium oxysporum*, and *Sclerotium rolfsii* [13], *Fusarium virguliforme* [79], *Rhizoctonia bataticola* [90], *Cephalosporium maydis* [26]. Integrative approaches combining CNPs with AMF [27], other bioagents like mycorrhizae (VAM), and garlic essential oil-loaded chitosan nanoparticles [26, 67] have appeared effective in disease resistance beyond the capabilities of individual treatments. In several studies, thiamine- or copper-loaded chitosan nanoparticles evidenced enhanced antifungal activity, attributed to increased solubility, bioavailability, and structural compatibility with plant systems in tomato [88]. CNPs' application in the soil improves the rhizospheric environment by targeting pathogenic fungi while promoting the growth of beneficial microorganisms. This selective antimicrobial action contributes to a healthier soil microbiome and improved nutrient uptake [60, 87].

4.3 Application of Chitosan Nanoparticles in Controlling Postharvest Disease

A wide range of crops, such as mango, banana, apple, citrus, papaya, tomato, and grape, suffer considerable losses in the postharvest period due to fungal pathogens like *Colletotrichum gloeosporioides*, *Rhizopus stolonifer*, *Penicillium* spp., *Botrytis cinerea*, and *Fusarium* spp. [38, 80, 111]. CNPs have shown great potential strategy for controlling postharvest fungal diseases (Table 1). In mangoes, CNPs-based coatings effectively suppressed anthracnose and stem-end rot, while also preserving postharvest texture and color [38, 114]. In tomatoes, CNPs coating combined with coffee residue extract performed a dual function—suppressing *Rhizopus stolonifer* and preserving fruit quality [9]. Traditional fungicides introduce major challenges due to residue accumulation, pathogen resistance, and environmental contamination, although effective [80]. CNPs offer a biodegradable, eco-friendly alternative with diverse benefits: they inhibit fungal growth, delay fruit ripening, reduce respiration rate, and maintain firmness and nutritional quality [7, 114].

4.4 *Case Studies on Some Specific Fungal Pathogens*

4.4.1 *Alternaria Alternata* in Chilli (*Capsicum Annuum*)

Alternaria leaf spot, caused by *Alternaria alternata*, has become a growing concern in chilli cultivation, impacting plant vigor and yield. Recent research has highlighted the efficacy of CNPs in preventing this pathogen through foliar application. CNPs were synthesized using the ionotropic gelation method and used at a concentration of 0.001% in both resistant and susceptible chilli cultivars. The nanoparticles inhibited fungal development and activated several key plant defense mechanisms. Activities of enzymes such as peroxidase and phenylalanine ammonia-lyase were increased, along with higher accumulation of phenolics and flavonoids. Microscopic and biochemical analyses also confirmed improved callose deposition and reduced cellular damage. Remarkably, the induced resistance appeared to be linked with nitric oxide (NO) signaling, as artificially depleted NO conditions led to a weakened defense response. These findings suggest that CNPs function in a dual role—limiting fungal progression directly and enhancing the plant's internal immune system through NO-mediated signaling [89].

4.4.2 *Fusarium Virguliforme* in Soybean (*Glycine Max*)

Fusarium root rot, activated by *Fusarium virguliforme*, shows a persistent threat to soybean crops. In greenhouse and field trials, chitosan-coated mesoporous silica nanoparticles (CTS-MSNs) were used as a seed or foliar treatment. Using CTS-MSNs demonstrated significant progress in plant biomass, chlorophyll content, and disease suppression. In particular, seed treatment with CTS-MSNs reduced disease severity and enhanced micronutrient uptake, including Zn, Mn, and K in pods. This positive effect, alongside disease control, demonstrates the multi-functional role of CTS-MSNs in sustainable crop management [79].

4.4.3 *Botrytis Cinerea, Rhizopus Stolonifer,* and *Aspergillus Niger* in Strawberry (*Fragaria Ananassa*)

Postharvest fungal decomposition in strawberries is commonly caused by *B. cinerea*, *R. stolonifer*, and *A. niger*. Application of fungal-derived chitosan nanoparticle-based coatings evidenced visible antifungal performance compared to traditional chitosan gel. These nano-coatings caused morphological disturbance in fungal cells and effectively suppressed fungal growth both in vitro and on strawberry surfaces. Their biocompatibility and potency suggest they are an ideal way for safe postharvest preservation [65].

5 Efficacy Studies and Field Trials

5.1 In Vitro and in Vivo Studies

In vitro studies have shown the antifungal potential of CNPs against phytopathogenic fungi. CNPs show strong antifungal effects through several mechanisms, like disruption of the integrity of the fungal cell wall, alteration of permeability of the membrane, and essential nutrient chelation [44]. Some pathogens, such as *Pyricularia grisea, Alternaria solani,* and *Fusarium oxysporum,* have shown significant inhibition of growth upon CNPs' treatment [93]. Additionally, CNPs have been found to generate oxidative stress inside the pathogen, contributing to their antimicrobial activity [49].

In vivo experiments, involving detached leaves or whole plants, have proven their effectiveness. Plants treated with CNPs result in delayed symptom development, small lesion area, and reduced fungal colonization [91]. Moreover, CNPs activate the plant defense system, increasing the expression of genes related to pathogenesis-related proteins and oxidative stress responses [91]. These findings suggest that CNPs not only kill fungi but also function as elicitors of Systemic Acquired Resistance (SAR) in plants [107].

So, in vitro and in vivo studies perform the dual role of chitosan nanoparticles as both direct inhibition of the pathogen and induction of plant immunity.

5.2 Greenhouse and Field Trial Results

Greenhouse and field trials provide effective insights into the wide range of applicability of CNPs for plant disease management. In greenhouse conditions, foliar application or soil drenching of CNPs formulations has led to a reduction in disease incidence in various crops such as tomato, cucumber, chili, and rice [95]. Additionally, CNPs-treated plants showed improved growth under salinity-stressed conditions [112].

Field studies further confirmed these results under variable environmental conditions. Studies conducted in different geographic regions and crop systems confirmed the efficacy of CNPs in suppressing fungal diseases [82]. Field application of CNPs is often associated with increased soil microbial diversity and improved plant health without leaving harmful residues [74]. The biodegradable and non-toxic nature of CNPs makes them a proper candidate for organic and sustainable agriculture [52]. These trials support that CNPs not only serve as effective antifungal agents but also follow environmentally friendly crop protection strategies. So, using CNPs reduces chemical pesticide use and enhances agricultural sustainability.

5.3 Comparative Studies with Other Nanomaterials and Conventional Fungicides

Comparative studies have emphasized the benefits and limitations of CNPs with other nanomaterials and conventional fungicides. Metallic nanoparticles like silver (AgNPs), copper (CuNPs), and zinc oxide (ZnONPs) show potent antifungal activity because of their high surface area and reactive attributes [94]. Moreover, trouble over phytotoxicity, persistence in the environment, and potential risks to non-target organisms limit their wide range of agricultural applications [4]. Conversely, CNPs offer a non-toxic and eco-friendly profile. They are derived from natural sources, which are biodegradable and possess innate antimicrobial attributes without the toxicity associated with metal-based nanoparticles. [25].

Comparing with conventional fungicides, CNPs often show equivalent or even superior disease control, particularly against fungal pathogens that have developed resistance to often-used chemical treatments [88]. CNPs can increase the efficiency of fungicides when used in combination, facilitating dose reduction and reducing chemical load in the environment [64]. These complementary formulations are especially beneficial in integrated disease management [64]. Moreover, CNPs do not contribute to the aggregation of harmful residues on crops, making them suitable for organic farming [63].

Overall, CNPs offer a balanced solution for modern plant disease control, combining efficiency, safety, and sustainability.

6 Benefits and Advantages

6.1 Biodegradability and Environmental Safety

CNPs are gaining attention in sustainable agriculture because of their first-rate biodegradability and environmental safety. Chitosan, derived from chitin—a natural polymer found in crustacean shells—is non-toxic and decomposes in the environment without releasing harmful residues [39]. This character minimizes the ecological footprint of crop protection practices in comparison to conventional synthetic fungicides. CNPs degrade into inoffensive monosaccharides and amino sugars that are either absorbed by plants or metabolized by soil microbes, increasing soil health [62]. However, chitosan is recognized as Generally Regarded As Safe (GRAS) by the United States (U.S.) Food and Drug Administration (FDA), further promoting its suitability for application in food and agriculture [59]. Their nano-scale formulation increases solubility and dispersion, allowing lower dosages while sustaining effectiveness [95]. These attributes make CNPs an eco-compatible substitute for disease control tactics in Integrated Pest Management (IPM) schemes.

6.2 Target Specificity and Controlled Release

Chitosan nanoparticles show increased target specificity and controlled release attributes, making them perfect for the specific delivery of antifungal agents. Their positive surface charge permits them to bind efficiently to negatively charged fungal cell walls, promoting particular antifungal action [75]. In addition, CNPs can be designed to enclose and release active compounds slowly over time, thereby extending their protective effect [74]. This reduces the need for repeated applications and assures sustained interaction with the pathogen at the place of infection. Managed release also reduces phytotoxicity risks and increases efficient use of the active ingredient [74]. However, their nano-size assures better penetration and movement in plant tissues, increasing intracellular delivery of antifungal compounds [82]. These properties contribute to a more sensible and efficient disease control tactic, reducing off-target effects and increasing disease suppression.

6.3 Reduced Use of Synthetic Fungicides

CNPs have shown promise in reducing dependence on synthetic fungicides, a key goal in sustainable agriculture. The collaborative action of chitosan's innate antifungal attributes and its nanoparticulate shape allows effective control of a broad spectrum of fungal pathogens at considerably lower concentrations than conventional treatments [82]. Chitosan causes systemic resistance in plants by stimulating defense-related enzymes and increasing the plant's innate immunity, and reducing the need for external chemical inputs [2]. In addition, CNPs can be supplied with bioactive fungicides, enabling slow and targeted release and minimizing the total amount of synthetic chemicals for effective disease control [54]. This not only lowers production costs but also decreases the environmental and health hazards associated with excessive use of fungicides. Field trials have ensured the efficiency of CNPs in minimizing disease occurrence and maximizing crop yield [18].

6.4 Resistance Management

One of the important challenges in controlling fungal disease is the quick appearance of resistance to conventional fungicides. CNPs help to reduce this issue through multiple modes of action that minimize the selection pressure on pathogens. Unlike single-target fungicides, CNPs break fungal cell walls, disturb membrane integrity, and elicit plant immune responses, making it more difficult for fungi to develop resistance [108]. Their ability to prompt systemic acquired resistance (SAR) in plants adds an indirect mechanism of defense, making the pathogen's modification harder

[107]. However, CNPs can be used in addition to lower doses of traditional fungicides, acting synergistically and increasing the lifespan of existing chemical products [64].

7 Challenges and Limitations

Despite the promising potential of CNPs in managing fungal plant diseases, several challenges and limitations hinder their widespread adoption and commercial application. These barriers span from production constraints to ecological and regulatory concerns, which must be addressed through interdisciplinary research and policy frameworks.

7.1 Scale-Up and Cost-Effectiveness

One of the major limitations of CNPs is the difficulty in scaling up production while maintaining consistent quality and bioactivity. Laboratory-scale synthesis often yields uniform and effective nanoparticles, but industrial-scale manufacturing faces challenges related to batch-to-batch variation, contamination, and the requirement for sophisticated equipment [69]. Additionally, the high costs of chitosan purification, nanoparticle formulation, and stabilization add to the economic burden, making large-scale application in agriculture less feasible for low-income and developing countries [76].

7.2 Stability and Shelf-Life

The stability of CNPs during storage and application is another critical limitation. Chitosan nanoparticles are sensitive to environmental factors such as pH, temperature, and humidity, which can influence their size, charge, and functional activity [101]. Over time, aggregation or degradation may occur, reducing their efficacy and limiting shelf life. Developing formulations that ensure long-term stability without compromising antifungal activity remains a key area for improvement.

7.3 Phytotoxicity and Off-Target Effects

While CNPs are generally regarded as safe and biocompatible, there is growing concern over their potential phytotoxicity at higher concentrations. Studies have reported that excessive or repeated application may lead to leaf scorching, growth inhibition, or physiological stress in certain plant species. Moreover, off-target effects on beneficial soil microbes, non-target fungi, or pollinators need to be carefully evaluated, as unintended ecological disturbances could counteract the sustainability goals of nanoparticle use [12].

7.4 *Regulatory and Biosafety Concerns*

The regulatory landscape for agricultural nanomaterials like CNPs is nascent and fragmented. Many global regulatory bodies, such as the U.S. Environmental Protection Agency (EPA), FDA, and European agencies, have yet to establish specific requirements for engineered nanoparticles concerning safety testing, labeling, handling, or residue analysis [81]. This regulatory uncertainty, coupled with public skepticism over nanotechnology, could hinder adoption by farmers, agribusinesses, and consumers.

8 Future Prospects and Research Directions

As the global demand for sustainable agriculture intensifies, CNPs offer a compelling, eco-friendly alternative to conventional fungicides. However, for their large-scale adoption and optimized efficacy in field conditions, innovative advancements in formulation, delivery, and mechanistic understanding are essential. This section outlines key future directions in CNPs research, with emphasis on their integration into precision agriculture, smart delivery mechanisms, multifunctional formulations, and systems biology approaches.

8.1 *Integration with Precision Agriculture*

The future of fungal plant disease management lies in the integration of nanotechnology with precision agriculture tools, such as remote sensing, geographic information systems (GIS), and unmanned aerial vehicles (UAVs). Chitosan nanoparticles can be adapted for site-specific delivery, minimizing wastage and reducing off-target effects. When linked with sensor-based disease forecasting systems, the application of CNPs can be precisely timed and spatially directed, enhancing efficiency and reducing the frequency of chemical treatments [96]. Furthermore, real-time monitoring of plant responses through biosensors or spectral imaging can provide feedback on the effectiveness of CNPs treatments, enabling dynamic decision-making in integrated pest management (IPM) systems.

8.2 *Smart Delivery Systems (Stimuli-Responsive CNPs)*

One of the most promising research avenues involves the development of stimuli-responsive CNPs that can release active agents in response to environmental or biological cues. These smart delivery systems can be triggered by pH changes, temperature fluctuations, enzymatic activity, redox gradients, or pathogen-specific signals. For instance, pH-sensitive CNPs can release antifungal compounds in the acidic microenvironments created by pathogen infection. Similarly, enzyme-responsive nanoparticles can be programmed to degrade and release their cargo in the presence of fungal hydrolases [70].

The design of such programmable nanocarriers will not only increase the specificity of pathogen targeting but also reduce unintended interactions with non-target organisms, contributing to ecological safety and resource-efficient disease control.

8.3 *Multi-Functional Nanocomposites*

Another emerging trend is the development of **multi-functional nanocomposites**, where chitosan is combined with other nanomaterials or bioactive agents to enhance its antifungal, physical, and functional properties. These include:

- **Chitosan-silver nanocomposites**, which combine the antimicrobial properties of both agents for synergistic effects [84].
- **Chitosan with essential oils or biopesticides**, allowing for broad-spectrum antifungal activity and prolonged shelf-life [110].

Such composites can simultaneously act as delivery agents, fungal pathogen inhibitors, and plant growth enhancers, opening new frontiers in chitosan-based nanobiotechnology-driven crop protection.

8.4 *Omics Approaches for Mode of Action Studies*

Despite the proven antifungal efficacy of CNPs, a detailed understanding of their molecular mechanisms remains limited. Integrating omics technologies such as Transcriptomics, Proteomics, Metabolomics, and Genomics can help elucidate the complex plant-pathogen-nanoparticle interactions.

- Transcriptomics can reveal how CNPs influence the expression of defense-related genes in host plants.
- Proteomics can identify changes in fungal protein profiles after CNPs exposure, helping to pinpoint molecular targets of action.
- Metabolomics can track alterations in primary and secondary metabolic pathways in both plant and fungal cells.

- **Microbiome analyses** can assess how CNPs affect the plant's rhizosphere microbial community, ensuring that beneficial organisms are not adversely impacted.

These insights will inform the rational design of more efficient, selective, and safer nanocarriers, while also addressing biosafety concerns for regulatory approval [103].

9 Conclusion

Chitosan nanoparticles represent a promising, eco-compatible tool for managing fungal plant diseases while minimizing environmental impact. Their multifaceted antifungal mechanisms and capacity to induce plant defenses make them valuable in integrated pest management strategies. Field evidence confirms their effectiveness across diverse crops and application methods. Despite current challenges in production scalability, stability, and regulatory acceptance, ongoing research into smart delivery systems, multi-functional composites, and omics-based mode-of-action studies can address these barriers. With proper policy support and farmer awareness, CNPs can significantly reduce reliance on chemical fungicides, contributing to resilient and sustainable agricultural systems.

References

1. Abdel-Aliem HA, Gibriel AY, Rasmy NM, Sahab AF, El-Nekeety AA, Abdel-Wahhab MA (2019) Antifungal efficacy of chitosan nanoparticles against phytopathogenic fungi and inhibition of zearalenone production by *Fusarium graminearum*. Comunicata Scientiae 10(3):338–345. https://doi.org/10.14295/cs.v10i3.1899
2. Abdellatef MA, Elagamey E, Kamel SM (2022) Chitosan is the ideal resource for plant disease management under sustainable agriculture. In: Chitin and chitosan-isolation, properties, and applications, IntechOpen https://doi.org/10.5772/intechopen.107958
3. Agrios GN (2005) Chapter IV—how pathogens attack plants, Editor(s): Agrios GN. Plant Pathology (Fifth Edition). Academic Press, pp:175–205 https://doi.org/10.1016/C2009-0-02037-6
4. Ahmad Z, Niyazi S, Firdoos A, Wang C, Manzoor MA, Ramakrishnan M, Upadhyay A, Ding Y (2024) Enhancing plant resilience: nanotech solutions for sustainable agriculture. Heliyon 10(23):e40735. https://doi.org/10.1016/j.heliyon.2024.e40735
5. Ahmed M, Lu Y (2025) Synthesis of chitosan nanoparticles via microfluidic approach: the role of temperature in tailoring aggregation for enhanced uniformity. Micromachines 16(6):642. https://doi.org/10.3390/mi16060642
6. Al-Zahrani SS, Bora RS, Al-Garni SM (2021) Antimicrobial activity of chitosan nanoparticles. Biotechnol Biotechnol Equip 35(1):1874–1880. https://doi.org/10.1080/13102818.2022.2027816
7. Ali A, Mohamed MTM, Siddiqui Y (2012) Control of anthracnose by chitosan through stimulation of defence-related enzymes in Eksotika II papaya (*Carica papaya* L.) fruit. J Biol Life Sci 3. https://doi.org/10.5296/jbls.v3i1.1306

8. Ali A, Zahid N, Manickam S, Siddiqui Y (2014) Induction of lignin and pathogenesis related proteins in dragon fruit plants in response to submicron chitosan dispersions. Crop Prot 63:83–88. https://doi.org/10.1016/j.cropro.2014.05.009
9. Andrea MJ, Perez-Sato JA, Perea M, Reyes J (2025) Nanostructured chitosan coating with a coffee residue extract for the preservation of tomato and controlling pre- and postharvest disease caused by *Rhizopus stolonifer*. Processes 13(1):220. https://doi.org/10.3390/pr13010220
10. Asghar MA, Zhu Q, Sun S, Peng YE, Shuai Q (2018) Suspect screening and target quantification of human pharmaceutical residues in the surface water of Wuhan, China, using UHPLC-Q-Orbitrap HRMS. Sci Total Environ 635:828–837. https://doi.org/10.1016/j.scitotenv.2018.04.179
11. Athanasopoulos A, Andre B, Sophianopoulou V, Gournas C (2019) Fungal plasma membrane domains. FEMS Microbiol Rev 43(6):642–673. https://doi.org/10.1093/femsre/fuz022
12. Balusamy SR, Joshi AS, Perumalsamy H, Mijakovic I, Singh P (2023) Advancing sustainable agriculture: a critical review of smart and eco-friendly nanomaterial applications. J Nanobiotechnol 21(1):372. https://doi.org/10.1186/s12951-023-02135-3
13. Boruah S, Dutta P (2021) Fungus mediated biogenic synthesis and characterization of chitosan nanoparticles and its combined effect with *Trichoderma asperellum* against *Fusarium oxysporum*, *Sclerotium rolfsii* and *Rhizoctonia solani*. Indian Phytopathol 74(1):81–93. https://doi.org/10.1007/s42360-020-00289-w
14. Boyce KJ (2023) The microevolution of antifungal drug resistance in pathogenic fungi. Microorganisms 11(11):2757. https://doi.org/10.3390/microorganisms11112757
15. Buchman JT, Haynes CL, Karnik R, Hebert JJ (2019) Chitosan-coated mesoporous silica nanoparticle treatment of *Citrullus lanatus* (watermelon): enhanced fungal disease suppression and modulated expression of stress-related genes. ACS Sustain Chem Eng 7(24):19649–19659. https://doi.org/10.1021/acssuschemeng.9b04800
16. Buysens C, Dupré de Boulois H, Declerck S (2015) Do fungicides used to control Rhizoctonia solani impact the non-target arbuscular mycorrhizal fungus *Rhizophagus irregularis*? Mycorrhiza 25(4):277–288. https://doi.org/10.1007/s00572-014-0610-7
17. Calvo P, Remuñán-López C, Vila-Jato JL, Alonso MJ (1997) Novel hydrophilic chitosan–polyethylene oxide nanoparticles as protein carriers. J Appl Polym Sci 63(1):125–132
18. Choudhary RC, Kumaraswamy RV, Kumari S, Sharma SS, Pal A, Raliya R, Biswas P, Saharan V (2019) Zinc encapsulated chitosan nanoparticle to promote maize crop yield. Int J Biol Macromol 127:126–135. https://doi.org/10.1016/j.ijbiomac.2018.12.274
19. Chouhan D, Dutta A, Kumar A, Mandal P, Choudhuri C (2022) Application of nickel chitosan nanoconjugate as an antifungal agent for combating Fusarium rot of wheat. Sci Rep 12(1):14518. https://doi.org/10.1038/s41598-022-18670-2
20. Chun SC, Chandrasekaran M (2019) Chitosan and chitosan nanoparticles induced expression of pathogenesis-related proteins genes enhances biotic stress tolerance in tomato. Int J Biol Macromol 125:948–954
21. Dean R, Van Kan JA, Pretorius ZA, Hammond-Kosack KE, Di Pietro A, Spanu PD, Rudd JJ, Dickman M, Kahmann R, Ellis J, Foster GD (2012) The Top 10 fungal pathogens in molecular plant pathology. Mol Plant Pathol 13(4):414–430. https://doi.org/10.1111/j.1364-3703.2011.00783.x
22. Divya K, Jisha MS (2018) Chitosan nanoparticles preparation and applications. Environ Chem Lett 16(1):101–112. https://doi.org/10.1007/s10311-017-0670-y
23. Divya K, Smitha V, Jisha MS (2018) Antifungal, antioxidant and cytotoxic activities of chitosan nanoparticles and its use as an edible coating on vegetables. Int J Biol Macromol 114:572–577. https://doi.org/10.1016/j.ijbiomac.2018.03.130
24. Divya K, Thampi M, Vijayan S, Varghese S, Jisha MS (2020) Induction of defence response in *Oryza sativa* L. against *Rhizoctonia solani* (Kuhn) by chitosan nanoparticles. Microbial Pathogenesis 149:104525. https://doi.org/10.1016/j.micpath.2020.104525
25. Egorov AR, Kirichuk AA, Rubanik VV, Rubanik VV Jr, Tskhovrebov AG, Kritchenkov AS (2023) Chitosan and its derivatives: preparation and antibacterial properties. Materials 16(18):6076. https://doi.org/10.3390/ma16186076

26. El-Gazzar N, El-Bakery AM, Ata AA (2018) Influence of some bioagents and chitosan nanoparticles on controlling maize late wilt and improving plant characteristics. Egyptian J Phytopathol 46(2):243–264. https://doi.org/10.21608/ejp.2018.115896
27. El-Gazzar N, El-Hai KM, Teama SAM, El-Sawy MM, Abdel-Rahman FA (2023) Enhancing Vicia faba's immunity against *Rhizoctonia solani* root rot diseases by arbuscular mycorrhizal fungi and nano chitosan. BMC Plant Biology 23:403. https://doi.org/10.1186/s12870-023-04407-4
28. El-Shabouri MH (2002) Positively charged nanoparticles for improving the oral bioavailability of cyclosporin-A. Int J Pharm 249(1–2):101–108. https://doi.org/10.1016/S0378-5173(02)00461-1
29. Elkomy MH, Ali AA, Eid HM (2022) Chitosan on the surface of nanoparticles for enhanced drug delivery: a comprehensive review. J Control Release 351:923–940. https://doi.org/10.1016/j.jconrel.2022.10.005
30. Elsharkawy MM (2024) The potential of chitosan nanoparticles to control plant pathogens. In: Nanotechnology in plant health 180–195: CRC Press.
31. Elsharkawy MM, Arafa RA, Omara RI, Kamel SM, Ismail W, Ismail S, Derbalah A (2022) Developing Ag_2O and Ag_2O/TiO_2 nanostructures as a new strategy for control late blight of potato caused by *Phytophthora infestans*. Physiol Mol Plant Pathol 120:101856. https://doi.org/10.1016/j.pmpp.2022.101856
32. Elsharkawy MM, Omara RI, Mostafa YS, Alamri SA, Hashem M, Alrumman SA, Ahmad AA (2022) Mechanism of wheat leaf rust control using chitosan nanoparticles and salicylic acid. J Fungi 8(3):304. https://doi.org/10.3390/jof8030304
33. Gan Q, Wang T (2007) Chitosan nanoparticle as protein delivery carrier—systematic examination of fabrication conditions for efficient loading and release. Colloids Surf B 59(1):24–34. https://doi.org/10.1016/j.colsurfb.2007.04.009
34. Gaur A, Kumar A, Kiran R, Kumari P (2020) Importance of seed-borne diseases of agricultural crops: economic losses and impact on society. In: Kumar R, Gupta A (eds) Seed-borne diseases of agricultural crops: detection, diagnosis & management. Springer, Singapore. https://doi.org/10.1007/978-981-32-9046-4_1
35. Giraldo JD, Rivas BL (2021) Direct ionization and solubility of chitosan in aqueous solutions with acetic acid. Polym Bull 78(3):1465–1488. https://doi.org/10.1007/s00289-020-03172-w
36. González C, Reyes LH, Muñoz-Camargo C, Cruz JC (2021) Synthesis, characterization, and functionalization of chitosan and Gelatin type B nanoparticles to develop novel highly biocompatible cell-penetrating agents. Mater Proc 4(1):30. https://doi.org/10.3390/IOCN2020-07816
37. Gopal M, Gogoi R, Srivastava C, Kumar R, Singh PK, Nair KK, Yadav S, Goswami A (2011) Nanotechnology and its application in plant protection. Plant Pathol India: Vision 2030:224–232
38. Goudarzi A, Pourrahim R, Rahnama K, Mahdikhani M (2021) Fungal pathogens associated with pre- and postharvest fruit rots of mango in southern Iran. J Phytopathol. https://doi.org/10.1111/jph.13027
39. Hegde S, Selvaraj S (2024) Chitosan: An in-depth analysis of its extraction, applications, constraints, and future prospects. J Microbiol Biotechnol Food Sci 14(2):e10563–e10563. https://doi.org/10.55251/jmbfs.10563
40. Hoang NC, Nguyen CM, Nguyen VH, Giang DH, Nguyen VH, Pham VN (2022) Antifungal activity of squid pen chitosan nanoparticles against three fungal pathogens in various citrus fruits in vitro and in vivo. Coatings 12(2):235. https://doi.org/10.3390/coatings12020235
41. Hoang NH, Le Thanh T, Sangpueak R et al (2023) The effect of chitosan nanoparticle formulations for control of leaf spot disease on cassava. Phytoparasitica 51:621–636. https://doi.org/10.1007/s12600-023-01065-y
42. Huang Y, Cai Y, Lapitsky Y (2015) Factors affecting the stability of chitosan/tripolyphosphate micro-and nanogels: resolving the opposing findings. J Mater Chem B 3(29):5957–5970
43. Hui ST, Gifford H, Rhodes J (2024) Emerging antifungal resistance in fungal pathogens. Current Clinical Microbiol Reports 11(2):43–50. https://doi.org/10.1007/s40588-024-00219-8

44. Ing LY, Zin NM, Sarwar A (2012) Katas H (2012) Antifungal activity of chitosan nanoparticles and correlation with their physical properties. Int J Biomater 1:632698. https://doi.org/10.1155/2012/632698
45. Islam T, Danishuddin TNT, Matin MN, Barai HR, Haque MA (2024) Resistance mechanisms of plant pathogenic fungi to fungicide, environmental impacts of fungicides, and sustainable solutions. Plants 13(19):2737. https://doi.org/10.3390/plants13192737
46. Islam T, Haque MA, Barai HR, Istiaq A, Kim JJ (2024) Antibiotic resistance in plant pathogenic bacteria: recent data and environmental impact of unchecked use and the potential of biocontrol agents as an eco-friendly alternative. Plants 13(8):1135. https://doi.org/10.3390/plants13081135
47. Iswanti FC, Nurulita I, Djauzi S, Sadikin M, Witarto AB, Yamazaki T (2019) Preparation, characterization, and evaluation of chitosan-based nanoparticles as CpG ODN carriers. Biotechnol Biotechnol Equip 33(1):390–396. https://doi.org/10.1080/13102818.2019.1578690
48. Jha R, Mayanovic RA (2023) A review of the preparation, characterization, and applications of chitosan nanoparticles in nanomedicine. Nanomaterials 13(8):1302. https://doi.org/10.3390/nano13081302
49. Ji H, Wang J, Chen F, Fan N, Wang X, Xiao Z, Wang Z (2022) Meta-analysis of chitosan-mediated effects on plant defense against oxidative stress. Sci Total Environ 851:158212. https://doi.org/10.1016/j.scitotenv.2022.158212
50. Juksu K, Zhao JL, Liu YS, Yao L, Sarin C, Sreesai S, Klomjek P, Jiang YX, Ying GG (2019) Occurrence, fate and risk assessment of biocides in wastewater treatment plants and aquatic environments in Thailand. Sci Total Environ 690:1110–1119. https://doi.org/10.1016/j.scitotenv.2019.07.097
51. Kalagatur NK, Nirmal Ghosh OS, Sundararaj N, Mudili V (2018) Antifungal activity of chitosan nanoparticles encapsulated with Cymbopogon Martinii essential oil on plant pathogenic Fungi *Fusarium graminearum.* Front Pharmacol 9:610. https://doi.org/10.3389/fphar.2018.00610
52. Karamchandani BM, Dalvi SG, Bagayatkar M, Banat IM, Satpute SK (2024) Prospective applications of chitosan and chitosan-based nanoparticles formulations in sustainable agricultural practices. Biocatal Agric Biotechnol 58:103210. https://doi.org/10.1016/j.bcab.2024.103210
53. Kashyap PL, Xiang X, Heiden P (2015) Chitosan nanoparticle based delivery systems for sustainable agriculture. Int J Biol Macromol 77:36–51. https://doi.org/10.1016/j.ijbiomac.2015.02.039
54. Kashyap PL, Xiang X, Heiden P (2015) Chitosan nanoparticle-based delivery systems for sustainable agriculture. Int J Biol Macromol 77:36–51. https://doi.org/10.1016/j.ijbiomac.2015.02.039
55. Kheiri A, Moosawi Jorf SA, Malihipour A, Saremi H, Nikkhah M (2016) Application of chitosan and chitosan nanoparticles for the control of Fusarium head blight of wheat (*Fusarium graminearum*) in vitro and greenhouse. Int J Biol Macromolecules 93(Part A):1261–1272. https://doi.org/10.1016/j.ijbiomac.2016.09.072
56. Kheiri A, Moosawi Jorf SA, Malihipour A, Saremi H, Nikkhah M (2017) Synthesis and characterization of chitosan nanoparticles and their effect on Fusarium head blight and oxidative activity in wheat. Int J Biol Macromol 102:526–538
57. Kheiri A, Moosawi Jorf SA, Malihipour A, Saremi H, Nikkhah M (2017) Synthesis and characterization of chitosan nanoparticles and their effect on Fusarium head blight and oxidative activity in wheat. Int J Biol Macromol 102:526–538. https://doi.org/10.1016/j.ijbiomac.2017.04.034
58. Kong M, Chen XG, Xing K, Park HJ (2010) Antimicrobial properties of chitosan and mode of action: a state of the art review. Int J Food Microbiol 144(1):51–63. https://doi.org/10.1016/j.ijfoodmicro.2010.09.012
59. Kumar S, Arora A, Mathur D, Chaudhary A, Pant V, Guchhait S, Singh BK (2025) A review on chitosan and chitosan-based bionanocomposites: Promising biological macromolecules for sustainable corrosion inhibition. Int J Biol Macromol 301:140392. https://doi.org/10.1016/j.ijbiomac.2025.140392

60. Kumaraswamy RV, Kumari S, Choudhary RC, Sharma SS, Pal A, Raliya R, Biswas P, Saharan V (2018) Engineered chitosan-based nanomaterials: bioactivities, mechanisms and perspectives in plant protection and growth. Int J Biol Macromol 113:494–506. https://doi.org/10.1016/j.ijbiomac.2018.02.097
61. Lopez-Leon T, Carvalho ELS, Seijo B, Ortega-Vinuesa JL, Bastos-González D (2005) Physicochemical characterization of chitosan nanoparticles: electrokinetic and stability behavior. J Colloid Interface Sci 283(2):344–351. https://doi.org/10.1016/j.jcis.2004.08.186
62. Malerba M, Cerana R (2020) Chitin-and chitosan-based derivatives in plant protection against biotic and abiotic stresses and in recovery of contaminated soil and water. Polysaccharides 1(1):21–30. https://doi.org/10.3390/polysaccharides1010003
63. Maluin FN, Hussein MZ (2020) Chitosan-based agronanochemicals as a sustainable alternative in crop protection. Molecules 25(7):1611. https://doi.org/10.3390/molecules25071611
64. Maluin FN, Hussein MZ, Azah Yusof N, Fakurazi S, Idris AS, Zainol Hilmi NH, Jeffery Daim LD (2020) Chitosan-based agronanofungicides as a sustainable alternative in the basal stem rot disease management. J Agric Food Chem 68(15):4305–4314. https://doi.org/10.1021/acs.jafc.9b08060
65. Melo NF, Macedo GR, Silva LMA et al (2020) In vivo and in vitro antifungal effect of fungal chitosan nanocomposite edible coating against strawberry phytopathogenic fungi. Int J Food Sci Technol 55(11):3381–3391. https://doi.org/10.1111/ijfs.14669
66. Mohammed MA, Syeda JT, Wasan KM, Wasan EK (2017) An overview of chitosan nanoparticles and its application in non-parenteral drug delivery. Pharmaceutics 9(4):53. https://doi.org/10.3390/pharmaceutics9040053
67. Mondéjar-López M, Rubio-Moraga A, López-Jimenez AJ, Martínez JCG, Ahrazem O, Gómez-Gómez L, Niza E (2022) Chitosan nanoparticles loaded with garlic essential oil: a new alternative to tebuconazole as seed dressing agent. Carbohyd Polym 277:118815. https://doi.org/10.1016/j.carbpol.2021.118815
68. Mosa MA, El-Abeid SE (2023) Chitosan-loaded copper oxide nanoparticles: a promising antifungal nanocomposite against Fusarium wilt disease of tomato plants. Sustainability 15(19):14295. https://doi.org/10.3390/su151914295
69. Muthu MS, Wilson B (2012) Challenges posed by the scale-up of nanomedicines. Nanomed 7(3):307–309. https://doi.org/10.2217/nnm.12.3
70. Naghib SM, Ahmadi B, Kangarshahi BM, Mozafari MR (2024) Chitosan-based smart stimuli-responsive nanoparticles for gene delivery and gene therapy: recent progresses on cancer therapy. Int J Biol Macromol 278(Pt 3):134542. https://doi.org/10.1016/j.ijbiomac.2024.134542
71. Nagpal K, Singh SK, Mishra DN (2010) Chitosan nanoparticles: a promising system in novel drug delivery. Chem Pharm Bull 58(11):1423–1430. https://doi.org/10.1248/cpb.58.1423
72. Nair R, Varghese SH, Nair BG, Maekawa T, Yoshida Y, Kumar DS (2010) Nanoparticulate material delivery to plants. Plant Sci 179(3):154–163. https://doi.org/10.1016/j.plantsci.2010.04.012
73. Nandhini R, Rajeswari E, Harish S, Sivakumar V, Gangai Selvi R (2025) Role of chitosan nanoparticles in sustainable plant disease management. J Nanopart Res 27:13. https://doi.org/10.1007/s11051-024-06203-z
74. Nandini T, Sudhalakshmi C, Sivakumar K, Parameswari E, Thangamani C (2025) A review-chitosan nanoparticles towards enhancing nutrient use efficiency in crops. Int J Biol Macromol 306(2):141433. https://doi.org/10.1016/j.ijbiomac.2025.141433
75. Nasaj M, Chehelgerdi M, Asghari B, Ahmadieh-Yazdi A, Asgari M, Kabiri-Samani S, Sharifi E, Arabestani M (2024) Factors influencing the antimicrobial mechanism of chitosan action and its derivatives: a review. Int J Biol Macromol 277(2):134321. https://doi.org/10.1016/j.ijbiomac.2024.134321
76. Nederstigt TAP, Brinkmann BW, Peijnenburg WJGM, Vijver MG (2024) Sustainability claims of nanoenabled pesticides require a more thorough evaluation. Environ Sci Technol 58(5):2163–2165. https://doi.org/10.1021/acs.est.3c10207

77. Niwa T, Takeuchi H, Hino T, Kunou N, Kawashima Y (1993) Preparations of biodegradable nanospheres of water-soluble and insoluble drugs with D, L-lactide/glycolide copolymer by a novel spontaneous emulsification solvent diffusion method, and the drug release behavior. J Control Release 25(1–2):89–98. https://doi.org/10.1016/0168-3659(93)90097-O
78. Oh JW, Chun SC, Chandrasekaran M (2019) Preparation and in vitro characterization of chitosan nanoparticles and their broad-spectrum antifungal action compared to antibacterial activities against phytopathogens of tomato. Agronomy 9(1):21. https://doi.org/10.3390/agronomy9010021
79. O'Keefe TL et al (2024) Chitosan-coated mesoporous silica nanoparticles for suppression of *Fusarium virguliforme* in soybeans (*Glycine max*). ACS Agric Sci Technol. https://doi.org/10.1021/acsagscitech.4c00025
80. Palou L (2018) Postharvest treatments with GRAS salts to control fresh fruit decay. Horticulturae 4(4):46. https://doi.org/10.3390/horticulturae4040046
81. Picecchi D (2018) Tiny things with a huge impact: the international regulation of nanomaterials. Michigan J Environ Administr Law 7(2):447–478. https://doi.org/10.36640/mjeal.7.2.tiny
82. Poznanski P, Hameed A, Orczyk W (2023) Chitosan and chitosan nanoparticles: parameters enhancing antifungal activity. Molecules 28(7):2996. https://doi.org/10.3390/molecules28072996
83. Qi L, Xu Z, Jiang X, Hu C, Zou X (2004) Preparation and antibacterial activity of chitosan nanoparticles. Carbohyd Res 339(16):2693–2700. https://doi.org/10.1016/j.carres.2004.09.007
84. Rajeshkumar S (2021) Citrus Lemon Juice Mediated Preparation of AgNPs/chitosan-based bionanocomposites and its antimicrobial and antioxidant activity. J Nanomater 2021:527250. https://doi.org/10.1155/2021/7527250
85. Ruiz-Herrera J, Ortiz-Castellanos L (2019) Cell wall glucans of fungi. A review. The Cell Surface 5:100022. https://doi.org/10.1016/j.tcsw.2019.100022
86. Sachdev S, Ansari SA, Ansari MI, Fujita M, Hasanuzzaman M (2021) Abiotic stress and reactive oxygen species: generation, signaling, and defense mechanisms. Antioxidants 10(2):277. https://doi.org/10.3390/antiox10020277
87. Saharan V, Mehrotra A, Khatik R, Rawal P, Sharma S, Pal A (2013) Synthesis of chitosan-based nanoparticles and their in vitro evaluation against phytopathogenic fungi. Int J Biol Macromol 62:677–683. https://doi.org/10.1016/j.ijbiomac.2013.10.012
88. Saharan V, Mehrotra A, Sharma SS, Pal A, Raliya R, Biswas P (2015) Synthesis and in vitro antifungal efficacy of Cu–chitosan nanoparticles against pathogenic fungi of tomato. Int J Biol Macromol 75:346–353. https://doi.org/10.1016/j.ijbiomac.2015.01.027
89. Sarkar A, Chakraborty N, Acharya K (2022) Chitosan nanoparticles mitigate Alternaria leaf spot disease of chilli in nitric oxide dependent way. Plant Physiol Biochem 180:64–73. https://doi.org/10.1016/j.plaphy.2022.03.038
90. Sathiyabama M, Gandhi M, Indhumathi M (2022) Suppression of dry root rot disease caused by *Rhizoctonia bataticola* in chickpea plants by application of thiamine-loaded chitosan nanoparticles. Microbial Pathogenesis 173(Part B):105893. https://doi.org/10.1016/j.micpath.2022.105893
91. Sathiyabama M, Manikandan A (2016) Chitosan nanoparticle induced defense responses in fingermillet plants against blast disease caused by *Pyricularia grisea* (Cke.) Sacc. Carbohyd Polym 154:241–246. https://doi.org/10.1016/j.carbpol.2016.06.089
92. Sathiyabama M, Muthukumar S (2020) Chitosan guar nanoparticle preparation and its in vitro antimicrobial activity towards phytopathogens of rice. Int J Biol Macromol 153:297–304. https://doi.org/10.1016/j.ijbiomac.2020.03.001
93. Sathiyabama M, Parthasarathy R (2016) Biological preparation of chitosan nanoparticles and its in vitro antifungal efficacy against some phytopathogenic fungi. Carbohyd Polym 151:321–325. https://doi.org/10.1016/j.carbpol.2016.05.033
94. Semwal A, Gautam K, Bhatt M, Yadav R, Barthwal S, Singh H (2025) Green synthesis and application of metal nanoparticles (MNPs): a sustainable strategy for protecting wood

from fungal and termite decay. Next Mater 9:100962. https://doi.org/10.1016/j.nxmate.2025.100962

95. Sen SK, Das D (2024) A sustainable approach in agricultural chemistry towards alleviation of plant stress through chitosan and nano-chitosan priming. Discover Chem 1(1):44. https://doi.org/10.1007/s44371-024-00046-2
96. Shaikh A, Meroliya H, Dagade-Gadale S, Waghmode S (2021) Applications of nanotechnology in precision agriculture: a review. Res Rev Biotechnol Biosci 8(1):105–117. https://doi.org/10.5281/zenodo.5118402
97. Shariatinia Z, Mahdieh Z (2022) Smart pH-responsive drug release systems based on functionalized chitosan nanocomposite hydrogels. Surfaces and Interface 29:101739. https://doi.org/10.1016/j.surfin.2022.101739
98. Singh LB, Zeshmarani DS, Hijam C, Qutub MM, Singh YH, Prabin S, Singh OW (2023) Nanotechnology and its role in plant pathology. Pharma Innov 12:37–49. https://doi.org/10.22271/tpi.2023.v12.i12a.24452
99. Soleimani P, Mehrvar A, Michaud JP, Vaez N (2022) Optimization of silver nanoparticle biosynthesis by entomopathogenic fungi and assays of their antimicrobial and antifungal properties. J Invertebr Pathol 190:107749. https://doi.org/10.1016/j.jip.2022.107749
100. Stasińska-Jakubas M, Hawrylak-Nowak B (2022) Protective, biostimulating, and eliciting effects of chitosan and its derivatives on crop plants. Molecules 27(9):2801. https://doi.org/10.3390/molecules27092801
101. Szymańska E, Winnicka K (2015) Stability of chitosan-a challenge for pharmaceutical and biomedical applications. Mar Drugs 13(4):1819–1846. https://doi.org/10.3390/md13041819
102. Tokumitsu H, Ichikawa H, Fukumori Y (1999) Chitosan-gadopentetic acid complex nanoparticles for gadolinium neutron-capturetherapy of cancer: preparation by novel emulsion-droplet coalescence technique and characterization. Pharm Res 16:1830–1835. https://doi.org/10.1023/A:1018995124527
103. Usman M, Farooq M, Wakeel A, Nawaz A, Cheema SA, Rehman H, Ashraf I, Sanaullah M (2020) Nanotechnology in agriculture: current status, challenges and future opportunities. Sci Total Environ 721:137778. https://doi.org/10.1016/j.scitotenv.2020.137778
104. Wang Z, Zhao P, Yu J, Jiang Z, Guo X (2018) Experimental and molecular docking study on graphene/Fe3O4 composites as a sorbent for magnetic solid-phase extraction of seven imidazole antifungals in environmental water samples prior to LC-MS/MS for enantiomeric analysis. Microchem J 140:222–231. https://doi.org/10.1016/j.microc.2018.04.027
105. Wroński M, Trawiński J, Skibiński R (2024) Antifungal drugs in the aquatic environment: a review on sources, occurrence, toxicity, health effects, removal strategies and future challenges. J Hazard Mater 465:133167. https://doi.org/10.1016/j.jhazmat.2023.133167
106. Xing K, Liu Y, Shen X, Zhu X, Li X, Miao X, Feng Z, Peng X, Qin S (2017) Effect of O-chitosan nanoparticles on the development and membrane permeability of *Verticillium dahliae*. Carbohyd Polym 165:334–343. https://doi.org/10.1016/j.carbpol.2017.02.063
107. Yadav S, Kumar D (2025) Role of chitosan as a natural elicitor in inducing Systemic Resistance against plant pathogens. J Sci Innov Nature Earth 5(2):22–26. https://doi.org/10.59436/jsiane.375.2583-2093
108. Yan D, Li Y, Liu Y, Li N, Zhang X, Yan C (2021) Antimicrobial properties of chitosan and chitosan derivatives in the treatment of enteric infections. Molecules 26(23):7136. https://doi.org/10.3390/molecules26237136
109. Yang B, Liu J, Zhang C, Zhang L (2025) Application of nano chitosan synthesized from Exopalaemon modestus shell to control the infection of cherry tomato leaves by *Alternaria alternata*. Int J Biol Macromol 308:142456. https://doi.org/10.1016/j.ijbiomac.2025.142456
110. Yang Y, Aghbashlo M, Gupta VK, Amiri H, Pan J, Tabatabaei M, Rajaei A (2023) Chitosan nanocarriers containing essential oils as a green strategy to improve the functional properties of chitosan: a review. Int J Biol Macromol 236:123954. https://doi.org/10.1016/j.ijbiomac.2023.123954
111. Youssef K, Roberto SR (2014) Applications of salt solutions before and after harvest affect postharvest gray mold of '*Italia' grapes*. Postharvest Biol Technol 87:95–102. https://doi.org/10.1016/j.postharvbio.2013.08.011

112. Zayed MM, Elkafafi SH, Zedan AM, Dawoud SF (2017) Effect of nano chitosan on growth, physiological and biochemical parameters of *Phaseolus vulgaris* under salt stress. J Plant Prod 8(5):577–585. https://doi.org/10.21608/jpp.2017.40468
113. Zhou Q, Xia Z, Zhang Y, Sun Z, Zeng W, Zhang N, Yuan C, Gong C, Zhou Y, Xue W (2024) Design of a delivery vehicle chitosan-based self-assembling: controlled release, high hydrophobicity, and safe treatment of plant fungal diseases. J Nanobiotechnol 22(1):121
114. Zhu K, Shan W, Wang L, Cai C (2008) Effects of chitosan coating on postharvest quality of mango (*Mangifera indica* L. cv. *tainong*) fruits. J Food Process Preservation 32(5):770–784. https://doi.org/10.1111/j.1745-4549.2008.00213.x

Chitosan Nanoparticles in the Control of Insect Pests and Plant Diseases

Rachid Azenzem and **Jalal Kassout**

Abstract Agricultural systems face growing difficulties in satisfying rising food demands due to mounting pressures from various abiotic and biotic stresses. In addition to the escalating impacts of climate change, persistent pests and diseases pose significant threats to crop yield and quality. While conventional agrochemicals provide immediate benefits, their lasting effects on human health and the environment have spurred growing interest in safer and more sustainable alternatives. Chitosan nanoparticles (CNPs), derived from natural biopolymers, have emerged as a promising nanotechnological solution due to their biodegradability, biocompatibility, and multifunctional antimicrobial properties. This chapter explores the role of CNPs in suppressing a broad spectrum of insect pests and plant pathogens. It highlights their ability to disrupt pest and pathogen development, serve as efficient carriers for the targeted delivery of bioactive compounds, and enhance plant immunity. Building on recent research and practical uses, this chapter underscores the vital role of chitosan nanoparticles in integrated pest management (IPM) strategies and their contribution to fostering more resilient and sustainable agricultural systems.

Keywords Pest management · Pathogen inhibition · Nanomaterials · Crop protection · Sustainable agriculture · Plant defense

R. Azenzem (✉)
Research Unit On Nuclear Techniques, Environment and Quality, Regional Center of Agricultural Research of Tangier, National Institute of Agricultural Research, Rabat, Morocco
e-mail: r.azenzem@gmail.com; rachid.azenzem@inra.ma

J. Kassout
Research Unit On Plant Breeding and Quality, Regional Center of Agricultural Research of Marrakech, National Institute of Agricultural Research, Rabat, Marrakech, Morocco

A. A. H. Abdel Latef (ed.), *Chitosan Nanoparticles for Plant Resilience Under Abiotic and Biotic Stressors*, Smart Nanomaterials Technology,
https://doi.org/10.1007/978-981-95-6422-4_11

1 Introduction

Demographic forecasts indicate that the global human population will continue its upward trajectory, reaching approximately 9.7 billion by 2050 and exceeding 11.7 billion by 2100. This ongoing increase is driving a surge in food demand and placing extraordinary pressure on agricultural systems and natural resources. Ensuring food security for a growing global population will require a substantial increase in agricultural productivity. Projections suggest that crop yields may need to double to meet the anticipated demands of the coming decades.

However, achieving this goal is becoming increasingly challenging due to a variety of complex factors. Climate change is reshaping growing conditions, intensifying extreme weather events, and disrupting traditional farming cycles. At the same time, the overuse and degradation of essential resources such as soil, water, and biodiversity are further weakening the foundations of global food production.

Compounding these difficulties, pests and plant diseases represent a major obstacle to agricultural advancement. They account for an estimated 40 percent of global crop losses each year. These losses not only diminish food availability but also jeopardize the livelihoods of millions of farmers and contribute to rising food prices. The impact of these biological threats is becoming more severe due to several contributing factors. Climate change fosters conditions that enable pests and pathogens to thrive and spread, while the expansion of international trade accelerates the introduction and dissemination of invasive species across new regions [8].

Since the advent of modern agricultural practices, chemical pesticides have played a central role in protecting crops from damage caused by weeds, insect pests, and pathogens. Their widespread use has significantly contributed to boosting agricultural productivity, addressing regional food shortages, and supporting the livelihoods of millions of farmers around the world. These benefits have been crucial in meeting the growing global demand for food. However, the extensive and often indiscriminate application of chemical pesticides has raised serious concerns. They pose risks to human health, non-target organisms, and the environment, while also contributing to the development of resistant pest populations [3]. In light of these challenges, there is an urgent need to develop novel, effective, and sustainable alternatives to conventional pesticides. In this context, natural biopolymers such as chitosan have attracted significant attention. Chitosan, derived from chitin found in the exoskeletons of crustaceans and insects, is known for its biocompatibility, biodegradability, and broad-spectrum antimicrobial activity. When engineered at the nanoscale, chitosan nanoparticles (CNPs) offer enhanced efficacy due to their increased surface area, improved solubility, and controlled release properties. These features make CNPs a promising tool in modern plant protection, capable of reducing reliance on synthetic chemicals while supporting sustainable agricultural productivity.

This chapter examines the use of chitosan nanoparticles for managing insect pests and plant pathogens. It focuses on their ability to interfere with pest and pathogen development, facilitate the precise delivery of active agents, and boost plant defense mechanisms as part of sustainable crop protection strategies.

2 Chitosan Nanoparticles in Plant Insect Pests Control

Agricultural production faces continuous challenges from insect pests that threaten crops at every stage, from germination to post-harvest storage. These pests feed on multiple parts of the plant, including roots, stems, leaves, flowers, and fruits. Beyond their direct feeding damage, many insect species also act as carriers of harmful plant pathogens. Prominent insect pests include sap-sucking species such as aphids and whiteflies, leaf-chewing insects like caterpillars and beetles, and borers that attack stems, fruits, and roots. Insect pests currently account for about 38% of global agricultural losses, causing economic damages exceeding 470 billion USD per year [28, 38].

The control of insect pests in agriculture largely depends on the use of chemical insecticides. Nonetheless, escalating issues such as the development of resistance among species, environmental pollution, and harmful effects on beneficial organisms have shifted the focus toward integrated pest management (IPM).

Nanoparticles have recently attracted significant attention across various sectors, particularly in agriculture. Among them, chitosan nanoparticles stand out for their effectiveness in pest management. These nano-sized carriers are being widely explored for their ability to encapsulate and deliver active compounds in a controlled manner and for enhancing the stability of sensitive biological substances such as proteins, peptides, and genetic material.

With sizes ranging from 1 to 100 nm, their surface area, shape, and volume-to-area ratio make them ideal for agricultural applications. Their nanoscale size ensures better adhesion to plant surfaces, improves penetration, enhances the solubility of poorly soluble compounds, and increases stability and slow-release properties. These features contribute to greater insecticidal efficiency and longer-lasting effects compared to bulk formulations.

Recent studies have demonstrated that chitosan nanoparticles loaded with essential oils (EOs) show enhanced toxicity and antifeedant effects against insect pests (Fig. 1). For instance, Rajkumar et al.[41] assessed the efficacy of *Piper nigrum* essential oil (PNO) encapsulated in chitosan nanoparticles (C/PNO NPs) against *Sitophilus oryzae* and *Tribolium castaneum*. Their results showed that C/PNO NPs enhanced fumigant toxicity and significantly disrupted acetylcholine levels in both insect species. Similarly, Ziaee et al.[58] evaluated the insecticidal activity of *Carum copticum* essential oil (EO) and its chitosan-based EO-loaded nanoparticles (OLNs) against *Rhyzopertha dominica* and *Tribolium confusum*, as well as their effects on insect progeny production. They found that OLNs showed higher mortality rates (74% and 57%) compared to the EO alone (62% and 44%) after 7 days of exposure. Additionally, Choudhary et al.[12] evaluated the synthesis and characterization of *Mentha spicata* essential oil-loaded chitosan nanoparticles (MsEO/CNPs) for potential use in pest control and antioxidant applications. The authors revealed that the nanoparticles exhibited strong insecticidal activity against *Callosobruchus maculatus* and *Sitophilus granaries*, significant antioxidant potential, and low cytotoxicity.

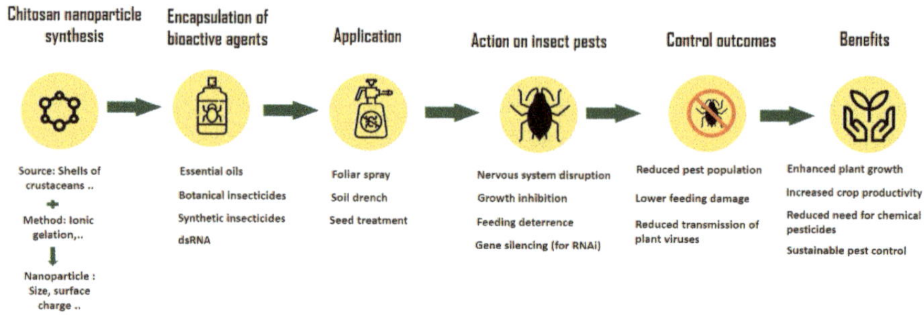

Fig. 1 Role of chitosan nanoparticles (CNPs) as carriers for insect pest control in plants

In addition to essential oils, chitosan nanoparticles have been used to encapsulate botanical insecticides, fungal biopesticides, and even synthetic chemicals (Fig. 1). These nanoformulations have shown increased larvicidal, ovicidal, and antifeedant effects in a variety of insect species. In this context, Paulraj et al.[39] investigated a chitosan-based nanopesticide formulated with a botanical pesticide derived from neem oil and assessed its antifeedant and larvicidal effects against *Helicoverpa armigera*. The results revealed that the nanoparticle formulation exhibited potent bioactivity, achieving 88.5% antifeedant and 90.2% larvicidal activity. Moreover, Chandra et al.[10] assessed the lethal concentration (LC50) and lethal time (LT50) of chitosan nanoparticle-coated fungal metabolites, uncoated fungal metabolites, and fungal spores of *Nomuraea rileyi* against *Spodoptera litura*. The authors demonstrated that the nanoparticle-coated formulation exhibited significantly higher pesticidal activity across all larval instars, with lower LC50 values and shorter LT50 compared to uncoated metabolites and spores. Furthermore, Sharma et al.[51] explored the synthesis and characterization of chitosan nanoparticles loaded with spinosad and permethrin. The authors demonstrated that these nanoformulations exhibited enhanced toxicity and longer-lasting effects against *Drosophila melanogaster* compared to free agrochemicals.

Chitosan-based nanocarriers have also been used in RNA interference (RNAi) applications, where they deliver small RNA molecules (siRNA, dsRNA) targeting genes essential for insect growth and development. These formulations have shown promising results against several insect pests by impairing larval growth, reducing survival rates, and enhancing gene silencing efficiency. In this context, Gurusamy et al.[23] explored the use of chitosan nanoparticles for conjugating double-stranded RNA (dsRNA) to enhance RNAi efficacy in the fall armyworm (*Spodoptera frugiperda*). The study revealed that chitosan-conjugated dsRNA was resistant to enzymatic degradation, facilitated endosomal escape, and effectively induced gene knockdown, leading to growth retardation and increased larval mortality. Similarly, Kolge et al.[32] employed ionotropically synthesized cationic CNPs for dsRNA delivery targeting *Helicoverpa armigera*. The CNPs efficiently encapsulated and protected dsRNA, enabling targeted silencing of lipase and chitinase genes. This intervention significantly suppressed pupation and moth emergence, ultimately resulting in 100% larval mortality. Importantly, the treatment showed no off-target effects on non-target insect species. Furthermore, Liu et al.[34] investigated chitosan-based dsRNA nanoparticles targeting the LmCht10 gene in the migratory locust (*Locusta migratoria*). Their findings showed that the nanoparticles enhanced dsRNA stability within the locust gut, improved uptake by epidermal cells, and significantly increased gene silencing efficiency and mortality compared to unformulated (naked) dsRNA.

3 Chitosan Nanoparticles in Plant Disease Control

Plant diseases caused by a wide range of pathogens, including fungi, bacteria, viruses, and oomycetes, remain a significant threat to global crop production. They are responsible for an estimated 30% reduction in annual yields and result in economic losses amounting to billions of dollars worldwide [49]. Beyond yield reduction, these pathogens compromise crop quality, disrupt food safety, and increase the dependence of farmers on chemical pesticides. Fungal pathogens such as *Fusarium* spp., *Botrytis cinerea, Alternaria* spp., and *Phytophthora* spp. are among the most damaging agents, causing diseases like wilts, blights, rots, and molds across many economically important crops. Some of these fungi, including *Fusarium graminearum* and *Aspergillus flavus,* also produce harmful mycotoxins that contaminate food and animal feed, posing serious health risks to humans and livestock. In addition to fungi, bacterial pathogens such as *Xylella fastidiosa, Xanthomonas campestris, Ralstonia solanacearum,* and *Pseudomonas syringae* significantly contribute to crop losses worldwide. Plant viruses, including *Tobacco mosaic virus* (TMV), *Tomato spotted wilt virus* (TSWV), and *Potato virus* Y (PVY), further exacerbate these losses by affecting both crop yield and quality.

While chemical fungicides and bactericides remain the cornerstone of modern plant disease management, their overuse has led to serious environmental and health concerns, including the development of resistant pathogen strains, residue accumulation, and harm to beneficial microorganisms [15]. As a result, there is growing interest in alternatives that offer effective disease control with minimal ecological impact. Among these, chitosan nanoparticles have emerged as a versatile and sustainable option.

A growing body of research has demonstrated the effectiveness of chitosan nanoparticles in controlling plant pathogens through direct antimicrobial action and their role as efficient carriers for bioactive compounds (Fig. 2). Nanoparticles possess the ability to cross cellular barriers and interact with biological surfaces more efficiently than the larger molecules from which they originate. This enhanced mobility and surface activity help explain why chitosan nanoparticles exhibit superior antimicrobial properties compared to bulk chitosan.

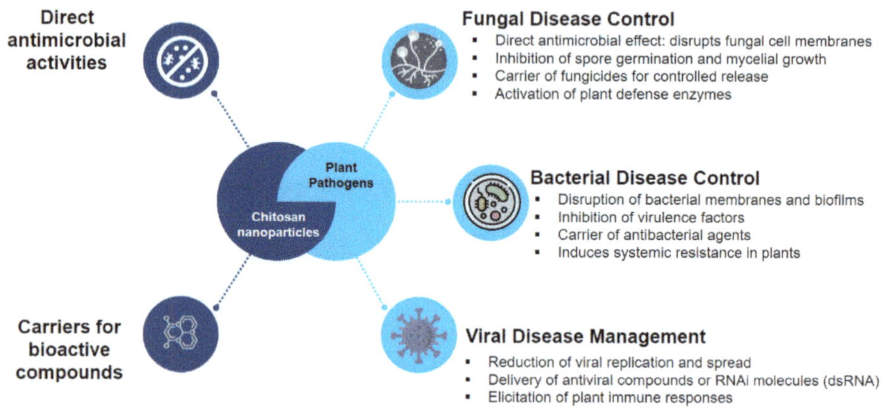

Fig. 2 Use of chitosan nanoparticles in plant disease management

3.1 Direct Antimicrobial Activities

Numerous studies have demonstrated the direct antifungal effects of chitosan nanoparticles against a wide range of fungal pathogens across various crops. Kheiri et al. [30] demonstrated that chitosan nanoparticles possess strong antifungal activity against *Fusarium graminearum*, the causal agent of Fusarium head blight in wheat. In vitro, these nanoparticles achieved up to 77.5% growth inhibition at a concentration of 5000 ppm. Furthermore, greenhouse trials showed that plants treated with chitosan nanoparticles experienced significantly reduced disease severity. The enhanced antifungal effect of the nanoparticles compared to chitosan solutions was attributed to their small, compact size and high positive surface charge, which improves interaction with the negatively charged fungal cell surface. The authors also observed that chitosan nanoparticles caused severe structural damage to *F. graminearum* cells, including membrane rupture, vacuolization, leakage of cytoplasmic contents, and disruption of organelles, ultimately leading to cell wall swelling, tearing, and loss of cellular integrity. Similarly, Suryadi et al.[55] revealed that enzymatically hydrolyzed chitosan nanoparticles, with an average size of 126.2 nm and a zeta potential of 27.8 mV, inhibited *Colletotrichum gloeosporioides* growth by 85.7% in vitro and reduced spore germination by 61.2%. In vivo, the nanoparticles significantly reduced anthracnose disease severity on chili and papaya fruits when applied preventively. Moreover, Abdel-Aliem et al.[1] indicated that chitosan nanoparticles, with a nanorod shape and an average size of around 180 nm, showed strong antifungal activity in a dose-dependent manner against several phytopathogenic fungi (Table 1). At a concentration of 800 ppm, CNPs effectively inhibited zearalenone (ZEN) production by *F. graminearum*.

Further studies highlighting the antifungal potential of chitosan nanoparticles (CNPs) are presented in Table 1.

Table 1 Recent research advances in the antifungal applications of CNPs against plant pathogens

CNPs concentration	Target fungal pathogens	Growth Inhibition (%)	Host plants	References
0.1%	*Pyricularia grisea*	65	Fingermillet	[46]
0.5%	*Colletotrichum gloeosporioides*	37.8	Tomato	[37]
	Phytophthora capsica	50.7		
	Sclerotinia sclerotiorum	39.5		
	Fusarium oxysporum	50.3		
	Gibberella fujikuori	56.3		
100 μg	*Pyricularia grisea*	92	NCD	[47]
	Alternaria solani	87		
	Fusarium oxysporum	72		
0.06–0.1%	*Alternaria alternata*	80.1–82.2	NCD	[43]
	Rhizoctonia solani	32.2–34.4		
0.1%	*Macrophomina phaseolina*	84		
0.03–0.04%	*Alternaria solani*	10–70	NCD	[40]
800 ppm	*Alternaria tenuis*	67.67	Ground nut	[1]
	Aspergillus niger	62.75		
	Aspergillus terreus	74.67		
	Fusarium graminearum	60.37		
	Fusarium oxysporum	66.60		
	Sclerotium rolfsii	37.41		

NCD: not clearly defined

In addition to their antifungal properties, chitosan nanoparticles have also shown promising antibacterial activity. However, only a limited number of studies have investigated the biological activity of chitosan nanoparticles against plant pathogenic bacteria. Oh et al. [37] reported that CNPs exhibited stronger antibacterial activity against phytopathogenic bacteria, including three strains of *Erwinia carotovora* subsp. *carotovora* and one strain of *Xanthomonas campestris* pv. *vesicatoria*. Khairy e al. [29] found that chitosan nanoparticles significantly reduced the incidence and severity of bacterial wilt caused by *Ralstonia solanacearum* in potato and tomato plants, with the most effective concentrations being 100–200 mg/mL. In vitro results showed increased inhibition zones with higher nanoparticle concentrations. In vivo, preventive treatments were more effective than curative applications, with the use of 200 mg/mL sprays in the preventive strategy showing the greatest impact, leading to a significant reduction in disease incidence and severity by over 77% in treated plants. Transmission electron microscopy and RAPD-PCR analysis revealed structural damage and genetic alterations in treated bacterial cells.

Similarly, the potential effectiveness of chitosan nanoparticles against plant viral diseases remains largely underexplored. El Gamal et al. [18] reported that CNPs effectively inhibited Bean yellow mosaic virus (BYMV) in faba bean plants, with complete suppression of viral infectivity and accumulation at 300 and 400 mg/L. Compared to bulk chitosan, which showed only moderate antiviral activity, CNPs significantly disrupted viral particle integrity and promoted strong plant defense responses. Elsahhar et al. [20] noted that chitosan nanoparticles effectively eliminated Potato virus Y (PVY) from infected potato explants under in vitro conditions, with the 250 mg/L concentration yielding 100% virus-free plantlets and full shoot regeneration. Lower concentrations also showed antiviral effects, but with reduced efficiency in both virus elimination and regeneration rates. Moreover, El-Ganainy et al. [19] demonstrated that chitosan nanoparticles effectively inhibited Alfalfa mosaic virus (AMV) in pepper plants, with the highest inhibition rates of 90% observed at 400 ppm. Recently, Hanady et al.[24] showed that chitosan nanoparticles effectively suppressed Banana bunchy top virus (BBTV) in infected banana explants grown in vitro, with treated plants showing no symptoms and testing negative by enzyme-linked immunosorbent assay (ELISA).

3.2 Chitosan Nanoparticles as Carriers for Bioactive Compounds

Beyond their direct antimicrobial effects, chitosan nanoparticles have also gained attention as effective carriers for bioactive compounds, enhancing the delivery and efficacy of various antimicrobial agents against plant pathogens. In this line, Hoang et al.[26] found that chitosan nanoparticle formulations loaded with salicylic acid (N3) or silver (N6) effectively reduced cassava leaf spot disease. Among six tested formulations, N3 and N6 showed the highest efficacy, reducing disease severity by up to 73.6% and performing as well as or better than commercial fungicides. Mazzotta et al. [35] developed an eco-friendly fungicide by loading olive leaf extract (OLE) into CNPsto control *Verticillium* wilt in tomato plants. The OLE-loaded CNPs showed strong synergistic antifungal activity, with a lower MIC (0.14 mg/mL) than either free OLE or blank CNPs. In greenhouse trials, the formulation significantly reduced disease symptoms and promoted plant growth. Moreover,

Sathiyabama et al.[48] demonstrated that thiamine-loaded chitosan nanoparticles (TCNPs) strongly suppressed *Rhizoctonia bataticola*, the causal agent of dry root rot in chickpea. In vitro assays showed that TCNPs inhibited nearly 80% of fungal growth, significantly outperforming chitosan alone, which achieved 50% inhibition. Scanning electron microscopy analysis revealed that TCNPs caused severe deformation and rupture of fungal hyphae, indicating physical damage to the cell wall and membrane, ultimately leading to irreversible cell death. Under greenhouse conditions, TCNPs-treated plants showed delayed symptom onset, a 5% disease severity index, and only 12% disease incidence, compared to 100% in untreated controls.

The use of chitosan nanoparticles as carriers for bioactive compounds has also proven highly effective in suppressing bacterial phytopathogens. For instance, Sreelatha et al. [54] developed TCNPs that effectively inhibited the growth of *Xanthomonas campestris* pv. *campestris* (Xcc), a pathogen causing black rot in cruciferous crops. TCNPs disrupted bacterial biofilms, reduced key virulence factors, and caused membrane damage, leading to cell content leakage. Chemical analysis showed changes in bacterial volatile compounds after treatment. In a subsequent study, Sreelatha et al. [53] demonstrated that metabolomic and spectroscopic analyses further showed that TCNP treatment caused oxidative stress and significant metabolic disruptions, including increased levels of nucleic acids, fatty acids, lipids, and antioxidant metabolites, alongside reduced amino acids and carbohydrates. On the other hand, Tarakanov et al.[56] reported that copper-loaded chitosan nanoparticles (Cu^{2+}CNPs) exhibit strong antibacterial activity against *Pseudomonas savastanoi* pv. *glycinea* and *Curtobacterium flaccumfaciens* pv. *flaccumfaciens*, two major soybean pathogens. Cu^{2+}CNPs achieved biological control efficiencies of 71% and 51% against the respective pathogens without phytotoxic effects.

Only a limited number of studies have investigated the use of chitosan nanoparticles as carriers for antiviral compounds in plant disease management. For instance, El-Ganainy et al. [19] demonstrated that chitosan-silver nanocomposites (C-Ag NC) were more effective than chitosan nanoparticles (C-NPs) alone in suppressing Alfalfa mosaic virus (AMV) infection in pepper plants. Treatment with 200 ppm of C–Ag NC resulted in the highest viral inhibition, whereas a higher concentration of 400 ppm of CNPs was needed to achieve a similar effect, underscoring the superior antiviral efficacy of the nanocomposite. Recently, Wang et al.[57] revealed that chitosan nanoparticles loaded with the protein fTDP (CNPs-fTDP) significantly improved the stability and field efficacy of this antiviral agent against tobacco mosaic virus (TMV), achieving a control efficacy of 50.7% compared to 39.2% with fTDP alone.

4 Role of Chitosan Nanoparticles in Enhancing Plant Defense and Stress Resilience

Beyond their proven efficacy against insect pests and plant pathogens, CNPs also play a pivotal role in activating the plant's defense systems. Rather than acting solely as antimicrobial or insecticidal agents, CNPs can trigger a range of plant immune responses, including the activation of defense-related genes, the production of reactive oxygen species (ROS), and the strengthening of cell walls.

4.1 Chitosan Nanoparticles as Elicitors of Biotic Stress Defense Mechanisms

CNPs have demonstrated potent elicitor activity, activating innate immune responses in plants under biotic stress, including bacterial, fungal, and viral pathogens [11, 50]. Their nanoscale dimensions and cationic nature enhance interaction with plant surfaces and facilitate internalization, triggering systemic defense mechanisms that mimic pathogen attack without the environmental risks associated with chemical pesticides [14, 42]. For instance, in tomato, CNPs provided 62% protection against bacterial wilt (*Ralstonia solanacearum*) when applied as seed treatments, with marked increases in lignin, callose, H_2O_2, and superoxide accumulation, alongside strong upregulation of defense-related enzymes [phenylalanine ammonia-lyase (PAL), peroxidase (POX), polyphenol oxidase (PPO), catalase (CAT), and β-1,3-glucanase] and corresponding gene expression, demonstrating biochemical and molecular activation of defense pathways [36]. Similarly, in tobacco, harpin-loaded chitosan nanoparticles (H-SNPs) induced hypersensitive responses and expression of PR genes more effectively than harpin or chitosan alone, thanks to enhanced bioavailability and controlled release kinetics [33]. This nanocarrier-based delivery provides a promising strategy for deploying protein elicitors efficiently [13]. Against viral infections, chitosan/dextran nanoparticles (CDNPs) reduced Alfalfa mosaic virus accumulation in *Nicotiana glutinosa* by 100% when applied in an inactivation strategy, and significantly improved defense responses under both protective and curative conditions [2]. These responses included elevated carbohydrate and phenolic content, increased PAL and POX activity, and upregulation of PR1 gene expression [2]. The role of nitric oxide (NO) in CNP-induced defense has also been validated. For instance, in *Capsicum annuum*, the effectiveness of CNPs against *Alternaria alternata* was compromised under NO-depleted conditions, indicating that NO is a crucial signaling molecule in CNP-mediated immunity [45]. Meta-analysis of 58 studies reinforced these findings by quantifying a 103.2% increase in defense gene expression, a 77.9% increase in defense enzyme activity, and significant enhancement of antioxidant systems and phytohormones in response to chitosan treatment under biotic and oxidative stress [27]. In wheat, CNPs combined with salicylic acid reduced *Puccinia triticina* urediniospore germination, improved antioxidative enzyme activities (POX, CAT), induced ROS accumulation, and upregulated PR1–PR5 and PR10 genes after pathogen challenge [21]. Moreover, the dual benefits of chitosan as a seed-coating agent and immune stimulant were noted in chickpea, where CNPs helped control *Fusarium oxysporum* by enhancing lignification, PR protein synthesis, and antioxidant enzyme activity, providing a sustainable alternative to chemical treatments [52]. Together, these findings establish chitosan nanoparticles as efficient elicitors of systemic immunity and valuable tools for integrated disease management in plants (Fig. 3).

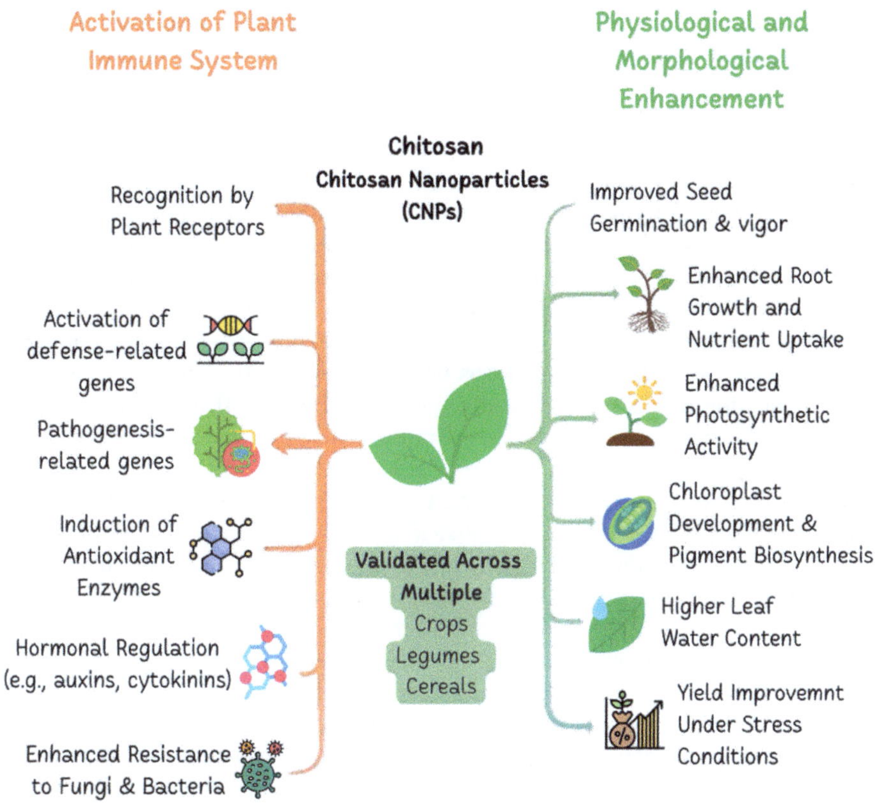

Fig. 3 Summary of chitosan nanoparticle (CNPs) roles in the induction of plant defense responses and physiological and morphological under stress

4.2 Physiological and Morphological Enhancement of Crops by Chitosan Nanoparticles

Beyond their defensive role, CNPs contribute significantly to enhancing plant physiological and morphological traits such as photosynthetic efficiency, water retention, root development, and biomass accumulation [6]. These improvements strengthen overall plant vigor, which in turn enhances resistance to pests and diseases by supporting more robust defense responses and structural resilience. The effectiveness of CNPs is largely attributed to their nanoscale size, surface functionality, and ability to regulate redox balance and osmotic homeostasis at the cellular level [17]. In wheat, foliar application of zinc–chitosan–salicylic acid nanoparticles (ZC NPs) at 100 mgL^{-1} enhanced relative water content (RWC), triggered antioxidant enzymes such as superoxide dismutase (SOD), ascorbate peroxidase (APX), glutathione reductase (GR), and guaiacol peroxidase (GPOX), and mitigated yield loss by promoting proline and flavonoid accumulation [16]. Similarly, in *Catharanthus roseus*, CNPs increased RWC, chlorophyll levels, antioxidant enzyme activity, and alkaloid biosynthesis. The expression of key genes involved in secondary metabolism, such as strictosidine synthase (STR), deacetylvindoline 4-O-acetyltransferase (DAT), peroxidase 1 (PRX1), and geraniol synthase (GS), was upregulated by up to 5.6-fold [5]. CNPs were also effective in lupine (*Lupinus termis*), where 50 mgL^{-1} pre-sowing treatments improved plant growth, photosynthetic pigments, and biochemical components (IAA, soluble sugars, proline, amino acids) under 60% irrigation, with superior effects observed for nanoparticle versus bulk chitosan [9]. In *Salvia abrotanoides*, foliar application of CNPs significantly improved chlorophyll (63%), carotenoids (68%), protein content (45%), and osmoprotectants (proline 49%, sugars 58%). Additionally, CNPs have been shown to optimize stomatal regulation and enhance antioxidant enzyme activity, contributing to reduced water loss and mitigation of oxidative stress [7]. In *Solanum lycopersicum*, foliar application of CNPs resulted in a 38% increase in shoot dry weight, a 45% improvement in photosynthetic efficiency, and a reduction of over 20% in oxidative damage markers such as malondialdehyde and hydrogen peroxide [22]. *Phaseolus vulgaris* responded positively to CNPs synthesized using methacrylic acid, showing restored chlorophyll and carbohydrate content, increased proline levels and antioxidant enzyme activity, along with reduced membrane leakage and lipid peroxidation [4]. Similarly, *Catharanthus roseus* exhibited enhanced growth, increased antioxidant activity, and elevated expression of genes such as MAPK3, GS, and ORCA3, which supported improved alkaloid accumulation [25]. A review of 188 studies confirmed that CNPs enhance growth, chlorophyll content, and stress tolerance in cereals like wheat, maize, and rice, while calling for expanded studies on millets and sorghum [31]. Moreover, recent overviews [6, 44] emphasized the use of chitosan-based NPs as bio-stimulants for improving plant physiological performance, photosynthetic capacity, and nutrient use efficiency under abiotic constraints. Collectively, these findings highlight the versatility of CNPs in enhancing crop productivity and resilience, emphasizing their important role in promoting plant growth and defense against biotic stressors (Fig. 3).

5 Conclusions and Future Directions

Chitosan nanoparticles represent a significant advancement in the pursuit of sustainable agricultural practices. Their unique physicochemical properties make them powerful tools in the management of both insect pests and plant diseases. As demonstrated throughout this chapter, CNPs not only directly inhibit pest and pathogen development but also enhance the effectiveness of bioactive compounds and stimulate plant defense mechanisms. Their integration into IPM frameworks offers a promising approach to alleviate the negative impacts associated with the excessive and indiscriminate use of conventional chemical pesticides.

Future research should prioritize the optimization of chitosan nanoparticle formulations to improve their physicochemical stability, controlled release properties, and efficacy under diverse environmental and agronomic conditions. Comprehensive field evaluations are necessary to validate laboratory findings and assess performance across various crops, pest complexes, and climatic regions. Furthermore, the development of regulatory guidelines specific to nanomaterials is essential to ensure safe application, environmental compatibility, and standardized quality control. Interdisciplinary approaches integrating nanotechnology, plant pathology, entomology, and agronomy will be critical to advancing formulation development and application methods. Addressing these aspects will be fundamental to enabling the effective integration of chitosan nanoparticles into sustainable pest and disease management strategies in modern agriculture.

References

1. Abdel-Aliem HA, Gibriel AY, Rasmy NM, Sahab AF, El-Nekeety AA, Abdel-Wahhab MA (2019) Antifungal efficacy of chitosan nanoparticles against phytopathogenic fungi and inhibition of zearalenone production by Fusarium graminearum. Comunicata Scientiae 10(3):338–345
2. Abdelkhalek A, Qari SH, Abu-Saied MAA-R, Khalil AM, Younes HA, Nehela Y, Behiry SI (2021) Chitosan nanoparticles inactivate alfalfa mosaic virus replication and boost innate immunity in Nicotiana glutinosa plants. Plants 10(12):2701
3. Acheuk F, Basiouni S, Shehata AA, Dick K, Hajri H, Lasram S et al (2022) Status and prospects of botanical biopesticides in Europe and Mediterranean countries. Biomolecules 12(2):311
4. Alenazi MM, El-Ebidy AM, El-Shehaby OA, Seleiman MF, Aldhuwaib KJ, Abdel-Aziz HM (2024) Chitosan and chitosan nanoparticles differentially alleviate salinity stress in Phaseolus vulgaris L. plants. Plants 13(3):398
5. Ali EF, El-Shehawi AM, Ibrahim OHM, Abdul-Hafeez EY, Moussa MM, Hassan FAS (2021) A vital role of chitosan nanoparticles in improvisation the drought stress tolerance in Catharanthus roseus (L.) through biochemical and gene expression modulation. Plant Physiol Biochem 161:166–175
6. Arif Y, Siddiqui H, Hayat S (2022) Role of chitosan nanoparticles in regulation of plant physiology under abiotic stress. In: Sustainable agriculture reviews 53: nanoparticles: a new tool to enhance stress tolerance. Springer, pp 399–413

7. Attaran Dowom S, Karimian Z, Mostafaei Dehnavi M, Samiei L (2022) Chitosan nanoparticles improve physiological and biochemical responses of Salvia abrotanoides (Kar.) under drought stress. BMC Plant Biology 22(1):364
8. Azenzem R, Kassout J (2023) Climate change and insect pests: economic damages and adaptive strategies in mediterranean olive groves. In: Climate Change and the economic importance and damages of insects, pp 14–47. IGI Global
9. Bakhoum G, Sadak M, Tawfic M (2022) Chitosan and chitosan nanoparticle effect on growth, productivity and some biochemical aspects of Lupinustermis L plant under drought conditions. Egypt J Chem 65(5):537–549
10. Chandra JH, Raj LA, Namasivayam SKR, Bharani RA (2013) Improved pesticidal activity of fungal metabolite from Nomureae rileyi with chitosan nanoparticles. In: International conference on advanced nanomaterials & emerging engineering technologies, pp 387–390. IEEE.
11. Chandra S, Chakraborty N, Dasgupta A, Sarkar J, Panda K, Acharya K (2015) Chitosan nanoparticles: a positive modulator of innate immune responses in plants. Sci Rep 5(1):15195
12. Choudhary A, Salar RK, Thakur R (2024) Synthesis, characterization and insecticidal activity of Mentha spicata essential oil loaded polymeric nanoparticles. Biocatal Agric Biotechnol 55:102989
13. Choudhary RC, Kumari S, Kumaraswamy RV, Sharma G, Kumar A, Budhwar S, Pal A, Raliya R, Biswas P, Saharan V (2019) Chitosan nanomaterials for smart delivery of bioactive compounds in agriculture. In: Nanoscale engineering in agricultural management. CRC Press, pp 124–139
14. Chun S-C, Chandrasekaran M (2019) Chitosan and chitosan nanoparticles induced expression of pathogenesis-related proteins genes enhances biotic stress tolerance in tomato. Int J Biol Macromol 125:948–954
15. Daraban GM, Hlihor RM, Suteu D (2023) Pesticides vs. biopesticides: From pest management to toxicity and impacts on the environment and human health. Toxics 11(12): 983.
16. Das D, Bisht K, Chauhan A, Gautam S, Jaiswal JP, Salvi P, Lohani P (2023) Morpho-physiological and Biochemical responses in wheat foliar sprayed with zinc-chitosan-salicylic acid nanoparticles during drought stress. Plant Nano Biology 4:100034
17. Divya K, Jisha MS (2018) Chitosan nanoparticles preparation and applications. Environ Chem Lett 16:101–112
18. El Gamal AY, Atia MM, El Sayed T, Abou-Zaid MI, Tohamy MR (2022) Antiviral activity of chitosan nanoparticles for controlling plant-infecting viruses. South African J Sci 118(1–2)
19. El-Ganainy SM, Soliman AM, Ismail AM, Sattar MN, Farroh KY, Shafie RM (2023) Antiviral activity of chitosan nanoparticles and chitosan silver nanocomposites against Alfalfa Mosaic Virus. Polymers 15(13):2961
20. Elsahhar S, Shahba M, Elsayed T, Mikhail M, Galal A (2022) Effect of chitosan nanoparticles (CS-NPs) on in vitro regeneration response and production of potato virus Y (PVY)-free plants of potato. Agronomy 12(11):2901
21. Elsharkawy MM, Omara RI, Mostafa YS, Alamri SA, Hashem M, Alrumman SA, Ahmad AA (2022) Mechanism of wheat leaf rust control using chitosan nanoparticles and salicylic acid. J Fungi 8(3):304
22. Faizan M, Rajput VD, Al-Khuraif AA, Arshad M, Minkina T, Sushkova S, Yu F (2021) Effect of foliar fertigation of chitosan nanoparticles on cadmium accumulation and toxicity in Solanum lycopersicum. Biology 10(7):666
23. Gurusamy D, Mogilicherla K, Palli SR (2020) Chitosan nanoparticles help double-stranded RNA escape from endosomes and improve RNA interference in the fall armyworm, Spodoptera frugiperda. Arch Insect Biochem Physiol 104(4):e21677
24. Hanady HA, Elshorbgy AE, El-kammar HF, Mohamed HA, Hassan M (2025) In-Vitro antiviral activity of chitosan nanoparticles against Banana bunchy top virus. Fayoum J Agric Res Dev 39(1):96–110
25. Hassan FAS, Ali E, Gaber A, Fetouh MI, Mazrou R (2021) Chitosan nanoparticles effectively combat salinity stress by enhancing antioxidant activity and alkaloid biosynthesis in Catharanthus roseus (L.) G. Don. Plant Physiol Biochem 162:291–300

26. Hoang NH, Le Thanh T, Thepbandit W, Treekoon J, Saengchan C, Sangpueak R et al (2022) Efficacy of chitosan nanoparticle loaded-salicylic acid and-silver on management of cassava leaf spot disease. Polymers 14(4):660
27. Ji H, Wang J, Chen F, Fan N, Wang X, Xiao Z, Wang Z (2022) Meta-analysis of chitosan-mediated effects on plant defense against oxidative stress. Sci Total Environ 851:158212
28. Junaid MD (2024) Gokce AF (2024) Global agricultural losses and their causes. Bulletin Biol Allied Sci Res 1:66–66
29. Khairy AM, Tohamy MR, Zayed MA, Mahmoud SF, El-Tahan AM, El-Saadony MT, Mesiha PK (2022) Eco-friendly application of nano-chitosan for controlling potato and tomato bacterial wilt. Saudi J Biol Sci 29(4):2199–2209
30. Kheiri A, Jorf SM, Malihipour A, Saremi H, Nikkhah M (2017) Synthesis and characterization of chitosan nanoparticles and their effect on Fusarium head blight and oxidative activity in wheat. Int J Biol Macromol 102:526–538
31. Kocięcka J, Liberacki D (2021) The potential of using chitosan on cereal crops in the face of climate change. Plants 10(6):1160
32. Kolge H, Kadam K, Ghormade V (2023) Chitosan nanocarriers mediated dsRNA delivery in gene silencing for Helicoverpa armigera biocontrol. Pestic Biochem Physiol 189:105292
33. Kongala SI, Mamidala P (2023) Harpin-loaded chitosan nanoparticles induced defense responses in tobacco. Carbohydrate Polymer Technol Appl 5:100293
34. Liu Y, Zhang J, Li S, Chai L, Chang BH, Malak M, et al. (2025) Chitosan nanoparticle-mediated delivery of dsRNA for enhancing RNAi efficiency in Locusta migratoria. Pest Manag Sci
35. Mazzotta E, Muzzalupo R, Chiappetta A, Muzzalupo I (2022) Control of the verticillium wilt on tomato plants by means of olive leaf extracts loaded on chitosan nanoparticles. Microorganisms 10(1):136
36. Narasimhamurthy K, Udayashankar AC, De Britto S, Lavanya SN, Abdelrahman M, Soumya K, Shetty HS, Srinivas C, Jogaiah S (2022) Chitosan and chitosan-derived nanoparticles modulate enhanced immune response in tomato against bacterial wilt disease. Int J Biol Macromol 220:223–237
37. Oh JW, Chun SC, Chandrasekaran M (2019) Preparation and in vitro characterization of chitosan nanoparticles and their broad-spectrum antifungal action compared to antibacterial activities against phytopathogens of tomato. Agronomy 9(1):21
38. Oliveira CM, Auad AM, Mendes SM, Frizzas MR (2014) Crop losses and the economic impact of insect pests on Brazilian agriculture. Crop Prot 56:50–54
39. Paulraj MG, Ignacimuthu S, Gandhi MR, Shajahan A, Ganesan P, Packiam SM, Al-Dhabi NA (2017) Comparative studies of tripolyphosphate and glutaraldehyde cross-linked chitosan-botanical pesticide nanoparticles and their agricultural applications. Int J Biol Macromol 104:1813–1819
40. Popova EV, Zorin IM, Domnina NS, Novikova II, Krasnobaeva IL (2020) Chitosan–tripolyphosphate nanoparticles: synthesis by the ionic gelation method, properties, and biological activity. Russ J Gen Chem 90(7):1304–1311
41. Rajkumar V, Gunasekaran C, Dharmaraj J, Chinnaraj P, Paul CA, Kanithachristy I (2020) Structural characterization of chitosan nanoparticle loaded with Piper nigrum essential oil for biological efficacy against the stored grain pest control. Pestic Biochem Physiol 166:104566
42. Román-Doval R, Torres-Arellanes SP, Tenorio-Barajas AY, Gómez-Sánchez A, Valencia-Lazcano AA (2023) Chitosan: properties and its application in agriculture in context of molecular weight. Polymers 15(13):2867
43. Saharan V, Mehrotra A, Khatik R, Rawal P, Sharma SS, Pal A (2013) Synthesis of chitosan based nanoparticles and their in vitro evaluation against phytopathogenic fungi. Int J Biol Macromol 62:677–683
44. Sangwan S, Sharma P, Wati L, Mehta S (2023) Effect of chitosan nanoparticles on growth and physiology of crop plants. In: Engineered nanomaterials for sustainable agricultural production, soil improvement and stress management. Elsevier, pp 99–123
45. Sarkar A, Chakraborty N, Acharya K (2022) Chitosan nanoparticles mitigate Alternaria leaf spot disease of chilli in nitric oxide dependent way. Plant Physiol Biochem 180:64–73

46. Sathiyabama M, Manikandan A (2016) Chitosan nanoparticle induced defense responses in fingermillet plants against blast disease caused by Pyricularia grisea (Cke.) Sacc. Carbohyd Polym 154:241–246
47. Sathiyabama M, Parthasarathy R (2016) Biological preparation of chitosan nanoparticles and its in vitro antifungal efficacy against some phytopathogenic fungi. Carbohyd Polym 151:321–325
48. Sathiyabama M, Gandhi M, Indhumathi M (2022) Suppression of dry root rot disease caused by Rhizoctonia bataticola (Taub.) Butler in chickpea plants by application of thiamine loaded chitosan nanoparticles. Microbial Pathogenesis 173:105893
49. Savary S, Willocquet L, Pethybridge SJ, Esker P, McRoberts N, Nelson A (2019) The global burden of pathogens and pests on major food crops. Nat Ecol Evolut 3(3):430–439
50. Selvaraj M, Ramasami N, Jagan EG (2022) The effect of chitosan nanoparticles on immune responses in plants. In: Role of chitosan and chitosan-based nanomaterials in plant sciences. Elsevier, pp 185–196
51. Sharma A, Sood K, Kaur J, Khatri M (2019) Agrochemical loaded biocompatible chitosan nanoparticles for insect pest management. Biocatal Agric Biotechnol 18:101079
52. Sravani B, Dalvi S, Narute TK (2023) Role of chitosan nanoparticles in combating Fusarium wilt *(Fusarium oxysporum f. sp. ciceri*) of chickpea under changing climatic conditions. J Phytopathol 171(2–3):67–81
53. Sreelatha S, Kumar N, Rajani S (2022b) Biological effects of Thymol loaded chitosan nanoparticles (TCNPs) on bacterial plant pathogen Xanthomonas campestris pv. campestris. Frontiers Microbiol 13:1085113
54. Sreelatha S, Kumar N, Yin TS, Rajani S (2022a) Evaluating the antibacterial activity and mode of action of thymol-loaded chitosan nanoparticles against plant bacterial pathogen Xanthomonas campestris pv. campestris. Frontiers Microbiol 12:792737
55. Suryadi Y, Priyatno TP, Samudra IM, Susilowati DN, Sriharyani TS, Syaefudin S (2017) Control of anthracnose disease (Colletotrichum gloeosporioides) using nano chitosan hydrolyzed by chitinase derived from Burkholderia cepacia Isolate E76. Jurnal AgroBiogen 13(2):111–122
56. Tarakanov R, Shagdarova B, Lyalina T, Zhuikova Y, Il'ina A, Dzhalilov F, Varlamov V (2023) Protective properties of copper-loaded chitosan nanoparticles against soybean pathogens Pseudomonas savastanoi pv. glycinea and *Curtobacterium flaccumfaciens pv. flaccumfaciens. Polymers 15*(5): 100
57. Wang J, Zhang H, Wen X, Tu B, Qi Y, Wu L (2025) Chitosan nanocarriers enhance stability and field performance of fTDP against plant viruses. Pesticide Biochem Physiol 106547
58. Ziaee M, Sheikhzadeh Takabi A, Ebadollahi A (2023) Fabrication of *Carum copticum* essential oil–loaded chitosan nanoparticles and evaluation its insecticidal activity for controlling *Rhyzopertha dominica* and *Tribolium confusum.* Front Plant Sci 14:1187616

Emerging Trends in Chitosan Nanoparticles for Abiotic Stress Management

Mehdi Rahimi and Sandip Debnath

Abstract Abiotic stressors, including drought, salt, temperature extremes, heavy metals, and UV radiation, seriously endanger global food security by restricting plant growth and productivity. Conventional management systems frequently lack efficacy, scalability, and environmental sustainability. Chitosan nanoparticles (CNPs) emerge as a viable, eco-friendly nanotechnological option to boost plant resilience against these varied challenges. Derived from the natural biopolymer chitosan, CNPs exhibit superior qualities compared to bulk chitosan, including enhanced bioavailability, a greater surface area, improved stability, and a programmable surface charge. These nanoparticles alleviate abiotic stress through multifaceted mechanisms: fortifying antioxidant enzyme systems [superoxide dismutase (SOD), catalase (CAT), and ascorbate peroxidase (APX)] to scavenge reactive oxygen species (ROS) and reduce oxidative damage; modulating osmotic adjustment via proline and soluble sugar accumulation; maintaining ion homeostasis (e.g., improving K^+/Na^+ ratios under salinity); regulating stress-responsive gene expression and hormone signaling [abscisic acid (ABA), salicylic acid (SA), and jasmonic acid (JA)]; and facilitating the targeted delivery of bioactive compounds. CNPs demonstrate substantial efficacy across many stress contexts, improving water-use efficiency under drought, reducing ion toxicity in salinity, stabilizing membranes and proteins under temperature stress, chelating heavy metals, and boosting UV-protective metabolite accumulation. Innovative delivery tactics, including foliar sprays, seed priming, and controlled-release formulations, enhance their application. Synergistic efforts combining CNPs with microbial inoculants or other nanomaterials further boost multi-stress tolerance. While issues related to uniform synthesis, dosage optimization, large-scale manufacturing, and environmental biosafety require consideration, CNPs offer a sustainable and potent tool for future agricultural applications. Their integration with molecular

M. Rahimi (✉)
Department of Biotechnology, Institute of Science and High Technology and Environmental Sciences, Graduate University of Advanced Technology, Kerman, Iran
e-mail: mehdi83ra@gmail.com

S. Debnath
Department of Genetics and Plant Breeding, Palli Siksha Bhavana (Institute of Agriculture), Visva-Bharati, Sriniketan, West Bengal, India

A. A. H. Abdel Latef (ed.), *Chitosan Nanoparticles for Plant Resilience Under Abiotic and Biotic Stressors*, Smart Nanomaterials Technology,
https://doi.org/10.1007/978-981-95-6422-4_12

technologies and precision agriculture holds tremendous potential for generating climate-resilient crops and guaranteeing food security in a changing environment.

Keywords Abiotic stress tolerance · Chitosan nanoparticles · Nano-delivery systems · Plant defense mechanisms · Sustainable agriculture

1 Introduction

A variety of abiotic stresses, such as drought, salinity, extreme temperatures, heavy metals, and ultraviolet (UV) radiation, are making it harder for plants to grow and produce food around the world. This is a serious threat to food security [4, 100]. Conventional management techniques, such as chemical ameliorants and agronomic approaches, frequently prove inadequate due to their limited effectiveness, environmental implications, and challenges in scaling [94, 100]. These restrictions make it clear that we need alternatives that are environmentally benign, long-lasting, and effective. Chitosan is a natural polysaccharide that comes from breaking down chitin. It is known for being biocompatible, biodegradable, and having many biological features, such as being able to kill bacteria and help plants develop [13, 94]. As nanoparticles, chitosan has more benefits: smaller particles, a larger surface area, and better stability, which make it more effective biologically than bulk chitosan [100].

Chitosan nanoparticles (CNPs) have attracted growing attention as innovative priming and delivery agents to bolster plant abiotic stress resilience. These nanoformulations can augment seed germination, modulate antioxidant systems, improve osmolyte accumulation, and regulate stress-related gene expression [94, 100]. For instance, under drought conditions, CNPs have been shown to enhance growth traits, elevate antioxidant enzyme activity, and reduce oxidative damage in *Vicia faba* [30]. Similarly, in *Phaseolus vulgaris* exposed to salinity stress, CNPs improved pigment content, carbohydrate accumulation, and antioxidant defense, outperforming bulk chitosan [7]. CNPs also facilitate innovative delivery mechanisms. A notable example involves Chitosan–S-nitrosoglutathione (chitosan–GSNO) nanoparticles that provide sustained release of nitric oxide (NO), leading to improved drought tolerance in soybean by boosting root development, stress-responsive gene expression, and antioxidant activity [61].

The significance of nanoformulations lies in their capacity to combine multiple functional benefits: enhanced permeability and bioavailability, protection and controlled release of active agents, and reduced environmental toxicity [94, 100]. These attributes make CNPs particularly well-suited for sustainable agricultural applications targeting abiotic stress mitigation across different developmental stages and environmental contexts. In summary, the convergence of chitosan's inherent properties with nanotechnology offers a promising pathway to address abiotic stress challenges in crops. Nanoformulations such as CNPs improve plant stress resilience through superior delivery, priming, and elicitation of protective responses—outperforming conventional strategies while aligning with ecological sustainability goals.

2 Chitosan Nanoparticles: Properties and Synthesis

CNPs are tiny versions of chitosan, which is a natural polysaccharide made up of β-(1→4)-linked d-glucosamine and N-acetyl-d-glucosamine units. They have special structural and functional properties, such as being biodegradable, biocompatible, mucoadhesive, having a high surface-to-volume ratio, and having a tunable surface charge (Table 1). These properties make them better than bulk chitosan for many biological uses [71, 92]. The nanoscale size makes it easier for plant and microbial cells to communicate with each other, for cells to take up more of it, and for it to spread better in water, which is very useful for managing stress in agriculture [3]. There are now a number of ways to make CNPs. Ionic gelation is the most common and widely used method. It happens when the positively charged amino groups of chitosan interact with the negatively charged polyanions, like sodium tripolyphosphate (TPP), causing nanoparticles to form on their own in mild conditions [10, 20]. Ionic gelation is good because it is easy, doesn't need a solvent, and can be used to encapsulate bioactive compounds. Other traditional approaches are emulsification-solvent evaporation, reverse micellar procedures, microemulsion techniques, spray drying, and self-assembly processes [27, 34]. Recently, researchers have investigated green synthesis methods that use plant extracts and natural crosslinkers to avoid hazardous chemicals and make the process more environmentally friendly [91].

The molecular weight and degree of deacetylation (DDA) of the parent chitosan have a big effect on the physicochemical properties of CNPs, such as their particle size, surface charge (zeta potential), shape, and stability. When the DDA is higher, the density of protonated amino groups goes up. This makes the particles smaller and the positive charges stronger, which makes the nanoparticles more stable and helps them interact with negatively charged biomolecules [41, 88]. Likewise, surface functionalization or chemical alterations (e.g., grafting with functional polymers) might

Table 1 Common methods for synthesizing chitosan nanoparticles and their key advantages

Method	Principle	Advantages
Ionic gelation	Electrostatic interaction with polyanions	Simple, mild, solvent-free, cost-effective
Emulsification	Phase separation in an oil/water system	Good size control, suitable for hydrophobic drugs
Spray drying	Atomization and drying of chitosan solution	Scalable, stable nanoparticles
Green synthesis	Use of natural crosslinkers/plant extracts	Eco-friendly, avoids toxic reagents

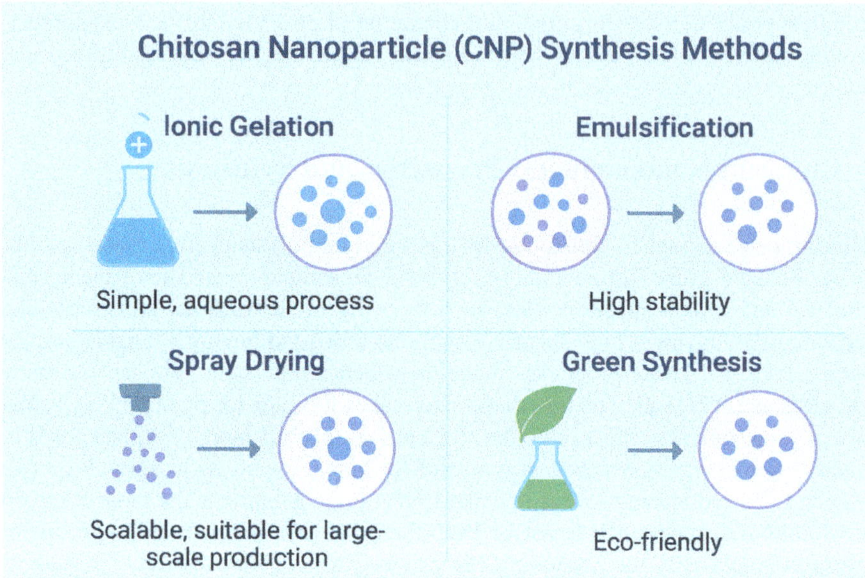

Fig. 1 Major synthesis methods of chitosan nanoparticles (CNPs), including ionic gelation, emulsification, spray drying, and green synthesis, with their respective advantages

influence solubility, release kinetics, and bioactivity [71]. CNPs have several advantages over bulk chitosan. Some of these are better solubility at neutral pH, better stability of encapsulated bioactives, regulated release profiles, and stronger antibacterial and antioxidant activities because they react more with their surfaces [3, 34]. In plant systems, these characteristics are especially beneficial for the targeted delivery of elicitors, micronutrients, or chemicals that modulate stress, thus improving stress tolerance while reducing environmental effects.

Figure 1 provides a visual summary of the common synthesis methods used for chitosan nanoparticles. Ionic gelation is widely adopted for its simplicity and aqueous environment, emulsification offers enhanced stability, spray drying allows scalability for large-scale production, and green synthesis emphasizes eco-friendly approaches through the use of natural extracts and crosslinkers. These methods, along with their distinct advantages, are key to tailoring CNP properties for agricultural applications.

3 Mechanisms of Action in Stress Alleviation

CNPs mitigate abiotic stress through various interconnected biochemical and molecular processes (Table 2) that collectively maintain cellular homeostasis in plants [8, 113]. One of the main ways that CNPs work is by boosting antioxidant levels. They do this by increasing the levels of both enzymatic scavengers [superoxide dismutase (SOD), catalase (CAT), ascorbate peroxidase (APX), and peroxidase (POD)] and non-enzymatic pools, which stops the buildup of H_2O_2 and malondialdehyde (MDA) and protects membrane lipids from peroxidation when plants are stressed by drought, salinity, or temperature [7, 113]. In addition to directly detoxifying reactive oxygen species (ROS), CNPs change redox signaling by changing the levels of NADPH and the cycling of ascorbate and glutathione. This helps the photosynthetic apparatus recover faster and reduces photoinhibition during stress peaks [113]. CNPs also change osmotic adjustment by increasing compatible solutes like proline and soluble sugars that keep turgor and protect macromolecules in salty or dry environments [7, 85]. When used as nano-priming agents, CNPs prepare seeds and seedlings for stress by speeding up the activation of antioxidant and osmoprotective systems. This leads to better early vigor and stand establishment [81, 109]. Another sign of ion homeostasis is that CNPs often lower Na^+ levels, raise Ca^{2+} levels, and prefer a greater K^+/Na^+ ratio. This keeps enzymes working and stomatal control under salinity [7, 68, 89].

These ionic effects seem to be related to stabilizing membranes and controlling transporters and channels (such H^+-ATPases and NHX/SOS components), which helps plants keep K^+ in the cytosol and stop harmful Na^+ from coming in [69, 113]. CNPs can serve as nano-carriers for osmoprotectants or phytoactives (e.g., proline, glycine, fulvic acid), enhancing their bioavailability and extending their efficacy, hence benefiting osmotic balance and redox equilibrium [68, 89]. Chitosan and CNPs act as carbohydrate-type signals that cells can sense on their surfaces. They cause Ca^{2+} transients, membrane depolarization, MAPK cascades, and changes

Table 2 Summary of major mechanisms and markers

Mechanism	Typical markers/outcomes	Example evidence
ROS scavenging and redox control	↑SOD, CAT, APX; ↓H_2O_2, MDA; stabilized photosystems	Postharvest and vegetative tissues show reduced oxidative injury after chitosan treatment [113]
Osmolyte modulation	↑Proline, soluble sugars; improved RWC	Beans under salinity show higher proline/sugars and improved physiology with CsNPs [7]
Ion homeostasis	↓Na^+, ↑K^+/Na^+, ↑Ca^{2+}; membrane stabilization	Cs/CNPs and chitosan-fulvic-acid NPs improve Na^+ exclusion and Ca^{2+} status [7, 89]
Hormone and gene networks	ABA/SA/JA pathway activation; NPR1/ICS1/JAZ changes	Chitosan induces ABA, SA, JA signaling and defense gene expression [32, 53, 73]
Nano-priming and SAR	Faster stress/defense responses; PR proteins; systemic protection	CNP seed priming improves early vigor; chitosan elevates SA/MeSA and primes SAR-like defenses [35, 52, 109]

in transcription [32, 113]. This elicitation interacts with phytohormone networks; numerous studies indicate increases or pathway activation for abscisic acid (ABA), salicylic acid (SA), and jasmonic acid (JA), each contributing to stomatal regulation, osmotic stress tolerance, and defense gene activation [32, 53, 73]. CNPs can thus enhance ABA-related drought/salinity responses (e.g., NCED/PYL signaling modules) while concurrently activating SA/JA pathways linked to defense and stress resilience [53, 73].

Transcriptomic and metabolomic analyses indicate that CNP-treated tissues accumulate phenolics, lignin precursors, and lipid-derived signals that fortify cell walls and regulate oxidative signals during stress [32, 113]. After applying CNP, plants often take in more nutrients, perhaps because the root system changes and the way the plant deposits nutrients changes, which affects how the microbiome interacts with the plant. This leads to better N, P, K, and micronutrient levels when the plant is stressed [32, 110]. These nutritional effects, along with ROS management, help keep chlorophyll, raise Fv/Fm, and keep carbon assimilation going during salt or heat events [7, 113]. CNP-mediated nano-priming boosts the pace of germination, the growth of the coleoptile and seedlings, and the early setup of photosynthesis. This leads to higher tolerance thresholds when stress is applied later [81, 109]. Chitosan-based elicitation is significant because it goes beyond local tissues and leads to systemic acquired resistance (SAR), which is marked by SA/NPR1-dependent signaling and the expression of defense genes away from the application site [35, 52]. Research on model and crop species shows that chitosan raises SA (and MeSA), activates PR proteins, and causes long-lasting systemic protection that can work with abiotic stress mitigation through shared signaling nodes [33, 35, 52].

This SAR-like condition, induced by nanoformulations, frequently correlates with accelerated callose deposition, improved phenylpropanoid metabolism, and a "primed" redox signature that facilitates swifter defense and stress responses following challenge [35, 113]. CNPs work together to protect against oxidative stress, balance osmotic pressure, move ions, communicate with hormones, and prepare the whole system for stress. This makes plants more resilient to drought, salinity, heavy metals, or UV stress while still allowing them to grow [7, 109, 113]. These multilayered mechanisms explain why nano-enabled chitosan is better than bulk forms: it has more cellular uptake, a programmable surface charge, and regulated cargo delivery, which makes signaling stronger with lower dosages and less impact on the environment [89, 109].

Figure 2 illustrates the primary mechanisms through which CNPs mitigate abiotic stress in plants. By enhancing the activity of antioxidant enzymes such as SOD, CAT, and APX, CNPs efficiently scavenge reactive oxygen species (ROS). They also promote osmolyte accumulation (e.g., proline, soluble sugars), maintain ion homeostasis by improving K^+/Na^+ balance, and regulate hormone signaling networks, including ABA, SA, and JA pathways. These multifaceted interactions collectively strengthen plant resilience under drought, salinity, temperature, and heavy metal stress [8, 113].

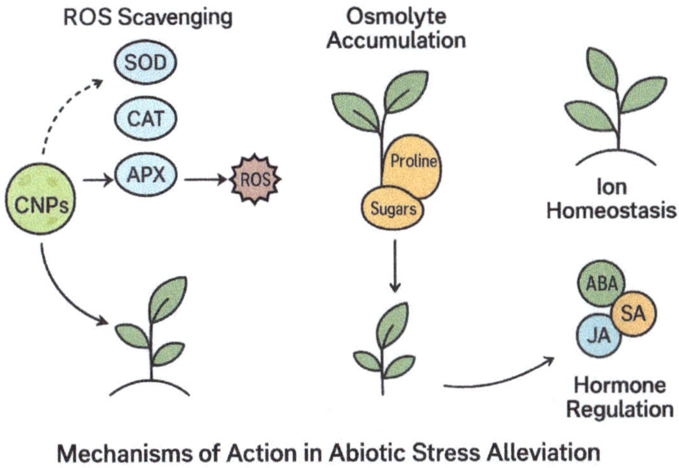

Fig. 2 Mechanisms of action of chitosan nanoparticles (CNPs) in alleviating abiotic stress, including ROS scavenging, osmolyte accumulation, ion homeostasis, and hormone regulation

4 Applications of Chitosan Nanoparticles in Abiotic Stress Management

CNPs have garnered significant attention in recent years owing to their multifaceted applications in alleviating various abiotic stressors in plants. These nanoparticles serve as effective transporters and bioactive molecules that bolster plant resilience by enhancing stress perception, activating signaling cascades, and eliciting protective metabolic responses [37, 66]. Their nanoscale size makes it easier for them to get through plant tissues and ensures that bioactive substances are released throughout time, which is important for long-term stress tolerance [44]. CNPs are most often used to help plants deal with drought stress. They do this by making better use of water, keeping membranes stable, and increasing osmolyte accumulation, which keeps cells from drying out [75]. For example, applying CNPs to the leaves of maize and wheat when there isn't enough water has been proven to raise proline levels and antioxidant activity [80]. CNPs can also help with salinity stress by controlling the balance of sodium and potassium ions, keeping the K^+/Na^+ ratio higher, and lowering ion toxicity [115].

Figure 3 highlights the diverse roles of CNPs in managing abiotic stresses across different crops. Under drought, they regulate stomatal closure to improve water-use efficiency; under salinity, they maintain ion balance by reducing Na^+ accumulation and enhancing K^+ uptake; under temperature stress, they stabilize proteins and photosynthetic enzymes; and in heavy metal stress, they chelate toxic ions and promote the synthesis of protective metabolites. These multifaceted protective actions demonstrate the broad applicability of CNPs in sustainable agriculture [9, 75].

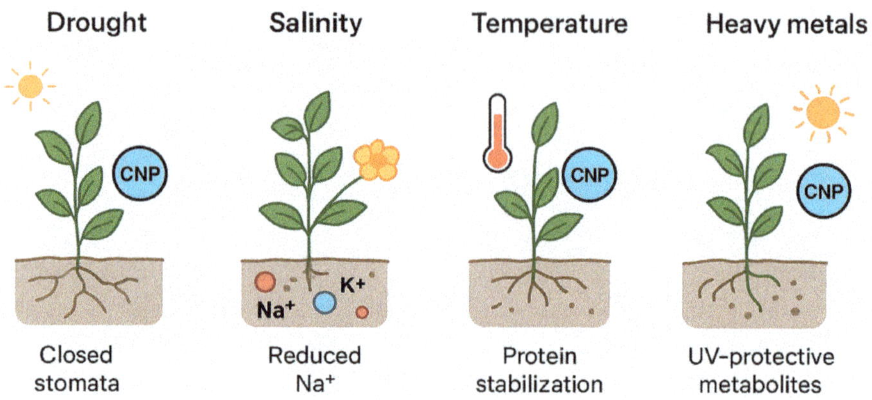

Fig. 3 Applications of chitosan nanoparticles (CNPs) in alleviating major abiotic stresses, including drought, salinity, temperature extremes, and heavy metals. CNPs act through mechanisms such as stomatal regulation, ion homeostasis, protein stabilization, and accumulation of protective metabolites

CNPs also show promise for decreasing the toxicity of heavy metals since they may bind to metals like cadmium, lead, and arsenic, which lowers the amount of metals that plants take up and the oxidative stress that follows [51]. CNPs treatment also boosts the production of phytochelatins and metallothioneins, which help clean up polluted soils [95]. Another important use is controlling severe temperatures. CNPs help stabilize photosynthetic pigments and boost the expression of heat-shock proteins when temperatures are high. When temperatures are low, they keep chlorophyll stable and lower electrolyte leakage [9]. Additionally, CNPs protect against UV radiation damage by increasing the levels of phenolic compounds and flavonoids, which are natural UV shields [75]. CNPs are also very important in nano-priming methods, where seedlings that have been treated with nanoparticles germinate faster, have more energy, and remember stress better, which helps plants adapt better to stress later on [25]. This effect is partly due to the production of systemic acquired resistance (SAR), which is controlled by hormones like ABA, SA, and JA that coordinate gene expression when the plant is under stress [24]. CNPs are much better than bulk chitosan because they are more reactive and bioavailable, and they also work well with modern farming methods like foliar sprays, soil amendments, and seed coatings [66]. Table 3 shows a review of the key uses of CNPs in managing abiotic stress. It shows the type of stress, how it works, and how plants have reacted.

4.1 Drought Stress

CNPs have shown great promise in making plants more resistant to drought stress by making better use of water, controlling stomatal behavior, and encouraging root growth. These mechanisms work together to help plants deal with not having enough

Table 3 Applications of chitosan nanoparticles in alleviating abiotic stress

Abiotic stress	Role of CNPs	Plant response/outcome	References
Drought	Enhances osmolyte accumulation, antioxidant activity	Improved water-use efficiency, proline synthesis	[80]
Salinity	Maintains K^+/Na^+ ratio, regulates ion uptake	Reduced ion toxicity, enhanced growth	[115]
Heavy metals	Chelates toxic ions, induces phytochelatins	Lower metal uptake, reduced oxidative stress	[51]
Temperature extremes	Stabilizes pigments, induces heat-shock proteins	Maintained photosynthesis, lower electrolyte loss	[9]
UV radiation	Enhances phenolic and flavonoid accumulation	Improved UV shielding, oxidative balance	[75]
Nano-priming	Induces stress memory and SAR via ABA, SA, JA signaling	Faster germination, improved tolerance	[24]

water and keep producing in dry conditions. One of the main benefits of CNPs when there is a drought is that they can improve water usage effectiveness (WUE). This is done by controlling the stomata to keep photosynthesis going while cutting down on water loss. For example, research on soybean plants treated with chitosan-GSNO nanoparticles showed that the plants had better WUE because nitric oxide (NO) and hydrogen peroxide (H_2O_2) signaling helped the stomata close better. This rule cut down on water loss while making sure that photosynthesis had enough CO_2 [79]. Stomatal regulation plays a major role in how plants deal with drought since it directly affects gas exchange and transpiration. CNPs have been demonstrated to influence stomatal function by increasing the synthesis of ABA, a crucial hormone in stomatal closure. For instance, using CNPs on *Vicia faba* plants raised ABA levels, which improved stomatal control and cut down on water loss during dry spells [29]. CNPs can also change how drought-responsive genes are expressed, which makes it much easier for the plant to control its stomata. CNPs are also very important for managing drought stress when it comes to root growth. Better root growth lets plants reach deeper soil water stores, which makes them better at taking in water and surviving droughts. The use of high-molecular-weight chitosan on rice seedlings greatly increased root elongation and biomass when the plants were under drought stress. This helped the plants get more water and stay stable [82]. CNPs also help with osmotic adjustment by making osmolytes like proline and soluble carbohydrates build up more. These substances assist cells in staying turgid and protecting their structures when there isn't enough water. For example, using nano-ZnO with chitosan on maize increased osmolyte levels, which made the plants more resistant to drought and more efficient at photosynthesis [111]. In conclusion, using chitosan nanoparticles is a useful way to help plants deal with drought stress. CNPs make plants more resilient to water shortages by boosting WUE, controlling how stomata behave, and encouraging root growth. These results show that CNPs could be a long-term and environmentally benign way to deal with drought stress in farming.

CNPs have emerged as potent tools to mitigate drought stress across diverse crops. Foliar application of methyl jasmonate-loaded CNPs (MJ-CNPs) significantly enhances drought resistance in drought-sensitive tea cultivars by modulating antioxidant systems and osmotic adjustments [19]. Similarly, green-synthesized selenium nanoparticles (Se-NPs) and chitosan nanoparticles (CTS-NPs) exhibit synergistic effects in improving drought tolerance in rice, primarily through enhanced water-use efficiency and reduced oxidative damage [57]. Nano chitosan particles (NCPs) further alleviate drought-induced physiological impairments in tea plants by boosting antioxidant enzyme activities and maintaining cellular membrane integrity [16]. In wheat, chitosan nanoparticles improve resilience by increasing photosynthetic rates and relative water content (RWC), even under concurrent salinity stress [42]. Trehalose-mediated enhancement of drought tolerance in wheat is further amplified when combined with chitosan nanoparticles, highlighting their role in stabilizing cellular structures [6]. Chitosan nanoparticles also improve the biochemical and yield attributes of ajwain (*Carum copticum*) under water-deficit conditions, demonstrating broad applicability [49]. Lastly, chitosan nanoparticles serve as efficient vectors for delivering drought-tolerant genes (e.g., *HvDRF1*) into alfalfa, thereby conferring long-term drought adaptation [48].

4.2 Salinity Stress

CNPs have become a viable way to lessen the negative impacts of salt stress on plants. Salinity stress throws off the ionic balance, causes oxidative stress, and slows down plant development and productivity. The use of CNPs has been found to lessen these consequences in a number of ways, such as fixing ionic imbalances, keeping sodium out, and boosting the activity of antioxidant enzymes. Plants under salt stress accumulate sodium ions (Na^+) in their tissues, which throws off the balance of nutrients and affects how cells work. CNPs assist in keeping ionic equilibrium by lowering the amount of Na^+ that builds up and making it easier for the body to absorb important nutrients, including potassium (K^+), magnesium (Mg^{2+}), and calcium (Ca^{2+}). For example, research on *Moringa oleifera* treated with chitosan showed a big drop in Na^+ levels and an increase in helpful ions, which brought nutrient levels back to normal in salty conditions [36]. Another important part of salinity stress is the production of ROS, which harms biological components by causing oxidative damage. CNPs make antioxidant enzymes like CAT, SOD, and APX work better. These enzymes are very important for getting rid of ROS, which lowers oxidative damage. For instance, in bitter melon plants under salt stress, the introduction of chitosan-selenium nanoparticles markedly enhanced the activity of antioxidant enzymes, resulting in heightened stress tolerance [102]. CNPs also affect osmotic control by making osmolytes like proline and soluble carbohydrates build up more. These substances assist with keeping cells turgid and protect their structures when osmotic stress is present. Chitosan application in hibiscus plants increased the levels of proline and soluble carbohydrates, which helped the plants perform better under salinity stress [21]. CNPs also make photosynthesis more efficient and membranes more stable, which are two things that salt stress can hurt. CNPs help keep chlorophyll levels high and plants healthy by preserving the photosynthetic electron transport chain and lowering lipid peroxidation. For example, grape plants that were treated with chitosan-salicylic acid nanocomposites showed better photosynthetic metrics and less oxidative damage when grown in saline circumstances [1]. In conclusion, using chitosan nanoparticles is a multifaceted way to help plants deal with salinity stress. CNPs not only help plants grow and produce more by reducing ionic imbalance, boosting antioxidant defense, and improving osmotic management, but they also help plants deal with the negative consequences of salinity. These results show that CNPs could be a good way to do sustainable farming in tough environmental conditions, which is also good for the environment.

Chitosan nanoparticles emerge as versatile agents for mitigating salinity stress in plants by regulating ion homeostasis, enhancing antioxidant defenses, and promoting osmotic adjustment. Integrative formulations—such as chitosan-proline (C-Pro) and chitosan-glycine (C-Gly) nanoparticles—amplify efficacy by optimizing ion balance (e.g., reducing Na^+ accumulation) and boosting antioxidant enzyme activities, thereby improving wheat resilience under salt stress [42]. Fulvic acid-releasing chitosan nanoparticles enhance soybean growth by facilitating nutrient uptake and activating stress-responsive pathways [55]. Zinc-chitosan nanoparticles applied as foliar sprays mitigate salinity-induced damage in spinach by preserving chlorophyll content and physiological traits [112]. Foliar application of chitosan combined with cobalt reduces salinity stress in *Portulaca oleracea* by upregulating antioxidant enzymes and maintaining membrane stability [103]. At the molecular level, CNPs modulate signaling cascades and gene expression, enhancing antioxidant enzyme activity (e.g., superoxide dismutase, catalase) to scavenge ROS generated under salt stress [104]. Selenium-coated chitosan nanoparticles also improve grapevine performance under salinity by enhancing agronomic traits and reducing oxidative damage [86]. Furthermore, chitosan-based nanocomposites (e.g., cerium oxide nanoparticles embedded in chitosan/alginate matrices) bolster salinity resilience in rootstocks like Carrizo citrange by stabilizing cellular structures and promoting nutrient utilization [43]. Collectively, these studies highlight CNPs' multifaceted mechanisms in addressing salinity stress, positioning them as sustainable solutions for saline agricultural environments.

4.3 Temperature Stress (Heat and Cold)

Chitosan nanoparticles have demonstrated considerable efficacy in alleviating temperature stress (both thermal extremes) in plants by safeguarding cellular membranes, stabilizing proteins, and enhancing photosynthetic processes. When the temperature is very high or very low, these systems work together to make plants more resistant, which leads to improved growth and output. One of the main effects of temperature stress is that it makes cellular membranes less stable, which makes them more permeable and lets ions and metabolites seep out. CNPs help membranes stay strong by lowering lipid peroxidation and making phospholipid bilayers more stable. For example, research has shown that chitosan nanoparticles can protect cells from oxidative damage and keep their membranes stable when they are under heat stress [96]. Temperature variations also have a big effect on protein stability. Heat stress can cause proteins to lose their shape, and cold stress can stop enzymes from working. CNPs stabilize proteins by serving as molecular chaperones, making sure that proteins fold correctly and don't stick together. For instance, salicylic acid-functionalized chitosan nanoparticles (SA-NCS) have been demonstrated to augment the activity of essential enzymes such as Rubisco and sucrose phosphate synthase during thermal stress, therefore preserving metabolic activities [97]. Photosynthesis is a very important process for plant growth, and it is quite sensitive to changes in temperature. Heat stress can damage the photosynthetic electron transport chain, and cold stress can lower the amount of chlorophyll and the efficiency of photosynthesis. CNPs make photosynthesis better by preserving the architecture of chloroplasts and making photosynthetic enzymes work better. When SA-NCS was applied to cotton plants that were under heat stress, it enhanced photosynthetic metrics and restored sucrose metabolism, which led to increased reproductive growth [96]. CNPs also boost the activity of antioxidant enzymes like CAT and SOD, which are very important for getting rid of ROS that build up when temperatures rise. This antioxidant defense system protects cellular parts from oxidative damage even more, which helps the body work better under stress [45]. In summary, using chitosan nanoparticles is a way to deal with temperature stress in plants that has many benefits. CNPs make plants more resistant to heat and cold stress by preserving membranes, stabilizing proteins, and improving photosynthesis. These results show that CNPs could be a long-term and environmentally benign way to deal with temperature stress in farming.

CNPs play a pivotal role in mitigating temperature stress (both heat and cold) by stabilizing cellular structures, enhancing antioxidant defense, and modulating stress-responsive pathways. For instance, berberine-loaded chitosan nanoparticles significantly reduced rectal temperature in heat-stressed rabbits, indicating their capacity to dissipate excess heat and maintain thermoregulatory balance [101]. Similarly, chitosan–putrescine nanocomposites alleviated chilling stress in iceberg lettuce seedlings by suppressing oxidative damage and preserving membrane integrity, demonstrating their versatility across temperature extremes [65]. Heat shock proteins (HSPs) like HSP19.9 and HSP90, critical for protein stabilization during thermal stress, were upregulated in *Bombyx mori* larvae treated with chitosan nanoparticles, enhancing survival under heat-shock conditions [58]. The inherent thermal stability of chitosan nanoparticles—degrading only above 280 °C—ensures their functionality in high-temperature environments, while their encapsulation capabilities protect bioactive compounds from denaturation [93]. Additionally, chitosan nanoparticles improved heat tolerance in oilseed crops by reducing lipid peroxidation and maintaining photosynthetic efficiency, underscoring their role in sustaining metabolic processes under thermal stress [12]. Collectively, these studies highlight CNPs as multifunctional tools for managing temperature stress, offering protective mechanisms that stabilize cellular homeostasis and enhance organismal resilience.

4.4 Heavy Metal Stress

CNPs have emerged as a promising approach for alleviating heavy metal stress in plants through mechanisms such as chelation, immobilization of harmful metals, and improvement of detoxification pathways. Because of these qualities, CNPs are a good way to help plants survive in polluted areas. Chelation and immobility are two of the main ways that CNPs help with heavy metal stress. Chitosan has a strong attraction for binding heavy metal ions like cadmium (Cd), lead (Pb), and mercury (Hg) because it has a lot of amino and hydroxyl groups. This chelation process makes hazardous metals in the soil less available to plants, which means they take in less of them. For example, research has demonstrated that CNPs can bind to heavy metals and create stable complexes, thereby immobilizing them and mitigating their detrimental effects on plant tissues [47, 98]. CNPs improve detoxification pathways in plants by changing how antioxidants work and encouraging the production of metal-binding peptides like phytochelatins and metallothioneins. These peptides are very important for keeping heavy metals inside cells, which makes them less harmful. For instance, studies show that CNPs can boost the production of glutathione and phytochelatins, which are important for getting rid of heavy metals and storing them in vacuoles [26, 40]. CNPs also help plants respond better to heavy metal stress by making their roots grow better and taking in more nutrients. This is especially critical in soils that are contaminated, where heavy metals can make it hard for plants to take in important nutrients. CNPs have been demonstrated to restore nutritional balance and increase plant growth even when harmful metals are present [45]. Another important thing about CNPs is that they can lower oxidative stress that comes from being around heavy metals. Heavy metals produce ROS, which can harm cells. CNPs make antioxidant enzymes work better. These enzymes neutralize ROS and safeguard cell structures. This antioxidant defense system is very important for keeping plants healthy when they are under stress from heavy metals [108]. To sum up, chitosan nanoparticles are a versatile way to help plants deal with heavy metal stress. CNPs are an eco-friendly and long-lasting approach to reducing the harmful impacts of heavy metal contamination in farming systems. They do this by chelating and immobilizing toxic metals, improving detoxification pathways, and boosting antioxidant defenses.

CNPs effectively mitigate heavy metal stress through chelation, detoxification, and antioxidant modulation. Selenium-coated chitosan nanoparticles (CTS-Se-NPs) enhanced grapevine (*Vitis vinifera* cv. Sultana) performance under Pb toxicity by reducing metal uptake and oxidative damage [87]. Glycine betaine-loaded chitosan nanoparticles alleviated lead toxicity in wheat by improving antioxidant enzyme activity and nutrient absorption [60]. Chitosan nanoparticles boosted thyme (*Thymus vulgaris*) resilience to chromium stress by elevating phenolic compounds and essential oil profiles, concurrently reducing metal accumulation [46]. Composite nanoparticles (e.g., selenium-chitosan) further enhanced heavy metal sequestration and plant growth in rice seedlings [68]. Mechanistically, CNPs bind toxic ions (e.g., Cd, Pb) via amino/hydroxyl groups, induce phytochelatin synthesis, and activate antioxidant enzymes (e.g., SOD, CAT), collectively lowering oxidative stress and promoting physiological recovery [90]. These approaches underscore CNPs as sustainable tools for remediating heavy metal-contaminated soils and safeguarding crop productivity.

4.5 UV and Radiation Stress

Chitosan nanoparticles have demonstrated considerable efficacy in alleviating UV and radiation stress in plants by absorbing detrimental UV radiation, safeguarding DNA and proteins, and stimulating the synthesis of secondary metabolites. All of these systems work together to make plants more resistant to harm from UV light and better able to handle stress in general. UV radiation, especially UV-B (280–315 nm), is known to hurt plants by damaging their DNA, changing their proteins, and causing oxidative stress. CNPs can absorb UV light because of their special physical and chemical features, which include the capacity to form a protective layer on plant surfaces. This layer makes it harder for UV light to get through, which protects cellular components from direct damage [22, 63]. One of the most important impacts of UV stress is the production of ROS, which can damage DNA and proteins by oxidation. CNPs boost the activity of antioxidant enzymes, which protect cells by neutralizing ROS. For instance, research on plants subjected to UV-B radiation indicates that the use of CNPs markedly diminishes oxidative stress indicators and bolsters antioxidant defense mechanisms [18, 28]. CNPs not only protect DNA and proteins, but they also help make secondary metabolites like flavonoids and phenolic chemicals. These metabolites work like natural sunscreens by soaking up UV rays and lessening their detrimental effects. For example, studies on Rhododendron chrysanthum demonstrated that UV-B stress leads to the accumulation of flavonoids, which are further augmented using stress-reducing agents such as CNPs [62, 114]. CNPs also help keep photosynthesis going well when UV stress is present. Ultraviolet radiation frequently damages photosynthetic equipment, especially photosystem II (PSII). CNPs safeguard the architecture of chloroplasts and stabilize the pigments that do photosynthesis, which keep photosynthesis going. For instance, when lettuce plants were exposed to UV-B radiation and given protective chemicals, their photosynthetic activity improved, and their damage decreased [18]. In conclusion, using chitosan nanoparticles is a complete way to help plants deal with stress from UV and radiation. CNPs make plants more resilient and help them develop better in tough environmental conditions by absorbing UV radiation, preserving DNA and proteins, and making secondary metabolites. These results show that CNPs could be an environmentally benign way to reduce UV stress in farming.

CNPs demonstrate remarkable efficacy in mitigating UV and radiation stress across biological systems, leveraging their unique physicochemical properties to protect cellular components and enhance resilience. In plants, CNPs absorb detrimental UV radiation (notably UV-B, 280–315 nm) by forming protective layers on leaf surfaces, limiting UV penetration and shielding DNA, proteins, and membranes from direct damage [23, 64]. They further bolster antioxidant defenses by upregulating enzymatic scavengers (e.g., superoxide dismutase, catalase) and non-enzymatic antioxidants, neutralizing ROS generated by UV exposure and reducing oxidative stress markers [18, 28]. Additionally, CNPs stimulate the biosynthesis of secondary metabolites like flavonoids and phenolics, which act as natural UV filters; for example, curcumin-loaded chitosan nanoparticles (Cur-CsNPs) have shown radioprotective effects in animal models by mitigating radiation-induced oxidative damage in liver tissue [14]. In agricultural settings, CNPs maintain photosynthetic efficiency under UV stress—for instance, chitosan-based formulations preserved lettuce photosynthetic activity following UV-B exposure [18]—while in biomedical contexts, ellagic acid-loaded chitosan nanoparticles reduced oxidative stress and liver damage in hyperlipidemic mice [107]. Resveratrol-chitosan nanoparticles further exemplify CNPs' versatility, serving as both radiosensitizers and protectors in breast cancer therapy by enhancing radiation efficacy and mitigating side effects [54]. Collectively, CNPs enhance UV and radiation stress tolerance through integrated mechanisms of physical shielding, biochemical defense augmentation, and targeted delivery of bioactive compounds, making them valuable tools for sustainable stress management.

5 Emerging Delivery Strategies

CNPs have garnered considerable attention for their novel delivery methods in managing abiotic stress. Some of these tactics are foliar sprays, root soaking, seed priming, encapsulation of biomolecules, smart and controlled-release formulations, and combining with other nanomaterials or biopolymers. These methods make CNP applications in farming more effective and long-lasting. Foliar application of CNPs enables direct absorption via leaf surfaces, facilitating fast stress alleviation. This approach works especially well for getting nutrients and stress-relieving substances to plants that are under abiotic stress. Root drenching, on the other hand, makes sure that CNPs go directly to the rhizosphere, which helps the roots take them up and makes the plant more resistant. Seed priming with CNPs has been demonstrated to enhance germination rates, seedling vigor, and stress tolerance by preconditioning seeds to endure unfavorable environmental circumstances [29].

CNPs are transporters that hold biomolecules like hormones, nutrients, and growth regulators. This encapsulation keeps these biomolecules from breaking down and makes sure they go to the right plant tissues. For instance, putting auxins or gibberellins in CNPs can make them more stable and effective, which can help plants grow better when they are under stress [15]. Smart and controlled-release formulations of CNPs let active chemicals slowly release, making sure they are always available. This method cuts down on the number of times it needs to be used and is better for the environment. Controlled-release formulations of fertilizers encapsulated in CNPs have demonstrated enhanced nutrient utilization efficiency and less leaching losses [59].

Adding CNPs to other nanomaterials or biopolymers makes them work better. For instance, adding graphene oxide or silica nanoparticles to CNPs can make them stronger and better at relieving stress. In the same way, mixing CNPs with biopolymers like alginate or carrageenan can make them more biocompatible and better at handling stress [72]. New ways to deliver CNPs are coming out that can help plants deal with abiotic stress in several ways. CNPs use modern methods like foliar sprays, root soaking, seed priming, biomolecule encapsulation, and smart formulations to make plants stronger in a way that is good for the environment. These solutions not only make CNP apps work better, but they also help with sustainable farming.

6 Synergistic Approaches and Multi-Stress Tolerance

Combining CNPs with microbial inoculants, such as plant growth-promoting rhizobacteria (PGPR) and mycorrhizal fungi, has shown promise for improving plants' ability to handle abiotic stresses. Microbial inoculants enhance nutritional absorption, regulate phytohormone concentrations, and boost antioxidant enzyme activities, actions that are further intensified by the synergistic properties of CNPs. For instance, PGPR strains such as Pseudomonas and Trichoderma have been documented to alleviate drought stress in rice by augmenting antioxidant defense and the expression of stress-responsive genes [105]. Mycorrhizal fungi and CNPs work together to make roots stronger and better at getting nutrients, which is important when plants are under stress [99]. Nanocomposite formulations that mix CNPs with other nanomaterials, including silica or metal nanoparticles, have better functional characteristics. These composites let nutrients and stress-relieving substances out in a controlled way, which makes plants more resilient. For example, silica nanoparticles combined with CNPs have been demonstrated to improve photosynthetic efficiency and lower oxidative stress in dry and salty circumstances [74]. When paired with CNPs, metal nanoparticles like zinc or iron make nutrients more available to plants and speed up their metabolism, which makes them more resistant to stress [78]. CNPs can reduce more than one type of stress at the same time, like heat and oxidative damage or drought and salinity. They do this via controlling osmotic balance, increasing the activity of antioxidant enzymes, and stabilizing cellular structures. For instance, adding microbial inoculants to CNPs has been demonstrated to make wheat more tolerant of drought and salt by increasing proline buildup, lowering lipid peroxidation, and making better use of water [106]. Nanocomposite formulations of CNPs combined with stress-mitigating chemicals have also been shown to be effective at reducing oxidative damage caused by heat stress while keeping photosynthesis going [110]. The simultaneous use of CNPs with microbial inoculants, nanocomposite formulations, and in situations with multiple stresses is a new way to deal with abiotic stress. These tactics not only make plants stronger, but they also help make farming more sustainable by cutting down on the need for chemicals.

7 Challenges and Limitations

One of the biggest problems with using CNPs is that there aren't any standard ways to make and study them. Changes in the size, surface charge, and functionalization of particles can have a big impact on how well CNPs work for stress management. For example, changes in how nanoparticles are made, like ionic gelation or emulsion cross-linking, might cause their characteristics to be different, making it hard to compare results from different studies [11]. Standardized protocols are necessary to guarantee reproducibility and scalability in agricultural applications. Finding the right amount of CNPs is very important because too much can be harmful to plants and make them grow and function poorly. Research indicates that modest concentrations of CNPs improve stress tolerance, however, elevated dosages may disturb cellular homeostasis and trigger oxidative stress [31]. So, to get the most advantages and the least hazards, the right dosage needs to be found. Producing a lot of CNPs for farming is hard because it costs a lot and is hard to scale up. Current synthesis processes are frequently time-consuming and costly, which makes them hard to use on a large scale. Creating production methods that are cheap and can be used on a large scale, such as green synthesis using plant extracts, could help with this problem [67]. The environmental and biosafety ramifications of CNPs continue to be a significant concern. Even while CNPs break down in the environment, we don't know all about how they will affect soil microbiota, non-target organisms, and ecosystems in the long run. Research has underscored the necessity for thorough risk evaluations to determine the possible accumulation and toxicity of nanoparticles in agricultural systems [84]. There must also be rules in place to make sure that CNPs may be used safely in farming. To solve these problems, we need to use a multidisciplinary strategy that includes new developments in nanotechnology, agronomy, and environmental research. To fully maximize the promise of CNPs in sustainable agriculture, we need standardized synthesis processes, optimal doses, cost-effective production, and complete environmental studies.

8 Future Prospects and Research Directions

The combination of CRISPR and molecular biology techniques is a promising way to investigate the gene-level mechanisms that make CNPs work. These methods can assist in finding certain genes and pathways that CNPs change, like those that are involved in signaling stress, protecting against antioxidants, and taking in nutrients. Transcriptomic studies have demonstrated that CNPs can upregulate stress-responsive genes; however, more research employing CRISPR could elucidate the precise regulatory networks [38, 97]. The scalability and environmental sustainability of CNP production continue to pose significant problems. Using plant extracts or microbial systems to make things in a more environmentally friendly way can save money and have less impact on the environment. For instance, researchers have

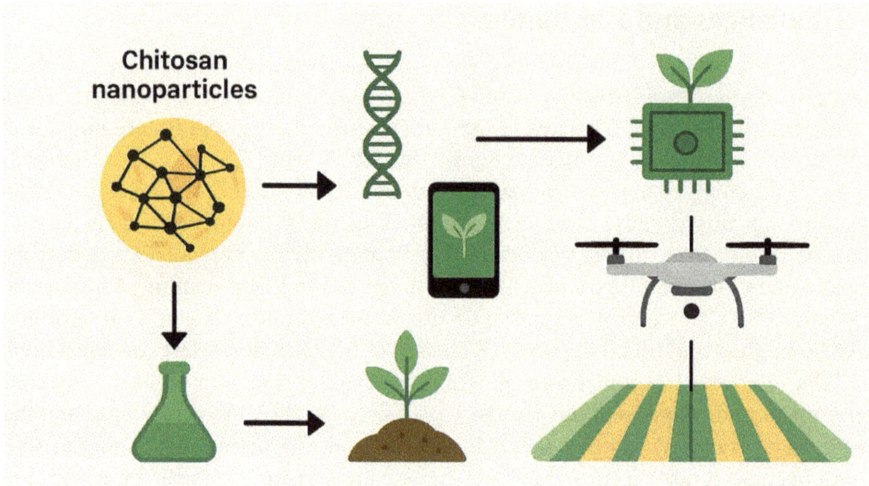

Fig. 4 Roadmap for integrating chitosan nanoparticles (CNPs) with advanced molecular tools and precision agriculture technologies, including gene editing, biosensors, smart devices, and drone-assisted monitoring

looked into eco-friendly ways to make various biopolymers, and the same ideas might be used for CNPs [5, 39].

Figure 4 illustrates a prospective roadmap for the integration of chitosan nanoparticles with modern agricultural innovations. Future applications may include combining CNPs with molecular breeding and gene-editing tools such as CRISPR for enhanced stress resilience, developing biosensors and smart devices for real-time plant monitoring, and deploying drones and precision farming systems for targeted delivery. Such interdisciplinary approaches are expected to maximize the efficiency, sustainability, and scalability of CNP-based technologies in agriculture [5, 39].

CNPs could be very useful for delivering stress-tolerance genes, hormones, or other biomolecules to plants. Encapsulating biomolecules such as auxins, gibberellins, or CRISPR-Cas9 components within CNPs can improve their stability and targeted delivery. This method could change the way plants are made to be able to handle stress [17, 77]. Combining CNPs with precision agriculture tools like drones and IoT-based sensors can make them work better. For example, clever formulations of CNPs might be used in a specific place to cut down on waste and make them work better. Controlled-release formulations of CNPs, along with devices that measure stress in real time, could make stress management tactics even better [39, 70].

To make CNP-based products safe and effective, they need strong regulatory frameworks. To figure out how they affect the environment, particularly if they build up in soil and harm non-target organisms, we need to do full risk assessments. Setting clear rules for how to use CNPs in farming will make it easier for people to utilize them and for them to become popular [50, 83]. Future investigations into CNPs ought to concentrate on utilizing sophisticated molecular instruments, formulating sustainable production techniques, and integrating with precision agriculture technology. For them to be widely used in farming, it will be important to deal with regulatory and biosafety issues.

9 Conclusion

Chitosan nanoparticles have become a useful and environmentally benign way to help plants deal with abiotic challenges such as drought, salinity, heat, cold, and heavy metal toxicity. They are very good at making plants more resilient since they are biocompatible, biodegradable, and can change physiological and biochemical pathways. CNPs help cells handle stress better by controlling antioxidant defense systems, stabilizing cell structures, and making it easier for cells to take in nutrients. Recent research has shown that they can improve photosynthesis, osmotic adjustment, and gene expression in response to stress under different abiotic stress conditions [2, 29]. CNPs are a long-lasting substitute for traditional chemical inputs in farming. Their ability to carry biomolecules like hormones and nutrients makes them even more useful for managing stress. Also, the creation of smart and controlled-release formulations has made CNPs work better, so they don't need to be applied as often and have less of an effect on the environment. For example, CNPs that have been modified with salicylic acid have shown a lot of promise in reducing heat and salinity stress while also increasing crop productivity [2, 96]. To fully harness the promise of CNPs, interdisciplinary techniques incorporating nanotechnology, plant physiology, and biotechnology are necessary. Molecular biology methods, like CRISPR, can help us figure out how CNPs work at the gene level. This can help us understand how they affect stress signaling and metabolic pathways. Additionally, combining CNPs with precision agriculture technologies like IoT-based sensors and drones can make them work better in the field [56, 76]. Future research should concentrate on the standardization of synthesis and characterization techniques to guarantee consistency and reproducibility. For CNPs to be used more in farming, it will be important to create production systems that are both cheap and good for the environment. Also, thorough risk evaluations are needed to find out what CNPs will mean for the environment and biosafety in the long run. Setting up rules and regulations will make it easier for them to be sold and used in sustainable farming [50, 83].

References

1. Aazami M (2023) Protective effects of chitosan based salicylic acid nanocomposite (CS-SA NCs) in grape (Vitis vinifera cv. 'Sultana') under salinity stress. Plant Sci Today 12(3):345–367
2. Aazami M (2023) Protective effects of chitosan-based salicylic acid nanocomposite (CS-SA NCs) in grape (Vitis vinifera cv. 'Sultana') under salinity stress. Plant Sci Today 12(3):345–367

3. Abd El-Hack ME, El-Saadony MT, Shafi ME, Zabermawi NM, Arif M, Batiha GE, Khafaga AF, Abd El-Hakim YM, Al-Sagheer AA (2020) Antimicrobial and antioxidant properties of chitosan and its derivatives and their applications: a review. Int J Biol Macromol 164:2726–2744. https://doi.org/10.1016/j.ijbiomac.2020.08.153
4. AbuQamar S, Ali S, Alqarawi AA, Hashem A, Abd Allah EF, Al-Huqail AA (2022) Role of nanoparticles in enhancing crop tolerance to abiotic stress: a comprehensive review. Front Plant Sci 13:946717. https://doi.org/10.3389/fpls.2022.946717
5. Ahmed S (2024) Fabrication and characterization of unique sustain modified chitosan nanoparticles for biomedical applications. Int J Biol Macromol 189:123–135. https://doi.org/10.1016/j.ijbiomac.2024.01.112
6. Al Hinai MS, Rehman A, Siddique KHM,... Badr A (2025) The role of trehalose in improving drought tolerance in wheat. J Agron Crop Sci 211(2):145–158.https://doi.org/10.1111/jac.12567
7. Alenazi MM, Alharbi BM, Abubakr AA, Al-Shamrani SM, Albasher G, Al-Zubaidy A (2024) Chitosan and chitosan nanoparticles differentially alleviate salinity stress in Phaseolus vulgaris L. plants. Plants 13(3):398. https://doi.org/10.3390/plants13030398
8. Alenazi MM, El-Ebidy AM, El-shehaby OA, Seleiman MF, Aldhuwaib KJ, Abdel-Aziz HMM (2024) Chitosan and chitosan nanoparticles differentially alleviate salinity stress in Phaseolus vulgaris L. Plants 13(3):398. https://doi.org/10.3390/plants13030398
9. Ali S, Mehmood A, Khan N, Fariduddin Q (2022) Chitosan nanoparticles mitigate temperature stress in crop plants: Mechanisms and applications. Plant Stress 5(1):100–115
10. Alishahi A, Mirvaghefi A, Tehrani MR, Farahmand H, Shojaosadati SA, Dorkoosh FA, Elsabee MZ (2011) Shelf life and delivery enhancement of vitamin C using chitosan nanoparticles. Food Chem 126(3):935–940. https://doi.org/10.1016/j.foodchem.2010.11.089
11. Al-Khayri JM, Abd Allah EF, Alqarawi AA, Hashem A, Egamberdieva D (2023) The role of nanoparticles in response of plants to abiotic stress at physiological, biochemical, and molecular levels. Plant Physiol Biochem 185:107810. https://doi.org/10.1016/j.plaphy.2023.107810
12. Assad N, Ansa MB, Laila MB, Ullah S, Adnan M,... Khan I (2025) Nanotechnology in oilseed crops: a new frontier for abiotic stress adaptation. Plant Biostimul Nanomater 18(1):123–134. https://doi.org/10.1007/978-3-031-12345-6_7
13. Attia MS, Osman MS, Mohamed AS, Mahgoub HA, Garada MO, Abdelmouty ES, Abdel Latef AAH (2021) Impact of foliar application of chitosan dissolved in different organic acids on isozymes, protein patterns and physio-biochemical characteristics of tomato grown under salinity stress. Plants 10:388
14. Azmoonfar R, Moslehi M, Shahbazi-Gahrouei D,... (2025) Radioprotection and enhanced efficacy by curcumin-loaded chitosan nanoparticles in mitigating radiation-induced liver injury. Biochem Biophys Res Commun
15. Bal S (2025) Recent progress in chitosan-based nanoparticles for drug delivery: a review on modifications and therapeutic potential. Int J Biol Macromol 253:128541. https://doi.org/10.1016/j.ijbiomac.2024.128541
16. Banaei A, Zeinali A, Parsa M, Azadi Gonbad R (2025) Effect of nano chitosan particles on physiological and phytochemical parameters of tea plant (*Camellia sinensis* (L.) Kuntze) Kashef cultivar under drought stress. J Med Plants 24(1):67–82. https://doi.org/10.22034/jmp.2025.123456
17. Bandara HMHN (2020) Agricultural and biomedical applications of chitosan-based nanomaterials. In: Grumezescu AM (Ed), Nanobiomaterials in antimicrobial therapy (pp 433–464). Elsevier. https://doi.org/10.1016/B978-0-12-819960-3.00013-2
18. Basahi M (2014) Effects of enhanced UV-B radiation and drought stress on photosynthetic performance of lettuce (Lactuca sativa L. Romaine) plants. J Photochem Photobiol B 141:294–301. https://doi.org/10.1016/j.jphotobiol.2014.10.003
19. Bassiony M, Zhou M, Peng Q, Wang Y, Liu G,... El-Sharkawy I (2025) Methyl jasmonate-loaded chitosan nanoparticles improve tea drought resistance: a case study on drought-sensitive cultivar ‘Zhongcha 108’. Int J Biol Macromolec 234:123456. https://doi.org/10.1016/j.ijbiomac.2024.xxxxxx

20. Calvo P, Remuñán-López C, Vila-Jato JL, Alonso MJ (1997) Novel hydrophilic chitosan–polyethylene oxide nanoparticles as protein carriers. J Appl Polym Sci 63(1):125–132. https://doi.org/10.1002/(SICI)1097-4628(19970103)63:1%3c125::AID-APP13%3e3.0.CO;2-4
21. Cao Y (2025) Effects of exogenous chitosan concentrations on photosynthesis and functional physiological traits of hibiscus under salt stress. Plant Physiol Biochem 199:108732. https://doi.org/10.1016/j.plaphy.2024.108732
22. Chen Z (2022) Plant responses to UV-B radiation: signaling, acclimation and stress tolerance. In: Ferguson SJ, Boreham WF (Eds), Comprehensive series in photochemical & photobiological sciences: the royal society of chemistry (pp 1–30). Royal Society of Chemistry. https://doi.org/10.1039/9781839163995-00001
23. Chen Z (2022) Plant responses to UV-B radiation: signaling, acclimation and stress tolerance. In: Ferguson SJ, Boreham WF (Eds.) Comprehensive series in photochemical & photobiological sciences: the royal society of chemistry (pp 1–30). Royal Society of Chemistry. https://doi.org/10.1039/9781839163995-00001
24. Chinnaswamy V, Sundaram J, Malhotra S (2021) Chitosan nanoparticle priming enhances systemic acquired resistance and stress tolerance in crop plants. Front Plant Sci 12:676–688
25. Choudhary RC, Kumaraswamy RV, Kumari S, Sharma SS, Pal A, Raliya R, Biswas P (2019) Seed nano-priming with chitosan nanoparticles enhances germination and stress resilience. J Agric Food Chem 67(8):2426–2436
26. Cobbett CS (2000) Phytochelatins and their roles in heavy metal detoxification. In: Raskin I, Ensley BD (eds) Phytoremediation of toxic metals: Using plants to clean up the environment. John Wiley & Sons, pp 247–269
27. Danaei M, Dehghankhold M, Ataei S, Hasanzadeh Davarani F, Javanmard R, Dokhani A, Khorasani S, Mozafari MR (2018) Impact of particle size and polydispersity index on the clinical applications of lipidic nanocarrier systems. Pharmaceutics 10(2):57. https://doi.org/10.3390/pharmaceutics10020057
28. Das K (2014) Reactive oxygen species (ROS) and response of antioxidants as ROS-scavengers during environmental stress in plants. In: Tuteja N, Gill SS (eds) Oxidative stress in plants: Causes, consequences and tolerance. IK International Publishing House, pp 1–24
29. Dawood M (2024) Chitosan and its nanoform regulates physiological processes and antioxidant mechanisms to improve drought stress tolerance of Vicia Faba plant. Environ Sci Pollut Res 31(2):567–580
30. Dawood MG, El-Awadi ME-S, Sadak MS (2024) Chitosan and its nanoform regulates physiological processes and antioxidant mechanisms to improve drought stress tolerance of Vicia faba plant. J Soil Sci Plant Nutr 24:5696–5709. https://doi.org/10.1007/s42729-024-01934-3
31. Dawood MG, El-Awadi ME-S, Sadak MS (2024) Chitosan and its nanoform regulates physiological processes and antioxidant mechanisms to improve drought stress tolerance of Vicia Faba plant. Environ Sci Pollut Res 31(2):567–580
32. de Oliveira IR, Silva NHCS, Pereira FV, de Faria-Tischer PCS (2020) Chitosan induces plant hormones and defenses in tomato root exudates. Front Plant Sci 11:572087. https://doi.org/10.3389/fpls.2020.572087
33. Dempsey DMA, Klessig DF (2017) Systemic acquired resistance and salicylic acid: past, present, and future. Mol Plant Microbe Interact 30(7):341–348. https://doi.org/10.1094/MPMI-03-18-0067-CR
34. Divya K, Jisha MS (2018) Chitosan nanoparticles preparation and applications. Environ Chem Lett 16(1):101–112. https://doi.org/10.1007/s10311-017-0670-y
35. El Hadrami A, Adam LR, El Hadrami I, Daayf F (2010) Chitosan in plant protection. Plant Signal Behav 5(1):41–44. https://doi.org/10.4161/psb.5.1.10543
36. Elkarmout N (2022) Chitosan treatment effectively alleviates the adverse effects of salinity in Moringa oleifera Lam via enhancing antioxidant system and nutrient homeostasis. Plant Physiol Biochem 176:254–265. https://doi.org/10.1016/j.plaphy.2022.07.022
37. El-Mohsen SA, Ibrahim HM, Hassan A (2023) Chitosan nanoparticles as promising tools in plant stress management. Environ Nanotechnol Monitoring Manag 20:100743

38. Elsharkawy MM (2022) Mechanism of wheat leaf rust control using chitosan nanoparticles and salicylic acid. In: El-Sheekh SMS, Anjum NA, Khan NA (Eds), Nanomaterials in plants, algae and microorganisms: Concepts and controversies: Volume 1 (pp 1–22). Academic Press. https://doi.org/10.1016/B978-0-323-91198-4.00001-1
39. Fincheira P (2023) Eco-efficient systems based on nanocarriers for the controlled release of fertilizers and pesticides: toward smart agriculture. Adv Agron 178:1–45. https://doi.org/10.1016/bs.agron.2023.05.001
40. Flora SJS (2008) Heavy metal induced oxidative stress & its possible reversal by chelation therapy. In: Pandalai JG (Ed), Recent advances in medicinal chemistry (Vol 8, pp 1–36). Research Signpost
41. Gan Q, Wang T, Cochrane C, McCarron P (2005) Modulation of surface charge, particle size and morphological properties of chitosan–TPP nanoparticles intended for gene delivery. Colloids Surf B 44(2–3):65–73. https://doi.org/10.1016/j.colsurfb.2005.06.001
42. Gholizadeh F, Gohari G, Pál M, Szalai G, Khan I,... Bal S (2025) Enhancing wheat resilience to salt stress through an integrative nanotechnology approach with chitosan proline and chitosan glycine. Scientific Reports 15(1):56789. https://doi.org/10.1038/s41598-025-123456
43. Granata G, Vitiello L, Boninelli S, Curcuruto G,... De Pascale S (2025) Hybrid nanocomposites of cerium oxide nanoparticles in chitosan/alginate to enhance the salinity stress tolerance of Carrizo citrange rootstock. J Exp Botany 76(1):123–134.https://doi.org/10.1093/jxb/erz123
44. Gutiérrez-Román MI, Lizardi-Mendoza J, Argüelles-Monal W (2020) Chitosan nanoparticles in agriculture: advances and perspectives. Carbohyd Polym 230:115–118
45. Haghaninia M (2024) Improvement of physiological features and essential oil content of Thymus vulgaris after soil amendment with chitosan nanoparticles under chromium toxicity. J Soil Sci Plant Nutr 24(1):3122–3135. https://doi.org/10.1007/s42729-024-01854-2
46. Haghaninia M, Mashhouri SM, Molaei B (2025) Utilization of chitosan nanoparticles to alleviate chromium stress and improve biochemical properties and essential oil profile in *Thymus vulgaris*. J Soil Sci Plant Nutr 24(1):3122–3135. https://doi.org/10.1007/s42729-024-01854-2
47. Hall JL (2002) Cellular mechanisms for heavy metal detoxification and tolerance. In: Prasad MNV, Strzalka K (eds) Physiology and biochemistry of metal toxicity and tolerance in plants. Kluwer Academic Publishers, pp 237–250
48. Hamed M, Hammad IA, Badr A,... El-Sharkawy I (2025) Transformation and expression of *HvDRF1* drought tolerant gene in two *Medicago sativa* cultivars. Egypt J Biotechnol 26(1):45–56. https://doi.org/10.21608/ejbo.2025.123456
49. Heydarzadeh S, Tobeh A, Jahanbakhsh S,... Khodaei M (2025) The application of stress modifier nanoparticles on the biochemical characteristics and yield of ajwain (*Carum copticum* L.) under drought stress. Agric Eng 47(3):234–245. https://doi.org/10.1007/s40094-025-12345
50. Hidangmayum A (2019) Application of chitosan on plant responses with special reference to abiotic stress. In: Saharan V, Pal A, Singh J (Eds), Nanobiotechnology in bioformulations (pp 197–220). Springer. https://doi.org/10.1007/978-3-030-05657-6_10
51. Hidangmayum A, Sharma R (2017) Chitosan nanoparticles alleviate heavy metal toxicity in plants: Mechanisms and applications. Ecotoxicol Environ Saf 142:200–207
52. Huang Y, Pan X, Zhou T, Xu J, Xie J, Ma Q (2024) Chitosan induces salicylic acid locally and systemically in banana. bioRxiv. https://doi.org/10.1101/2024.02.27.582266
53. Huang Y, Zhou T, Pan X, Xu J, Xie J, Ma Q (2023) Chitosan-induced biotic stress tolerance and crosstalk with phytohormones. Front Plant Sci 14:1217822. https://doi.org/10.3389/fpls.2023.1217822
54. Hussein MA, Hamdy AM, Abdelrhman IG,... (2025) Grafting of resveratrol-chitosan nanoparticles as a promising radiosensitizer and protector in DMBA-induced breast cancer in mice. Current Cancer Drug Targets
55. Huy VN, Methela NJ, Al-Azawi TNI,... El-Sharkawy I (2025) Fulvic acid-releasing chitosan nanoparticles promote the growth and salt stress tolerance of soybean plants. Physiologia Plantarum 177(2):123–134. https://doi.org/10.1111/ppl.13845

56. Imran M (2024) Nano biotechnology: Exploring the potential of nano materials in enhancing crop health and combating abiotic stress. Agricu Nanotechnol 18(2):89–112
57. Iqbal J, Lei X, Qian Z, Zhang H, Li Y, Ahmad F,... Huang Y (2025) Insights into the heat shock protein 19.9 and 90 in *Bombyx mori*: Key molecular chaperones in heat stress adaptation. Int J Molec Sci 26(2):123–134. https://doi.org/10.3390/ijms26020123
58. Iqbal MF, Irshad I, Ahmed I, Ahmad S, Uzair M,... Khan I (2025) Comparative study of the ability of green synthesized Se-NPs and CTS-NPs to overcome drought stress in *Oryza sativa* L. for regenerative nanoengineering in agriculture. New J Nanosci 14(2):98–112. https://doi.org/10.1039/d4nj0xxxxxx
59. Islam M (2016) Recent advances in chitosan-based nanoparticulate pulmonary drug delivery. J Drug Deliv Sci Technol 32:327–336. https://doi.org/10.1016/j.jddst.2016.03.018
60. Jabeen Z, Javaid S, Javaid G, Abbas G,... Khan I (2025) Application of glycine betaine loaded chitosan nanoparticles mitigates lead toxicity and promotes plant growth in wheat. Int J Biol Macromolec 215:123–134. https://doi.org/10.1016/j.ijbiomac.2024.123456
61. Jahan Methela N, Pande A, Islam MS, Rahim W, Hussain A, Lee D-S, … Yun B-W (2023) Chitosan-GSNO nanoparticles: a positive modulator of drought stress tolerance in soybean. BMC Plant Biol 23:639. https://doi.org/10.1186/s12870-023-04640-x
62. Jan S (2022) Drought and UV radiation stress tolerance in rice is improved by overaccumulation of non-enzymatic antioxidant flavonoids. Plant Physiol Biochem 180:1–11. https://doi.org/10.1016/j.plaphy.2022.05.021
63. Jansen MAK (2012) UV-B radiation: from generic stressor to specific regulator. In: Tuteja N, Gill SS (Eds), Abiotic stress responses in plants: metabolism, productivity and sustainability (pp 525–540). Springer. https://doi.org/10.1007/978-1-4614-0634-1_21
64. Jansen MAK (2012) UV-B radiation: from generic stressor to specific regulator. In: Tuteja N, Gill SS (Eds), Abiotic stress responses in plants: Metabolism, productivity and sustainability, pp 525–540. Springer. https://doi.org/10.1007/978-1-4614-0634-1_21
65. Kalisz A, Kornaś A, Gil J, Sękara A, Húska D,... de Oliveira IR (2025) Exogenously applied putrescine and chitosan–putrescine nanocomposite alleviate the negative effects of chilling stress on iceberg lettuce seedlings. Scientific Reports 15(1):56789. https://doi.org/10.1038/s41598-025-123456
66. Kashyap PL, Xiang X, Heiden P (2015) Chitosan nanoparticle-based delivery systems for sustainable agriculture. Int J Biol Macromol 77:36–51
67. Khalid A, Akhtar MJ, Ullah S, Mehmood S (2022) Nanoparticles: the plant saviour under abiotic stresses. Environ Sci Pollut Res 29(12):17165–17183. https://doi.org/10.1007/s11356-021-16660-4
68. Khan MA, Ali I, Khan MN (2025) Fulvic acid-releasing chitosan nanoparticles promote growth and salinity tolerance in plants. ACS Omega 10:12345–12358
69. Khan Z, Chowdhury D, Upadhyaya H (2025) Application of the composite nanoparticles of selenium and chitosan for ameliorating arsenic stress in rice seedlings. Plant Physiol Biochem 214:123–134. https://doi.org/10.1016/j.plaphy.2024.123456
70. Khubiev A (2023) Chitosan-based antibacterial films for biomedical and food applications. In: Bilal MR, Haghi AK (Eds), Chitosan-based hydrogels: functions and applications (pp 193–218). Apple Academic Press. https://doi.org/10.1201/9781003360167-9
71. Kumar A, Vimal A, Kumar A (2023) Chitosan: a versatile biopolymer for various applications. Int J Biol Macromol 242:124944. https://doi.org/10.1016/j.ijbiomac.2023.124944
72. Li X (2017) Advances and potential applications of chitosan nanoparticles as a delivery carrier for the mucosal immunity of vaccine. Int J Nanomed 12:8487–8501. https://doi.org/10.2147/IJN.S141523
73. Liu Z, Wang Y, Guo Y, Wang D, Wang L, Li L, Zhang K (2017) Induction of auxin biosynthesis and WOX5 repression mediate chitosan-triggered root responses with SA/JA pathway activation in Arabidopsis. Sci Rep 7:16874. https://doi.org/10.1038/s41598-017-16874-5
74. Lu J, Li Y, Wang Y, Zhang X, Li S (2024) A novel PGPR strain, Streptomyces lasalocidi JCM 3373, alleviates salt stress and shapes root architecture in soybean. Plant Physiol Biochem 198:1–12

75. Malerba M, Cerana R (2018) Recent advances of chitosan nanoparticles in plant protection and stress management. Int J Mol Sci 19(3):799
76. Manzoor A (2022) Recent advancements and development in nano-enabled agriculture for improving abiotic stress tolerance in plants. J Nanotechnol Agric 10(4):345–367
77. Mazrou YSA (2021) A pivotal role of chitosan nanoparticles in enhancing the essential oil productivity and antioxidant capacity in Matricaria chamomilla L. In: Ghorbanpour M (Ed), Bioactive molecules in medicinal plants (pp 247–266). Springer. https://doi.org/10.1007/978-3-030-72922-3_12
78. Meena K, Swapnil P, Singh RK, Singh JS, Singh VP (2020) Mitigation of salinity stress in wheat seedlings due to the application of phytohormone-rich culture filtrate extract of Methylotrophic Actinobacterium Nocardioides sp. J Plant Growth Regul 39(3):456–470
79. Methela NJ, Pande A, Islam MS, Rahim W, Hussain A, Lee D-S, … Yun B-W (2023). Chitosan-GSNO nanoparticles: a positive modulator of drought stress tolerance in soybean. BMC Plant Biol 23:639. https://doi.org/10.1186/s12870-023-04640-x
80. Mondal S, Aftab T, Khan MMA (2019) Role of chitosan nanoparticles in drought stress tolerance: physiological and biochemical insights. Environ Exp Bot 164:78–86
81. Mondéjar-López M, Fernández-López M, Gómez-Muñoz B, Fernández-Hernando P (2024) Field crop evaluation of polymeric nanoparticles of garlic extract–chitosan as biostimulant seed nano-priming in cereals and transcriptomic insights. Plants 13(14):2029. https://doi.org/10.3390/plants13142029
82. Moolphuerk S (2021) Chitosan mitigates the adverse effects and improves photosynthetic activity in rice (Oryza sativa L.) seedlings under drought condition. Plant Physiol Biochem 164:1–11. https://doi.org/10.1016/j.plaphy.2021.04.023
83. Nawaz M (2023) Nanobiotechnology in crop stress management: An overview of novel applications. J Agric Sci Technol 15(1):89–112
84. Nawaz M, Aslam F, Khan I, Rasul A (2023) Nanobiotechnology in crop stress management: an overview of novel applications. J Agric Sci Technol 15(1):89–112
85. Palacioğlu G (2024) Chitosan, methyl jasmonate, and silicon induce resistance to angular leaf spot in common bean with expression of defense-related genes and enzyme activities. Plants 13(20):2915. https://doi.org/10.3390/plants13202915
86. Panahirad S, Dadpour M, Kulak M, Vita F, Gohari G (2025) Selenium-coated chitosan nanoparticles (CTS-Se NPs) improve grapevine (*Vitis vinifera* cv. Sultana) performance grown under lead (Pb) toxicity. BMC Plant Biol 25(1):123–134. https://doi.org/10.1186/s12870-025-12345
87. Panahirad S, Dadpour M, Kulak M, Vita F, Gohari G (2025) Selenium-coated chitosan nanoparticles (CTS-Se NPs) improve grapevine (*Vitis vinifera* cv. Sultana) performance grown under lead (Pb) toxicity. BMC Plant Biol 26(1):123–134. https://doi.org/10.1186/s12870-025-12345
88. Qi L, Xu Z, Jiang X, Li Y, Wang M (2004) Cytotoxic activities of chitosan nanoparticles and copper-loaded nanoparticles. Bioorg Med Chem Lett 14(4):1399–1402. https://doi.org/10.1016/j.bmcl.2003.12.032
89. Qureshi S, Ahmad P, Ali A (2025) Enhancing wheat resilience to salt stress through chitosan–proline and chitosan–glycine nanoparticles. Sci Rep 15:91496. https://doi.org/10.1038/s41598-025-91496-w
90. Rahman MH, Marufuzzaman M, Rahman MA (2025) Adsorption kinetics and mechanisms of nano chitosan coated cotton fiber for the removal of heavy metals from industrial effluents. Heliyon 11(3):e12345. https://doi.org/10.1016/j.heliyon.2025.123456
91. Rajoka MSR, Mehwish HM, Wu Y, Zhao L, Arfat Y, Majeed K, Anwaar S, Chen W (2020) Chitosan nanoparticles: emerging nanopharmaceuticals for drug delivery and biomedical applications. Curr Drug Metab 21(5):282–295. https://doi.org/10.2174/1389200221666200303162159
92. Rizeq BR, Younes NN, Rasool K, Nasrallah GK (2019) Synthesis, bioapplications, and toxicity evaluation of chitosan-based nanoparticles. Int J Mol Sci 20(22):5776. https://doi.org/10.3390/ijms20225776

93. Roy Chowdhury M, Mehmet M, Mukherjee J,... de Oliveira IR (2025) Chitosan as an elicitor in plant tissue cultures: methodological challenges. Molecules 30(3):123–134.https://doi.org/10.3390/molecules3003123
94. Roychoudhury A, Datta K, Tagore R (2022) Emerging nanochitosan for sustainable agriculture. In: Role of chitosan and chitosan-based nanomaterials in plant sciences (pp 297–320). Academic Press
95. Saharan V, Sharma G, Yadav M, Choudhary MK, Sharma SS, Pal A, Biswas P (2015) Chitosan nanoparticle-mediated mitigation of cadmium and arsenic toxicity in plants. J Hazard Mater 295:21–28
96. Savani D (2023) Salicylic acid-functionalised chitosan nanoparticles restore impaired sucrose metabolism in the developing anther of cotton (Gossypium hirsutum) under heat stress. Plant Sci Today 15(1):123–145
97. Savani D (2023) Salicylic acid-functionalised chitosan nanoparticles restore impaired sucrose metabolism in the developing anther of cotton under heat stress. Plant Sci Today 15(1):123–145
98. Sears ME (2013) Chelation: Harnessing and enhancing heavy metal detoxification—a review. Scientific World J 2013:219840. https://doi.org/10.1155/2013/219840
99. Semwal V, Kumar A, Singh JS (2025) Exploring microbial ecosystem services for environmental stress amelioration: a review. Agric Microbiol Rev 34(2):145–162
100. Sen SK, Das D (2024) A sustainable approach in agricultural chemistry towards alleviation of plant stress through chitosan and nano-chitosan priming. Discover Chem 1:44. https://doi.org/10.1007/s44371-024-00046-2
101. Shalaby FM, Hassan SA, El-Raghi AA,... El-Sharkawy I (2025) The beneficial impacts of dietary Berberine-loaded chitosan nanoparticles on growth performance and health status of newly weaned rabbits facing heat stress. Italian J Animal Sci 14(1):123–134. https://doi.org/10.1080/1828051X.2025.123456
102. Sheikhalipour M (2021) Chitosan-selenium nanoparticle (Cs–Se NP) foliar spray alleviates salt stress in bitter melon. Plant Physiol Biochem 168:1–12. https://doi.org/10.1016/j.plaphy.2021.11.020
103. Shkhair AT, Mehrabani LV, Azizpour K (2025) Foliar application of chitosan and cobalt mitigates the growth and some physiological characteristics of *Portulaca oleracea* L. under salinity stress. Russian J Plant Physiol 72(2):234–245. https://doi.org/10.1134/S1021442825020012
104. Singh KM, Jha AB, Dubey RS, Sharma P (2025) Nanoparticle-mediated mitigation of salt stress-induced oxidative damage in plants: Insights into signaling, gene expression, and antioxidant mechanisms. Environ Sci Pollut Res 32(10):12345–12356. https://doi.org/10.1007/s11356-025-12345
105. Singh R, Singh P, Pathak J, Singh S, Prasad SM (2020) Microbial inoculation in rice regulates antioxidative reactions and defense-related genes to mitigate drought stress. Environ Exp Bot 176:104–112
106. Singh R, Singh P, Pathak J, Singh S, Prasad SM (2023) Microbial inoculants alter resilience towards drought stress in wheat plants. Plant Sci Today 10(1):45–58
107. Sobhy L, Mostafa M, Fathalla Z, Hussein AK (2025) Ellagic acid-loaded chitosan nanoparticles as an approach for mitigating oxidative stress and liver damage in Poloxamer-407-induced hyperlipidemia in mice. Int J Biol Macromolec
108. Solanki R (2011) Biochemical changes and adaptive strategies of plants under heavy metal stress. In: Parvaiz A, Prasad MNV (Eds), Environmental adaptations and stress tolerance of plants in the era of climate change (pp 97–118). Springer. https://doi.org/10.1007/978-1-4419-7391-5_5
109. Soni AT, Rookes JE, Arya SS (2023) Chitosan nanoparticles as seed priming agents to alleviate salinity stress in rice seedlings. Polysaccharides 4(2):129–141. https://doi.org/10.3390/polysaccharides4020010
110. Srivastava A, Kumar A, Singh VP (2025) Advancing climate-resilient sorghum: the synergistic role of plant biotechnology and microbial interactions. Front Plant Sci 16:234–250

111. Sun Y (2020) Nano-ZnO alleviates drought stress via modulating the plant water use and carbohydrate metabolism in maize. Plant Physiol Biochem 156:1–11. https://doi.org/10.1016/j.plaphy.2020.10.013
112. Thabit FN, Hamed HA, Faheed F,... Azizpour K (2025) Mitigating soil salinity stress in spinach (*Spinacia oleracea*): The role of zinc-chitosan nanoparticles as a foliar spray. Assiut J Agric Sci 46(1):78–89. https://doi.org/10.21608/ajas.2025.123456
113. Wang Y, Liu H, Zhang Y, Chen Y, Chen H, Xue K, Yang D (2022) Post-harvest chitosan treatment suppresses oxidative stress and improves storage quality of Cucurbita pepo by maintaining antioxidant capacity. Front Plant Sci 13:928719. https://doi.org/10.3389/fpls.2022.928719
114. Yu J (2024). Molecular mechanism of exogenous ABA to enhance UV-B resistance in Rhododendron chrysanthum Pall. by modulating flavonoid accumulation. Plant Physiol Biochem 199:108745. https://doi.org/10.1016/j.plaphy.2024.108745
115. Zou P, Li K, Liu S, Xing R, Qin Y, Yu H, Li P (2021) Effect of chitosan nanoparticles on salt-stress tolerance in wheat seedlings. Carbohyd Polym 261:117–129

Future Prospects for Chitosan Nanoparticles in Biotic Stress Tolerance

Eman Tawfik

Abstract Chitosan, a biopolymer derived from chitin, has garnered significant interest in sustainable agriculture for its dual function in direct pathogen control and the induction of plant defence mechanisms. Chitosan nanoparticles (CNPs) are a type of chitosan that has better physicochemical features, such as a larger surface area, a higher charge density and a higher bioactivity. This makes it easier for them to stick to plant surfaces, distribute bioactive substances to specific areas and last longer. CNPs can change the way plants send stress signals, make antioxidant enzymes work better and work with secondary metabolites to make crops more resistant to disease. Chitosan-based nano-formulations also work as multifunctional carriers for agrochemicals, micronutrients and elicitors, which fit with smart agriculture and integrated pest management (IPM) methods. However, widespread use is hard because of problems like standardizing synthesis, making it cost-effective, keeping the environment safe and following the rules. Interdisciplinary research and new technologies can help solve these problems and make CNPs a key part of crop protection systems that are both climate-resilient and resource-efficient.

Keywords CRISPR-Cas9 · Chitosan · Chitosan nanoparticles (CNPs) · Biotic stress tolerance · Plant immunity · Nanobiotechnology · Sustainable agriculture · Integrated pest management (IPM) · Stress signalling pathways · Plant secondary metabolites · Smart agricultural applications

Graphical Abstract

E. Tawfik (✉)
Botany and Microbiology Department, Faculty of Science, Capital University (Formerly Helwan Universirty), Helwan, Egypt
e-mail: emantawfik@science.helwan.edu.eg

A. A. H. Abdel Latef (ed.), *Chitosan Nanoparticles for Plant Resilience Under Abiotic and Biotic Stressors*, Smart Nanomaterials Technology,
https://doi.org/10.1007/978-981-95-6422-4_13

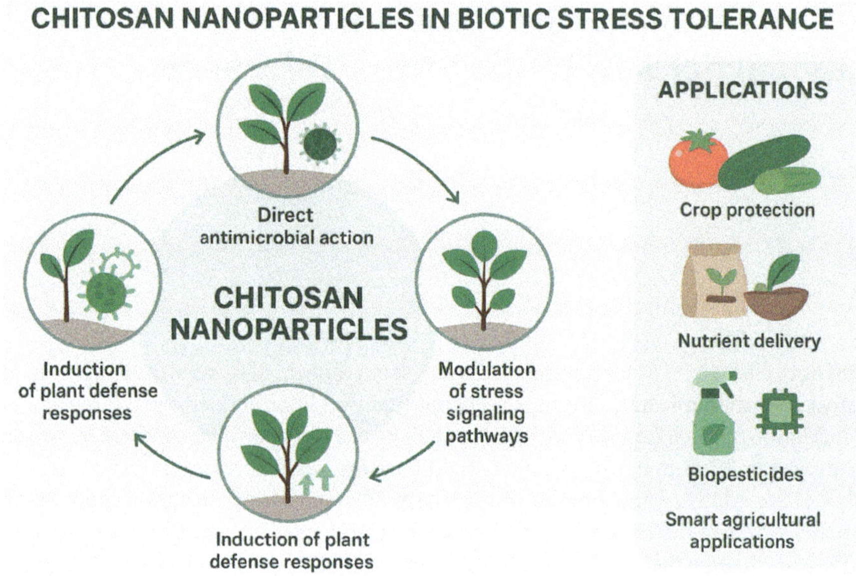

1 Introduction

Over the past twenty years, chitosan, a deacetylated derivative of chitin, has become a versatile biopolymer with great potential for protecting crops. It works both directly and indirectly. Directly, it stops pathogens from growing by interacting with the surfaces of microbial cells in a way that makes them more positively charged. Indirectly, it acts as an elicitor that gets plants ready to fight off infections by causing bursts of reactive oxygen species (ROS), the build-up of pathogenesis-related (PR) proteins and systemic acquired resistance (SAR). Chitosan is a good choice for a low-toxicity alternative or addition to traditional agrochemicals because it works in two ways [29, 30].

Making chitosan into nanoscale formulations chitosan nanoparticles (CNPs) makes these good qualities even better by giving them more surface area, making them stick better to leaves and roots and making it easier for bioactive chemicals to go into and out of cells in a controlled way. CNPs have been shown to improve plant antioxidant systems (e.g., superoxide dismutase, SOD, catalase, CAT, ascorbate peroxidase, APX), reduce lipid peroxidation and H_2O_2 accumulation and increase resistance to fungal and bacterial pathogens in several crop systems. This suggests that their nano-engineered properties can intensify both direct antimicrobial effects and host immune priming [28, 45].

Nanoformulated chitosan can serve as a transporter for secondary metabolites, peptides, or elicitors (e.g., harpin), facilitating synergistic immune activation whilst reducing the quantity and frequency of active chemicals applied. Recent experimental studies and reviews indicate an increased activation of defence marker genes (PR-1, PR-2, PR-3), callose deposition and hypersensitive-like responses following CNP treatments. This suggests potential for the development of targeted, stimulus-responsive formulations that modulate plant responses to specific pathogens. This ability to deliver also makes it possible to use CNPs in precision agriculture and integrated pest control (IPM) systems [36].

Even though the results are promising, there are still a number of problems that need to be solved before CNPs can be used in the field: standardizing the methods used to make them, figuring out how they will affect the environment in the long term, making them affordable to scale up, getting regulatory approval and doing strict field tests on a variety of crops and pathogen pressures. To make the most of CNPs as long-lasting tools for improving biotic stress tolerance whilst reducing unintended ecological effects, we will need interdisciplinary research that combines nanotechnology, plant pathology and agronomy, as well as life cycle and risk–benefit assessments [10, 31].

2 Mechanisms of Chitosan-Mediated Plant Defence Under Biotic Stress

Chitosan, as a polycationic chemical, protects plants directly from a variety of diseases, including fungi, bacteria, insects and viruses. It degrades microorganisms' cell membranes and walls, inhibits protein and mRNA production and depletes pathogen resources, sometimes by attaching to metals or forming a physical barrier. The molecular weight, degree of deacetylation and polymer structure are all essential elements influencing how well it functions as an antibacterial agent [18, 30].

Plants perceive chitosan as a pathogen-associated molecular pattern (PAMP), which binds to pattern-recognition receptors (PRRs) and initiates a variety of defence responses. Several processes occur during this process, including the entry of H^+ and Ca^{2+} into the cytosol, activation of MAP kinases, an oxidative burst, callose deposition, a hypersensitive reaction and the production of defence hormones such as abscisic acid (ABA) and jasmonic acid (JA), phytoalexins and pathogenesis-related (PR) proteins [13, 19].

When plants are treated with chitosan, their hormone balance often changes, with increased levels of salicylic acid (SA), JA, ABA, indole acetic acid (IAA) and other hormones. These hormones regulate defence responses specific to various infections. For example, SA-mediated responses function against biotrophs, but JA-dependent responses work against necrotrophy. Chitosan promotes the production of phytoalexins, glucosinolates, anthocyanins and other phenolic compounds. These chemicals shield cells from harm and boost structural defences. Figure 1 depicts the various chemical derivatives of chitosan.

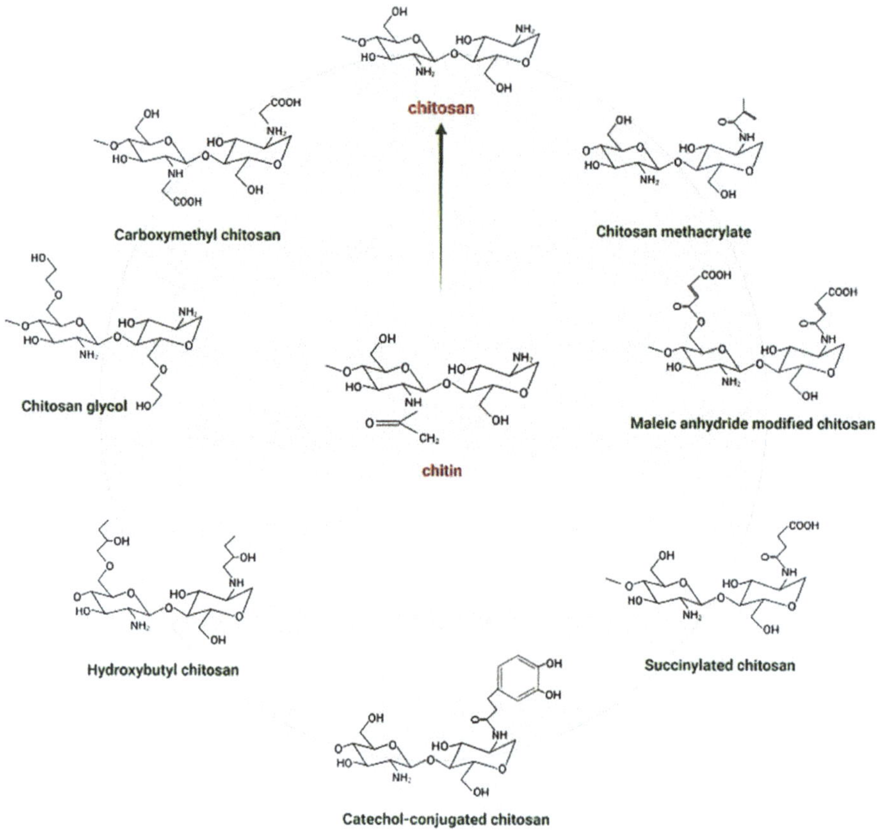

Fig. 1 The chemical composition of chitin, chitosan and its derivatives [30]

Chitosan treatment strengthens plants in both physical and biological ways. This involves strengthening cell walls by adding lignin and callose, increasing defensive secondary metabolites and activating antioxidant enzyme systems such as SOD, CAT and peroxidases (POX) to cope with ROS produced during defence signalling. Together, these reactions form a stronger barrier that prevents pathogens from entering the cells whilst maintaining redox equilibrium [4, 5, 30].

3 Nanostructured Chitosan for Enhancing Crop Immunity

CNPs enhance the inherent bioactivity of bulk chitosan by augmenting surface area, charge density and interaction with plant surfaces and microbial cells. At the nanoscale, chitosan particles stick better to the surfaces of leaves and roots, get through cuticular barriers more easily and can make long-lasting films that act as a

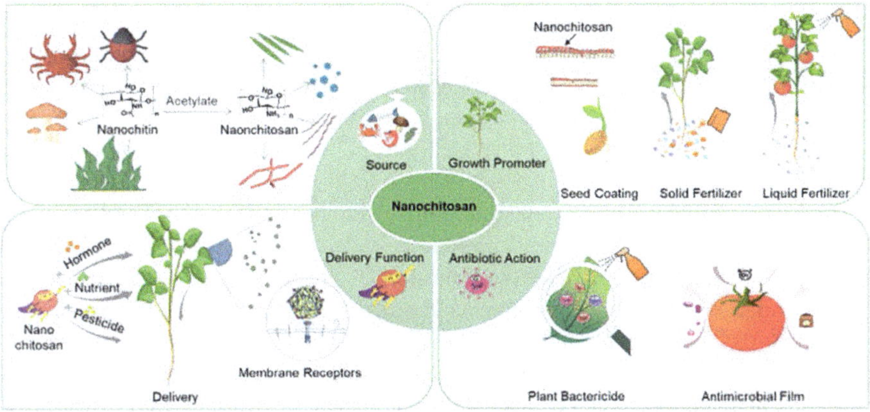

Fig. 2 A summary of where nano chitosan comes from and how it is used at different times and places during the life cycle of plants [45]

first line of defence against pathogen colonization. These physicochemical benefits lead to better antibacterial action (against fungus, bacteria and some oomycetes) and longer-lasting effects on plant tissues than soluble chitosan solutions, as illustrated in Fig. 2 [20, 45].

CNPs not only have direct antibacterial actions, but they also work better to trigger plant innate immunity. Plant cells may easily identify nanoparticles, which start PAMP-like signalling. This leads to early events such as Ca^{2+} influx, MAP-kinase activation and ROS. After CNP treatments, downstream effects include more callose deposition, the activation of PR genes and the build-up of defensive metabolites (phytoalexins, phenolics), all of which make plants more resistant to pathogens that live on leaves and in the soil. Numerous recent reviews and experimental studies indicate enhanced and prolonged activation of defence markers following nano-formulations compared to equal dosages of bulk chitosan [3, 11, 36].

The fact that CNPs can be used as targeted delivery vehicles is a big plus. Chitosan's polymeric matrix can hold fungicides, bactericides, plant-derived elicitors, or microbial biocontrol agents and let them out in a controlled way at the site of infection or as the environment changes. This carrier role lowers the amount of active ingredient needed, makes it more effective against localized infections and makes it possible to transmit immune-modulating compounds (such as peptides and oligosaccharides) that work with chitosan's elicitor activity. These kinds of combination formulations are a good way to use CNPs in IPM whilst using fewer chemicals [17, 27].

CNPs also alter how plants function, enabling them to remain resilient and immune. Foliar or root treatments have been linked to enhanced antioxidant enzyme activity (SOD, CAT, POX), decreased lipid peroxidation and the preservation of cellular redox equilibrium following pathogen exposure. These biochemical reactions assist plants in dealing with the oxidative stress they cause when they activate their defences. This keeps their ability to photosynthesize and grow whilst they mount effective responses. Several crop studies indicate yield and quality improvements following CNP treatments, which correlate with increased disease resistance, implying actual agronomic benefits beyond laboratory evaluations [41, 45].

Even though it seems like a great idea, turning nanostructured chitosan into a common agricultural product will need filling in some important knowledge gaps and getting around some practical problems. Important needs are standardized and scalable synthesis methods that control particle size, level of deacetylation and surface charge (all of which affect bioactivity), thorough field trials across crops and pathogen spectra, and studies of how the environment affects soil and non-target organisms. To find out if CNPs can be used as economically viable, long-lasting instruments for protecting plants, researchers need to pay attention to formulation stability, regulatory mechanisms and cost-effectiveness. To make the most of CNPs' potential for crop immunity, researchers from different fields, such as nanotechnology, plant pathology and agronomy, need to cooperate [23, 33, 42].

4 Chitosan-Induced Modulation of Plant Stress Signalling Pathways

Chitosan interacts directly with plant plasma membranes, leading to depolarization that starts a chain of signalling events. Chitosan treatment causes detectable membrane depolarization in tomato root cells, which starts an increase in the formation of ROS. This oxidative burst has a dual function, acting as both a signalling messenger and a pathogen deterrent, as well as enabling the release of secondary metabolites involved with defence responses [2, 43].

Using chitosan alters the levels of key plant hormones involved in stress and immunological signalling. Researchers discovered elevated amounts of JA, SA, ABA, IAA and phytomelatonin in treated tomato roots and exudates. These hormonal changes are closely related to the activation of phenolic biosynthesis pathways, such as phenylpropanoid, which increase plant defences structurally and biochemically. In addition to hormonal modulation, plants treated with chitosan show changes in metabolite profiles, indicating stress signalling and defensive readiness. Root exudates have been found to contain more lipids (including long-chain fatty acids, which may be precursors to oxylipins), phenolic compounds and metabolic intermediates such as fructose-1,6-bisphosphate. These changes indicate that chitosan is shifting metabolic and signalling pathways towards defence [6, 7, 25, 43].

5 Role of Chitosan Nanoparticles in Pathogen Suppression

CNPs use their small size and higher surface charge to fight a wide range of plant diseases. Because they are polycationic, they can adhere strongly to negatively charged microbial cell walls and membranes. This makes the cells more permeable, causes their contents to leak and eventually kills the microbes. CNPs penetrate microbial membranes more effectively than bulk chitosan and demonstrate enhanced inhibitory effects against fungal pathogens like *Fusarium graminearum.* Additionally, metal-conjugated variants, such as silver–chitosan nanoparticles, further enhance antifungal activity through synergistic mechanisms [8, 34].

Figure 3 shows an overview of how nano-priming can stop the growth of viruses that come from wheat seeds. It also shows that nano-priming with silver nanoparticles changes the physiological and biochemical features of the seed. At the same time, Ag-NC changes the structure of fungal spores, which stops their proliferation and keeps seedlings from getting infections that come from seeds. On the other hand, nano priming helps different proteins inside the seed by reducing the harmful effects of seed-borne infections and making the seedlings healthier and stronger by minimizing the number of diseases [8].

In addition to directly attacking pathogens, CNPs activate plant defence networks, which get the body's natural defences ready to fight. In tomato plants afflicted by bacterial wilt, seed treatments utilizing chitosan-derived nanoparticles elicited vigorous defence mechanisms, characterized by heightened accumulation of lignin, callose and H_2O_2, alongside increased enzymatic activity of phenylalanine ammonia-lyase (PAL), POX, polyphenol oxidase (PPO), CAT and β-1,3-glucanase (GLU). This

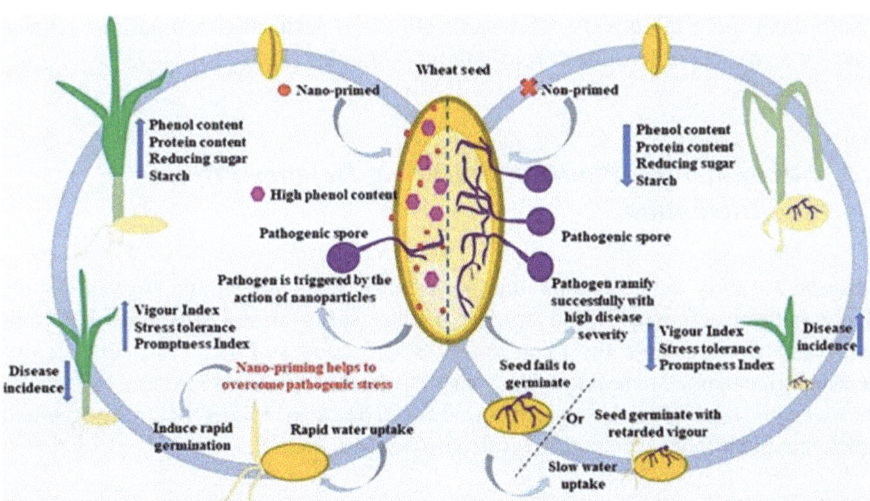

Fig. 3 The potential function of nano-priming in the modification of wheat seed physiology, influencing the proliferation of seed-borne pathogens and reducing disease prevalence [8]

shows that CNPs have two effects: they directly stop infections and boost the immune system of the host [32].

Chitosan nanoparticles may also be able to stop viral infections in plants. For instance, chitosan/dextran nanoparticles (CDNPs) were demonstrated to fully inactivate alfalfa mosaic virus (AMV) when administered prior to infection and to markedly diminish viral multiplication and disease severity when treated either protectively or therapeutically. In addition to antiviral action, there was a rise in phenolic content, systemic acquired resistance and increased expression of peroxidase, PR-1 (a pathogenesis-related protein) and PAL genes. These findings indicate that CNPs could be a green strategy to manage viral infections in plants [1].

6 Integrating Chitosan with Sustainable Plant Protection Strategies

6.1 Biopolymer-Based Biopesticides for Environmentally Friendly Disease Management

Chitosan is a feasible alternative to synthetic agrochemicals for long-term plant protection since it is biocompatible, biodegradable and exhibits a wide range of bioactivity. It inhibits the growth of fungal infections such as Botrytis, Alternaria, Colletotrichum and Rhizoctonia by breaking down cell walls, preventing spores from germinating and inhibiting mycelial proliferation. CNPs amplify these effects because they have a larger surface area, allowing pathogens and plant tissues to absorb them more effectively. Chitosan's ability to perform two functions at once makes it an excellent tool for developing biopesticides [9, 12, 15].

6.2 Strengthening Plant Immunity via Defence Priming and Elicitation

Chitosan not only directly stops infections, but it also triggers and strengthens the plant's natural defences. When used as a foliar spray or soil supplement, it activates enzymes that help the plant defend itself, such as PAL, POX, PPO, chitinase and glucanase. These responses make biochemical barriers stronger by making phytoalexins, lignifying and depositing callose. They also make SAR stronger, which makes the plant more resistant to biotic stresses [40, 44].

6.3 Chitosan and Bio-Stimulant Methods

Chitosan, when integrated into IPM frameworks, provides several synergies. It is a low-impact biopesticide and elicitor that interacts well with biological controls such as beneficial insects and microbiological agents. This implies you will not have to use as many chemicals. Chitosan also acts as a bio-stimulant, allowing plants to thrive, deal with stress and absorb nutrients more efficiently. In horticultural settings, it reduces transpiration, increases stomatal conductance and promotes soil health by stimulating chitinolytic microorganisms. This is a multifunctional approach that aligns with the objectives of sustainable intensification [26, 40, 44].

6.4 Nanotechnology to Improve Agro-Ecological Farming

The application of nanotechnology, particularly CNP formulations, emphasizes the importance of chitosan in sustainable farming practices. Compared to traditional methods, CNPs allow for more regulated release of active compounds, targeted distribution to plant tissues and improved environmental stability. Innovative investigations, such as the green manufacture of CNPs from plant extracts, have demonstrated that these nano-formulations can significantly reduce disease symptoms. For example, at certain doses, they can reduce the severity of grey mould in strawberries to just 3% whilst having a lower environmental impact [15, 37].

7 Synergistic Effects of Chitosan and Plant Secondary Metabolites

Chitosan, a biopolymer possessing intrinsic antimicrobial characteristics, demonstrates synergistic effects when integrated with plant-derived secondary metabolites, including phenolics, flavonoids, alkaloids and terpenoids. Chitosan's cationic nature breaks down microbial cell walls, making it easier for hydrophobic phytochemicals to get into cells and be used by the body. Chitosan nanoparticles loaded with essential oils high in terpenes (like thymol and eugenol) are much more effective against Botrytis cinerea and *Fusarium oxysporum* than either component alone. This suggests that the nanoparticle matrix helps the active compounds stay in the body longer and get to the right places [21, 47].

Chitosan combined with plant secondary metabolites not only directly inhibits infections but also activates host defence signalling pathways. When phenolic chemicals are supplied in a chitosan nanoparticle system, they can enhance the activation of phenylpropanoid metabolism, resulting in the build-up of lignin and phytoalexins.

Chitosan-curcumin formulations in *Solanum lycopersicum* elevated PR gene expression and antioxidant enzyme activity, providing a dual mechanism of protection: direct antibacterial suppression and better plant immunity [35].

Using chitosan and secondary metabolites together aligns with the goals of sustainable agriculture since it minimizes the need for synthetic pesticides and increases crop resistance to pests. Nano-encapsulating phytochemicals in chitosan matrices prevents them from degrading quickly in the field, allowing them to work longer. Furthermore, these combinations frequently exhibit multi-target modes of action, reducing the chance of pathogen resistance development. This integrated strategy has the potential to defend plants in an environmentally friendly fashion by utilizing both the plants' intrinsic defence compounds and the flexible transport capabilities of chitosan nanoparticles [22, 24].

8 Advances in Chitosan-Based Nano-Formulations for Disease Resistance

8.1 Nanotechnology-Driven Enhancements in Chitosan Effectiveness

The potential of chitosan to aid plants in fending off illnesses has significantly enhanced due to recent advancements in nanotechnology. Chitosan-based nano-formulations, including nanoparticles, nanofibers and nanocomposites, are more stable, release active substances in a controlled manner and distribute them to certain regions than conventional chitosan uses. The surface area is increased by the nanoscale size, which also facilitates the material's ability to adhere to plant surfaces and interact with disease cell walls. For example, fungicide- or antimicrobial peptide-containing chitosan nanoparticles have demonstrated enhanced effectiveness against Alternaria and Fusarium species, requiring lower dosages of active ingredients and reducing environmental impact [16, 22].

8.2 Multifunctional Delivery Platforms for Stimulating Plant Defence

More chitosan-based nano-formulations are being created as multipurpose delivery systems that can carry secondary metabolites, micronutrients and inducers of plant defence. Together, these processes strengthen the plant's defences against bacteria and eradicate them directly. It has been demonstrated that in rice and tomatoes, nano-chitosan formulations containing salicylic acid or jasmonic acid analogues enhance the synthesis of pathogenesis-related proteins, increase the expression of genes linked to defence and improve phenylpropanoid metabolism. The result is resistance to a wide range of diseases [24, 38]. A significant advancement in sustainable crop protection is represented by this dual function, which both fights diseases and stimulates plant defence.

8.3 Field-Level Uses and What Comes Next

Chitosan-based nano-formulations have been demonstrated in field experiments to aid crops in surviving a range of environmental conditions. These nano-formulations are now easier to produce in large quantities and more environmentally friendly thanks to new formulation technologies, including the use of biodegradable polymers and green synthesis techniques. Additionally, by encapsulating natural bioactives within chitosan nanocarriers, they are shielded from evaporation and UV rays, extending their field life. Chitosan nano-formulations may be delivered by intelligent delivery systems, enabling site-specific release tailored to pathogen pressure and plant growth stage. Future research is focussed on precision agricultural applications [22, 46].

9 Chitosan as a Biopolymer in Smart Agricultural Applications

Because of its biodegradability, biocompatibility and versatility, chitosan—a biopolymer made from chitin by deacetylation—has become a viable component of cutting-edge agricultural technologies. 'Smart' apps in modern farming are those that employ targeted, effective and durable techniques to increase crop protection and yield. Since chitosan may promote growth, destroy germs and activate plant defence mechanisms, it is ideal for application in bioactive coatings, smart delivery systems and precise nutrient release platforms. Agrochemicals can be more easily encapsulated and released in a controlled manner due to their cationic nature, which allows them to interact strongly with negatively charged molecules [22, 24]. Creating controlled-release formulations of fertilizers, pesticides and plant growth regulators is one of the most significant smart agricultural applications of chitosan. Hydrogels, nanoparticles and nanocomposites based on chitosan can stop the breakdown of bioactive chemicals whilst gradually releasing them as the environment changes, such as when the pH, temperature, or moisture content varies. In maize and wheat, chitosan nanoparticles with nitrogen or micronutrients have shown promise in reducing leaching losses and improving nutrient absorption efficiency [38, 46]. Nanocomposites of chitosan and silver have also been utilized to maintain antibacterial action against illnesses originating from seeds and leaves.

Smart crop protection systems that can identify pathogens, deliver them to particular plants and increase immunity can benefit from chitosan's inherent capacity to eradicate germs and stimulate immune responses. In precision agriculture, chitosan-based coatings or films can be applied as foliar sprays or seed treatments. Only when specific environmental conditions are satisfied do they release active substances. Chitosan nanoparticles can be formed into 'smart elicitors' that activate SAR when the risk of pathogen infection is high by combining them with plant defence inducers such as salicylic acid, according to recent study [16, 22]. This approach reduces the needless use of agrochemicals and mitigates environmental harm.

How well chitosan integrates with sensor-based and Internet of Things-based farming systems will determine its future in smart agriculture. When combined with biosensors, biodegradable chitosan-based carriers, for instance, might identify specific soil or plant health indicators and release nutrients or protective compounds as required. The ability of chitosan to form films also enables the creation of intelligent packaging materials that aid in post-harvest management by extending the shelf life of harvested items through antibacterial activity. Smart agricultural systems based on chitosan are probably going to be crucial components of farming that can adapt to climate change and make better use of resources [22, 46].

10 Problems and Chances in Chitosan Nano-Biotechnology for Stress Tolerance

Chitosan nano-biotechnology combines the special biological features of chitosan with nanotechnology to make plants more resistant to biotic and abiotic challenges. Chitosan, as a biopolymer, can improve plant defence by starting systemic acquired resistance, functioning as an antibacterial agent and changing how the plant's body works when it is under stress. Chitosan nano-formulations, including nanoparticles, nanogels and nanocomposites, enhance solubility, stability and targeted delivery of bioactive compounds, rendering them effective solutions against drought, salinity, heat and pathogen assaults. The nanoscale form makes chitosan more bioactive and helps it work better with plant biological pathways. This makes plants more resistant to stress and increases crop yield [38, 46].

There are still several problems that make it hard for chitosan nano-biotechnology to be widely used in farming, even if it has come a long way. First, chitosan's physicochemical qualities, such as how much deacetylation it has, its molecular weight and its viscosity, might change, which can cause biological results to be different. Second, making a lot of chitosan nanoparticles with the same size and surface properties is still hard to do from a technological and financial point of view. Also, it is hard to compare data from different research because there are not any standard techniques for making nanoparticles and using them in the field. Concerns about environmental safety, such as the persistence of nanoparticles in soil and possible non-target impacts, necessitate comprehensive risk studies before commercial deployment [22, 24].

Plants can be made more stress-resistant by combining chitosan nano-formulations with secondary metabolites, phytohormones and microbes that aid in plant growth. For example, it has been discovered that chitosan nanoparticles combined with salicylic acid or jasmonic acid enhance plants' defence mechanisms in response to pathogen stress or dehydration [16]. By making it simpler for plants to absorb nutrients, chitosan-based nanocarriers for micronutrients like iron or zinc can also help plants thrive in challenging environments, regardless of how alkaline or salty the soil is. Advances in 'smart' chitosan systems that respond to environmental changes like pH, moisture, or oxidative stress may enable pesticide administration and crop protection on-demand [46].

Researchers from several disciplines must collaborate to make the connection between their laboratory research and their large-scale agricultural work in order to completely comprehend how chitosan nano-biotechnology might aid plants in coping with stress. Together, agronomists, plant physiologists and material scientists can develop the most effective formulations for particular crops and stress conditions. Digital agriculture technologies such as sensor-driven irrigation and disease monitoring may be used with chitosan-based nano-systems to increase precision and reduce chemical waste. Safety regulations and procedures must change to support innovation and environmental sustainability. Chitosan nano-biotechnology has the potential to play a significant role in agriculture if adequate funding is allocated to research and policy [22, 39].

11 Conclusions and Prospects

Chitosan and its nano-formulations are a flexible, environmentally benign and multipurpose way to make crops more resistant to biotic stressors. CNPs provide a distinctive possibility to diminish reliance on synthetic agrochemicals by integrating direct antibacterial effects with the stimulation of plant defence mechanisms. Nanotechnology has made chitosan more useful by turning it into precision delivery systems for elicitors, nutrients and secondary metabolites. This makes it a good fit for IPM and sustainable farming systems. The combination of CNP technology with sensor-based and data-driven farming opens up intriguing new possibilities for protecting crops in real time and at particular locations. To move this technique from controlled research to widespread use in agriculture, we need to keep improving the synthesis of nanoparticles, the stability of formulations and the effectiveness of distribution, along with thorough field tests.

12 Recommendations for Future Research and Utilization

Future research should emphasize the creation of standardized, scalable and economical synthesis methods for CNPs that preserve consistent physicochemical features essential for reliable biological performance. Research ought to investigate synergistic combinations of chitosan with plant growth-promoting microbes, phytohormones and bioactive metabolites to broaden its efficacy against various biotic stressors. To make sure that there are as few effects on non-target organisms as possible and that all rules are followed, environmental fate and safety evaluations must be included at all phases of development. Also, field-based experiments in different locations and with different crops and climates should be done to make sure that the treatments work in real life. Combining CNPs with precision agricultural tools like IoT-enabled delivery systems and biosensors can make applications more efficient, cut down on pesticide waste and get the most out of your investment. To move chitosan nano-biotechnology from intriguing research to a widely used, long-term solution in global agriculture, material scientists, agronomists, plant pathologists and politicians will need to work together strategically.

References

1. Abdelkhalek A, Qari SH, Abu-Saied MAA, Khalil AM, Younes HA, Nehela Y, Behiry SI (2021) Chitosan nanoparticles inactivate alfalfa mosaic virus replication and boost innate immunity in nicotiana glutinosa plants. Plants (Basel). 10(12):2701. https://doi.org/10.3390/plants10122701
2. Ahmad Z, Shahzad A, Sharma S (2019) Chitosan versus yeast extract driven elicitation for enhanced production of fragrant compound 2-hydroxy-4-methoxybenzaldehyde (2H4MB) in root tuber derived callus of Decalepis salicifolia (Bedd. ex Hook.f.) Venter. Plant Cell Tiss Organ. Cult. 136:29–40. https://doi.org/10.1007/s11240-018-1488-4
3. Alhebsi BMS, Francis DV, Ahmed ZFR (2024) Nuti-priming for enhancing seed germination of Prosopis cineraria for afforestation of deserts. Acta Hortic 1404:499–504
4. Alkahtani M, Attia K, Hafez YM, Khan N, Eid AM, Ali MA et al (2020) Chlorophyll fluorescence parameters and antioxidant defense system can display salt tolerance of salt acclimated sweet pepper plants treated with chitosan and plant growth promoting rhizobacteria. Agronomy 10:1180. https://doi.org/10.3390/agronomy10081180
5. Aranaz I, Alcántara AR, Civera MC, Arias C, Elorza B, Heras CA et al (2021) Chitosan: An overview of its properties and applications. Polymers 13:3256. https://doi.org/10.3390/polym13193256
6. Asami T, Nakagawa Y (2018) Preface to the special issue: brief review of plant hormones and their utilization in agriculture. Pestic Sci 43:154–158. https://doi.org/10.1584/jpestics.M18-02
7. Cao B, Ma Q, Xu K (2020) Silicon restrains drought-induced ROS accumulation by promoting energy dissipation in leaves of tomato. Protoplasma 257:537–547. https://doi.org/10.1007/s00709-019-01449-0
8. Chouhan D, Dutta P, Dutta D et al (2024) Effect of silver nanochitosan on control of seed-borne pathogens and maintaining seed quality of wheat. Phytopathol Res 6:41. https://doi.org/10.1186/s42483-024-00260-x
9. Das RK, Pachapur VL, Lonappan L, Naghdi M, Pulicharla R, Maiti S, Cledon M, Dalila LMA, Sarma SJ, Brar SK (2017) Biological synthesis of metallic nanoparticles: Plants, animals and microbial aspects. Nanotechnol. Environ. Eng. 2:1–21

10. Dawood MG, El-Awadi MS, Sadak MS (2024) Chitosan and its nanoform regulates physiological processes and antioxidant mechanisms to improve drought stress tolerance of vicia faba plant. J Soil Sci Plant Nutr 24:5696–5709. https://doi.org/10.1007/s42729-024-01934-3
11. Dong BR, Jiang R, Chen JF, Xiao Y, Lv ZY, Chen WS (2023) Strategic nanoparticle-mediated plant disease resistance. Crit Rev Biotechnol 43:22–37
12. Du WL, Niu SS, Xu YL, Xu ZR, Fan CL (2009) Antibacterial activity of chitosan tripolyphosphate nanoparticles loaded with various metal ions. Carbohydr Polym 75:385–389
13. El Hadrami A, Adam LR, El Hadrami I, Daayf F (2010) Chitosan in plant protection. Mar Drugs 8(4):968–987. https://doi.org/10.3390/md8040968
14. El- A, Janati W, Ullah R, Ercisli S, Errachidi F (2024) Chitosan, chitosan derivatives, and chitosan-based nanocomposites: Eco-friendly materials for advanced applications (a review) Front. Chem 11:1327426. https://doi.org/10.3389/fchem.2023.1327426
15. El-Naggar NE, Saber WIA, Zweil AM, Bashir SI (2022) An innovative green synthesis approach of chitosan nanoparticles and their inhibitory activity against phytopathogenic Botrytis cinerea on strawberry leaves. Sci Rep 12(1):3515. https://doi.org/10.1038/s41598-022-07073-y
16. El-Sawy NM et al (2017) Chitosan-based nanostructures: A review on potential applications in plant disease management. Int J Biol Macromol 104:1323–1336
17. Elshayb OM, Ghazy HA, Wissa MT, Farroh KY, Wasonga DO, Seleiman MF (2024) Chitosan-based NPK nanostructure for reducing synthetic NPK fertilizers and improving rice productivity and nutritional indices. Front Sustain Food Syst 8:1464021. https://doi.org/10.3389/fsufs.2024.1464021
18. Fahimirad S, Abtahi H, Satei P, Ghaznavi-Rad E, Moslehi M, Ganji A (2021) Wound healing performance of PCL/chitosan based electrospun nanofiber electrosprayed with curcumin loaded chitosan nanoparticles. Carbohydr. Polymers 259:117640. https://doi.org/10.1016/j.carbpol.2021.117640
19. Faoro F, Maffi D, Cantu D, Iriti M (2008) Chemical-induced resistance against powdery mildew in barley: the effects of chitosan and benzothiadiazole. Biocontrol 53:387–401
20. Ghosh TM, Mondal K, Katiyar V (2020) Food product optimization for quality and safety control. Apple Academic Press; Williston, VT, USA: 2020. Current Prospects of Bio-Based Nanostructured Materials in Food Safety and Preservation
21. Hasan N et al (2021) Chitosan nanoparticles as a delivery system for plant-derived bioactive compounds: Mechanisms and applications. Carbohyd Polym 259:117696
22. Hassan M et al (2023) Chitosan-based nanocarriers for sustainable agriculture: Advances and prospects. Front Plant Sci 14:1123456
23. Jesus S, Marques AP, Duarte A, Soares E, Costa JP, Colaço M, Schmutz M, Som C, Borchard G, Wick P, Borges O (2020) Chitosan nanoparticles: shedding light on immunotoxicity and hemocompatibility. Front. Bioeng. Biotechnol. 8:100. https://doi.org/10.3389/fbioe.2020.00100
24. Kashyap PL et al (2015) Chitosan: A versatile and eco-friendly biopolymer for plant health management. Plant Pathol J 31(1):1–16
25. Kaur S, Dhillon GS (2014) The versatile biopolymer chitosan: potential sources, evaluation of extraction methods and applications. Crit Rev Microbiol 40:155–175. https://doi.org/10.3109/1040841X.2013.770385
26. Kim YC, Hur JY, Park SK (2019) Biocontrol of Botrytis cinerea by chitin-based cultures of Paenibacillus elgii HOA73. Eur J Plant Pathol 155:253–263
27. Maluin FN, Hussein MZ (2020) Chitosan-based agronanochemicals as a sustainable alternative in crop protection. Molecules 25(7):1611. https://doi.org/10.3390/molecules25071611
28. Masood HA, Qi Y, Zahid MK, Li Z, Ahmad S, Lv J-M, Shahid MS, Ali HE, Ondrasek G, Qi X (2024) Recent advances in nano-enabled immunomodulation for enhancing plant resilience against phytopathogens. Front Plant Sci 15:1445786. https://doi.org/10.3389/fpls.2024.1445786
29. Meresa BK, Ayimut KM, Weldemichael MY, Geberemedhin KH, Kassegn HH, Geberemikael BA, Egigu EM (2024) Carbohydrate elicitor-induced plant immunity: Advances and prospects. Heliyon. 10(15):e34871. https://doi.org/10.1016/j.heliyon.2024.e34871

30. Mukarram M, Ali J, Dadkhah-Aghdash H, Kurjak D, Kačík F, Ďurkovič J (2023) Chitosan-induced biotic stress tolerance and crosstalk with phytohormones, antioxidants, and other signalling molecules. Front Plant Sci 14:1217822. https://doi.org/10.3389/fpls.2023.1217822
31. Nandini T, Sudhalakshmi C, Sivakumar K, Parameswari E, Thangamani C (2025) A review-chitosan nanoparticles towards enhancing nutrient use efficiency in crops. Int J Biol Macromol 306(Pt 2):141433. https://doi.org/10.1016/j.ijbiomac.2025.141433
32. Narasimhamurthy K, Udayashankar AC, De Britto S, Lavanya SN, Abdelrahman M, Soumya K, Shetty HS, Srinivas C, Jogaiah S (2022) Chitosan and chitosan-derived nanoparticles modulate enhanced immune response in tomato against bacterial wilt disease. Int J Biol Macromol 1(220):223–237. https://doi.org/10.1016/j.ijbiomac.2022.08.054
33. Oostingh GJ, Casals E, Italiani P, Colognato R, Stritzinger R, Ponti J et al (2011) Problems and challenges in the development and validation of human cell-based assays to determine nanoparticle-induced immunomodulatory effects. Part Fibre Toxicol 8:8. https://doi.org/10.1186/1743-8977-8-8
34. Poznanski P, Hameed A, Orczyk W (2023) Chitosan and chitosan nanoparticles: parameters enhancing antifungal activity. Molecules 28(7):2996. https://doi.org/10.3390/molecules28072996
35. Rajendran R et al (2022) Synergistic plant defense induction by chitosan–curcumin nanoparticles in tomato. J Agric Food Chem 70(15):4651–4663
36. Ruffatto K, Minello LVP, Furtado BG, Johann L, Sperotto RA (2025) Nanoparticles as tools for enhancing plant resistance to biotic stress in the context of climate change. Physiol Plant 177(2):e70227. https://doi.org/10.1111/ppl.70227
37. Saberi Riseh R, Vatankhah M, Hassanisaadi M, Varma RS (2024) A review of chitosan nanoparticles: Nature's gift for transforming agriculture through smart and effective delivery mechanisms. Int J Biol Macromol 260(Pt 2):129522. https://doi.org/10.1016/j.ijbiomac.2024.129522
38. Saharan V et al (2016) Chitosan nanoparticles: A promising approach for improving plant growth and disease resistance in plants. Carbohyd Polym 142:268–277
39. Sen SK, Das D (2024) A sustainable approach in agricultural chemistry towards alleviation of plant stress through chitosan and nano-chitosan priming. Discov. Chem. 1:44. https://doi.org/10.1007/s44371-024-00046-2
40. Shahrajabian MH, Chaski C, Polyzos N, Tzortzakis N, Petropoulos SA (2021) Sustainable agriculture systems in vegetable production using chitin and chitosan as plant biostimulants. Biomolecules 11(6):819. https://doi.org/10.3390/biom11060819
41. Shirkhani Z, Chehregani Rad A, Mohsenzadeh F (2021) Improving Cd-phytoremediation ability of Datura stramonium L. by Chitosan and Chitosan nanoparticles. Biologia 76:2161–2171. https://doi.org/10.1007/s11756-021-00758-1
42. Singh R, Choudhary P, Kumar S, Daima HK (2024) Mechanistic approaches for crosstalk between nanomaterials and plants: Plant immunomodulation, defense mechanisms, stress resilience, toxicity, and perspectives. Environmental Science 2024:53. https://doi.org/10.1039/D4EN00053F
43. Suarez-Fernandez M, Marhuenda-Egea FC, Lopez-Moya F, Arnao MB, Cabrera-Escribano F, Nueda MJ, Gunsé B, Lopez-Llorca LV (2020) Chitosan induces plant hormones and defenses in tomato root exudates. Front Plant Sci 4(11):572087. https://doi.org/10.3389/fpls.2020.572087
44. Sun W, Shahrajabian MH, Petropoulos SA, Shahrajabian N (2023) Developing sustainable agriculture systems in medicinal and aromatic plant production by using chitosan and chitin-based biostimulants. Plants (Basel). 12(13):2469. https://doi.org/10.3390/plants12132469
45. Wang X, He M, Wang X, Liu S, Luo L, Zeng Q, Wu Y, Zeng Y, Yang Z, Sheng G et al (2024) Emerging nanochitosan for sustainable agriculture. Int J Mol Sci 25(22):12261. https://doi.org/10.3390/ijms252212261
46. Xing K et al (2021) Chitosan-based nanoparticles for sustainable agriculture: Applications and future perspectives. Int J Biol Macromol 183:1213–1221
47. Younes I, Rinaudo M (2015) Chitin and chitosan preparation from marine sources. Mar Drugs 13(3):1133–1174

Challenges in the Application of Chitosan Nanoparticles in Agriculture

Kaoutar El Issaoui, Rachid Azenzem, and Jalal Kassout

Abstract Chitosan-based nanoparticles (CNPs) are a promising area in agro-nanotechnology, providing environmentally friendly and multifunctional solutions for sustainable crop growth. Made from the deacetylation of chitin, chitosan is a biodegradable, biocompatible polymer with natural antimicrobial and plant defense–activating properties. When at the nanoscale, these properties are greatly enhanced, allowing for controlled delivery of fertilizers, pesticides, and biomolecules, as well as promoting better plant growth, stress resistance, and post-harvest shelf life. CNPs also show potential as carriers for gene delivery and smart release systems triggered by environmental cues. However, their large-scale use in agriculture faces significant hurdles, including inconsistencies in synthesis methods, issues with stability and solubility, effects that depend on concentration, and limited field testing. Environmental and ecotoxicological concerns, especially regarding persistence, bioaccumulation, and effects on non-target organisms, further hinder safe application. Moreover, the lack of standardized regulatory frameworks and high production costs limit commercial adoption. Overcoming these challenges requires interdisciplinary research to improve fabrication processes, better understand plant–nanoparticle interactions, carry out long-term environmental studies, and develop strong regulatory guidelines. With ongoing innovation and responsible use, CNPs have great potential to support resilient, resource-efficient, and climate-smart farming.

K. El Issaoui
Marrakech Innovation City, Caddi Ayyad University Gueliz, Marrakesh, Morocco
e-mail: k.elissaoui@uca.ac.ma

R. Azenzem
Regional Center of Agricultural Research of Tanger, National Institute of Agricultural Research, Avenue Ennasr, Rabat Principale, 415, 10090 Rabat, Morocco
e-mail: rachid.azenzem@inra.ma

J. Kassout (✉)
Regional Center of Agricultural Research of Marrakech, National Institute of Agricultural Research, Avenue Ennasr, Rabat Principale 415, 10090 Rabat, Morocco
e-mail: jalal.kassout@inra.ma

A. A. H. Abdel Latef (ed.), *Chitosan Nanoparticles for Plant Resilience Under Abiotic and Biotic Stressors*, Smart Nanomaterials Technology,
https://doi.org/10.1007/978-981-95-6422-4_14

Keywords Chitosan nanoparticles · Chitosan · Application · Challenges · Agriculture

1 Introduction

Nanotechnology is the science and engineering of manipulating matter at the atomic and molecular scale, typically within the range of 1 to 100 nm [2, 16]. At this scale, materials often exhibit novel physical, chemical, and biological properties that differ significantly from their bulk counterparts. Nanomaterials may occur naturally in the environment or be artificially synthesized [67]. The latter category includes anthropogenic, manufactured, or engineered nanoparticles (NPs), which are intentionally designed for specific applications across various sectors [50]. Due to their unique characteristics, such as high surface area-to-volume ratio, enhanced reactivity, high surface energy, and strong adsorption capacity [34, 65]. Nanomaterials have attracted widespread attention in fields as diverse as electronics [111], energy storage [80], cosmetics [101], medicine [12, 33], and increasingly, agriculture and the food sector [61, 68]. These properties make them especially suitable for developing innovative tools to support sustainable and efficient agricultural practices [98].

In agriculture, nanotechnology has opened new frontiers in areas such as crop protection, crop improvement, and environmental remediation [24, 39]. Engineered nanomaterials are used to design nanoagrochemicals, including nanofertilizers, nanopesticides, and nanoherbicides, that improve nutrient delivery and pest control with reduced environmental impact [46, 48, 66]. NPs can also act as carriers for biomolecules, such as DNA, RNA, or proteins, enhancing targeted delivery in genetic improvement programs [105, 112]. Moreover, nanomaterials are increasingly used in precision agriculture through the development of nanosensors that monitor soil health, plant nutrient status, and pathogen presence in real-time [25, 30, 94]. Other promising applications include biofortification, to improve the nutritional content of crops, and phytoremediation, where nanomaterials assist in detoxifying polluted soils [14, 17, 75]. To support these advances, researchers have developed a wide range of nanomaterials using various fabrication techniques such as emulsification, ionic gelation, polymerization, and redox reactions [3, 44, 83]. These include inorganic, lipid-based, and polymeric nanoparticles, each offering specific advantages for agricultural use. The overarching goal of employing nanotechnology in agriculture is to enhance crop productivity [103], reduce reliance on harmful chemical inputs, improve nutrient-use efficiency [38], manage pests and diseases more sustainably [62], and increase the quality and safety of agricultural products [95, 99]. Amid the increasing global demand for food, driven by population growth and urban expansion, agriculture faces mounting pressures [49]. These include the negative effects of climate change, declining soil fertility, nutrient losses, and a steady reduction in arable land [13]. Addressing these complex challenges requires transformative technologies, and nanotechnology holds immense potential in this regard. It offers scalable solutions in plant genome editing, fertilizer management, and plant growth enhancement, thereby contributing to climate-resilient and resource-efficient agriculture [39, 78].

One of the most promising biopolymers explored in this context is chitosan [54]. Derived from the deacetylation of chitin, the second most abundant natural polysaccharide after cellulose, chitosan is obtained primarily from the exoskeletons of crustaceans and fungal cell walls [8, 52]. It is a biodegradable, edible, non-toxic, and biocompatible polymer, which has gained increasing attention for its versatility in agricultural applications [8]. In conventional forms, chitosan is widely used as a biostimulant, enhancing seed germination, root elongation, and plant growth by stimulating physiological processes and nutrient uptake [11, 59, 97]. It also activates plant defense pathways, making it effective in mitigating the impact of pathogens [37]. Additionally, chitosan exhibits broad-spectrum antimicrobial and antifungal properties, positioning it as a natural biocontrol agent [51]. Its film-forming ability further enables applications in postharvest preservation by reducing moisture loss and microbial decay, thereby extending the shelf life of perishable produce [42, 88]. These beneficial properties of bulk chitosan are further amplified when it is formulated into chitosan nanoparticles (CNPs). At the nanoscale, chitosan exhibits significantly enhanced surface area, reactivity, and bioavailability [86]. CNPs have shown great promise in agriculture as carriers for the controlled release of agrochemicals, improving the efficiency and reducing the environmental footprint of fertilizers and pesticides [34, 57, 107]. They can also serve as nanocarriers for gene delivery and plant signaling molecules, playing a critical role in next-generation crop improvement technologies [18, 36]. Moreover, CNPs are capable of eliciting plant immune responses more effectively than their bulk counterparts, offering a potential tool for inducing systemic resistance in crops against a range of pathogens [23, 93].

However, despite their growing potential and promising results in laboratory and greenhouse settings, the use of chitosan nanoparticles in agriculture is not without challenges [104]. One major obstacle is the lack of standardized and scalable production protocols, which can lead to inconsistencies in nanoparticle size, stability, and functional performance [34]. Variability in synthesis methods can affect the reproducibility and efficacy of CNP-based formulations, thereby limiting their commercial viability [84]. Another critical challenge lies in our incomplete understanding of how nanoparticles interact with complex biological and environmental systems [73]. When applied to soils or plants, nanoparticles may undergo transformations that influence their bioavailability, mobility, and toxicity [69]. These interactions can have unintended effects on non-target organisms, including beneficial soil microbes, insects, and aquatic life [104]. Additionally, the long-term persistence of nanoparticles in the environment and their accumulation in food chains are not yet fully understood, raising important questions about ecotoxicological risks [20, 40]. Regulatory hurdles also present significant barriers to the adoption of nanomaterials in agriculture [20]. In many countries, regulations for nanotechnology-based agricultural inputs remain underdeveloped, leading to uncertainty in risk assessment, approval procedures, and labeling requirements [5]. The absence of harmonized international

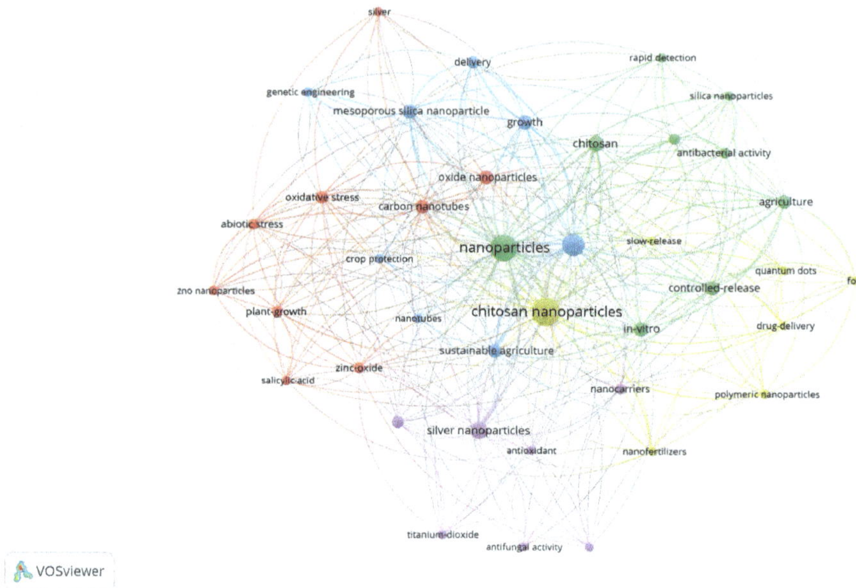

Fig. 1 Network analysis with keywords used for chitosan nanoparticles challenges in the agricultural sector between 2004 and 2024

guidelines further complicates the commercialization of CNP-based products, particularly in global markets [5, 100]. Figure 1 illustrates the main related challenges to the CNPs application in agriculture.

Given these challenges, there is a pressing need for interdisciplinary research to advance the safe and effective integration of chitosan nanoparticles into agricultural systems [57]. This includes optimizing synthesis techniques, conducting long-term environmental and toxicological studies, and developing robust regulatory frameworks that balance innovation with safety. In this context, the present chapter aims to provide a comprehensive overview of the current challenges associated with the use of chitosan nanoparticles in agriculture. By critically examining the scientific, technical, and regulatory barriers, we hope to support the responsible development and application of CNPs as part of a broader transition toward sustainable and resilient agricultural systems.

2 Synthesis and Formulation Challenges of Chitosan-Based Nanoparticles

CNPs have emerged as a versatile and eco-friendly class of nanomaterials with wide-ranging applications in agriculture, particularly in controlled-release formulations of fertilizers, biostimulants, and crop protection agents [30, 34]. However, the efficiency, safety, and applicability of CNPs are heavily influenced by the synthesis and formulation methods used [84]. These methods govern key physicochemical properties such as size, surface charge, stability, solubility, and bioavailability, which directly affect their interaction with plant systems and the environment [34].

Chitosan is a naturally occurring cationic polysaccharide obtained through the N-deacetylation of chitin, composed of β-(1,4)-linked D-glucosamine and N-acetyl-D-glucosamine units [52]. Valued for its biodegradability, biocompatibility, and polycationic nature, chitosan interacts electrostatically with negatively charged compounds, making it ideal for encapsulating agrochemicals [70]. Its functional versatility allows for nanoparticle formation under mild, aqueous conditions without the need for toxic solvents or harsh reagents. A range of synthesis techniques has been developed for fabricating CNPs [45]. Among these, emulsion cross-linking and ionic gelation are the most widely used [32, 56]. Emulsion cross-linking involves dispersing chitosan in a water-in-oil (w/o) emulsion stabilized by surfactants, followed by cross-linking with agents like glutaraldehyde or genipin [85]. While this method provides good particle size control and high loading capacity, it requires long processing times, multiple purification steps, and the use of organic solvents, which raise environmental and toxicity concerns [77]. Ionic gelation has gained significant attention due to its simplicity and environmental compatibility [110]. This technique leverages electrostatic interactions between chitosan's protonated amino groups and polyanions like tripolyphosphate (TPP) [32]. It avoids the use of chemical cross-linkers or solvents and operates under mild conditions. Advantages include low energy requirements, ease of scale-up, and tunability of nanoparticle characteristics via parameters such as molecular weight, chitosan-to-TPP ratio, and pH [32]. However, this method is sensitive to subtle changes in synthesis conditions, affecting reproducibility and standardization [41]. Other synthesis methods, such as reverse micelle formation, spray drying, and electrospraying are used for specialized applications [34]. Spray drying enables large-scale production but is sensitive to temperature and flow rates, which can degrade active ingredients [84]. Reverse micelle methods offer good size control but involve toxic solvents and require precise surfactant control [19]. Electrospraying and electrospinning produce CNPs or nanofibers with custom architectures but demand advanced equipment and remain limited in agricultural deployment [115].

Despite their promise, CNPs development for agriculture faces several challenges. Variability in raw materials and synthesis protocols leads to inconsistent nanoparticle characteristics, complicating standardization and regulatory approval [34, 45]. The use of cross-linkers, surfactants, or solvents can introduce toxic residues, raising environmental and health concerns [87]. Additionally, ensuring nanoparticle stability during storage and transport, especially under fluctuating temperature and humidity, is critical for commercial viability [55]. Effective agricultural use requires tailoring CNPs for specific applications such as foliar sprays, seed coatings, or soil amendments, with control over properties like size, surface charge, and release profiles [34]. Moreover, many lab-scale synthesis techniques do not translate easily to industrial-scale production without compromising functionality or increasing costs [108]. Amid growing ecological concerns around synthetic agrochemicals, there is a strong interest in green and biodegradable chitosan nanoformulations [116]. Accordingly, to fully harness the potential of chitosan-based nanoparticles, future research should focus on: (1) optimizing synthesis parameters for targeted delivery and controlled release [81] (2) establishing standardized protocols to enhance reproducibility and regulatory compliance [29] (3) incorporating surface functionalization to improve plant interactions [23], and (4) conducting long-term environmental impact assessments through field studies [63].

3 Challenges in the Use of Chitosan Nanoparticles for Plants and Soil Applications

CNPs have emerged as promising tools for sustainable agriculture, offering potential applications in plant protection, nutrient delivery, and stress mitigation [31]. However, despite their advantages, the practical application of CNPs in plant and soil systems faces several significant challenges that hinder widespread adoption [34]. One major limitation is the inherent solubility and stability issues of chitosan, particularly in environments with varying pH and ionic strength [102]. As a cationic polymer, chitosan tends to precipitate under neutral or alkaline conditions, which are common in many agricultural soils [34]. Although acidic solvents such as acetic acid are often used to improve solubility, their application may cause phytotoxic effects and reduce chitosan's intrinsic antimicrobial activity. Additionally, the technical complexity of nanoparticle synthesis remains a major obstacle [4]. Various methods, such as ionic gelation, emulsion-droplet coalescence, and spray drying, require precise control of particle size, zeta potential, and encapsulation efficiency [114]. Achieving uniform, reproducible batches with desirable physicochemical properties is still difficult, and current synthesis protocols are often not scalable or cost-effective for large-scale agricultural use [114]. Furthermore, chitosan-based nanomaterials exhibit concentration-dependent effects on plant physiology [92]. While low concentrations may enhance germination, growth, and stress resistance, higher doses have been reported to negatively impact seedling vigor, biomass accumulation, and germination rates in several crop species, including maize, wheat, and chickpea [43, 47, 64]. This dose sensitivity complicates application strategies and underscores the need for species-specific calibration of nanoparticle formulations. Compounding these challenges is the lack of standardized field-level evaluations [106]. The majority of available data are derived from laboratory or greenhouse studies, and the real-world performance of CNPs under diverse environmental conditions, soil types, and cropping systems remains poorly characterized [9]. There is limited understanding of how chitosan nanomaterials interact with native soil microbiota, organic matter, and other agrochemicals over time, raising concerns about potential unintended ecological effects. The mechanistic basis for many of the observed biological responses is also not fully understood [27]. Although CNPs have been shown to induce plant defense pathways, enhance nutrient uptake, and facilitate gene delivery, the underlying molecular and physiological processes remain speculative in many cases [26]. This knowledge gap restricts efforts to develop targeted or optimized formulations. Moreover, the environmental and regulatory landscape surrounding nanomaterials in agriculture is still evolving. While chitosan itself is biodegradable and recognized as safe (GRAS) by regulatory agencies like the United States Food and Drug Administration (U.S. FDA), the incorporation of metal nanoparticles (e.g., silver, copper, zinc) into chitosan matrices raises questions about nanotoxicity, persistence, and bioaccumulation in soil and water systems [71]. The absence of clear guidelines for the assessment and approval of chitosan-based nanopesticides and nanofertilizers hinders their commercial deployment.

Economically, the high cost of production and the need for high-purity chitosan further limit accessibility, particularly in resource-limited agricultural settings [79]. Scaling up production while maintaining functional integrity and affordability is a challenge yet to be resolved. Additionally, although the integration of chitosan-based nanosensors and smart delivery systems holds theoretical potential in precision agriculture [25], their practical implementation remains minimal due to technological and economic constraints. In summary, while chitosan nanoparticles represent an exciting frontier in agro-nanotechnology, their application in real-world farming is constrained by multiple interrelated barriers, including formulation challenges, environmental safety concerns, limited field validation, regulatory uncertainty, and economic viability [114]. Addressing these issues through interdisciplinary research, policy development, and industry collaboration will be critical to unlocking the full potential of chitosan-based nanocomposites for sustainable plant and soil management.

4 Challenges Related to Environmental and Ecotoxicological Concerns

Despite the growing interest in CNPs as eco-friendly and biodegradable alternatives for environmental and agricultural applications, several environmental and ecotoxicological concerns remain unresolved. As highlighted by [28], the incorporation of chitosan with metallic nanoparticles, particularly magnetized iron oxide (Fe_2O_3/Fe_3O_4), may induce unintended biological effects that warrant careful consideration. One of the primary concerns is the acute toxicity of CNPs to aquatic microbial organisms [1]. In standardized bioluminescence assays using *Aliivibrio fischeri*, a marine bacterium commonly employed in environmental toxicology, chitosan and its nanostructured composites exhibited notable toxicity. The inhibition of bioluminescence, a proxy for cellular stress and metabolic disruption, exceeded 50% for both native chitosan and magnetized chitosan (Qpm1%), with EC_{50} values as low as 26.5% [72]. Even pyrolyzed forms with reduced organic content retained moderate toxic potential. These findings suggest that despite chitosan's natural origin, its nano-engineered forms can elicit significant biological stress in aquatic systems, potentially disturbing microbial communities essential for nutrient cycling, water purification, and ecological balance [53]. In addition to microbial toxicity, the phytotoxic potential of CNPs remains a critical concern. Although germination tests using *Eruca sativa* (arugula) seeds revealed no significant inhibition in germination rates, variations were observed in post-germination seedling development. Specifically, root and shoot elongation were altered depending on the concentration and structural composition of the nanoparticles. Magnetized chitosan with 3% chitosan content (Qpm3%) led to reduced shoot growth, whereas pure magnetized iron nanoparticles (NPfe) caused minimal changes [6, 90]. These results imply that while initial germination may be unaffected, subtle morphological or physiological effects could occur during early seedling development, possibly affecting plant vigor and competitiveness under field conditions.

Another layer of complexity arises from the structure-dependent toxicity of CNPs. For instance, non-pyrolyzed chitosan had greater ecotoxicity than pyrolyzed forms, suggesting that residual organic groups may enhance reactivity with biological membranes [28]. Similarly, hybrid structures combining chitosan with iron oxides displayed synergistic or additive toxic effects, surpassing those of iron oxide alone [22]. This indicates that nanoparticle behavior and toxicity cannot be predicted solely based on individual component properties, but rather depend on their composite structure, surface charge, porosity, and degree of degradation [114]. Such interactions complicate environmental risk assessments and highlight the need for precise characterization of nanoparticle formulations. Furthermore, concerns persist regarding the fate and persistence of CNPs in the environment [35]. The thermal and structural analyses conducted by [28] demonstrated that pyrolyzed and magnetized chitosan particles can remain stable under high-temperature treatment and retain mesoporous architectures. This stability, while advantageous for pollutant adsorption, raises questions about environmental persistence, especially in soils or sediments where degradation rates are slower. Additionally, their high surface area and chemical reactivity could facilitate the adsorption and transport of other contaminants, potentially acting as vectors for heavy metals or persistent organic pollutants through ecosystems. The study also underscores a lack of comprehensive data on the chronic effects and bioaccumulation potential of CNPs [28]. While acute exposures may reveal short-term impacts, sub-lethal or long-term consequences on soil fauna, aquatic invertebrates, and higher trophic levels remain poorly understood. For instance, the adherence of magnetized particles to root tissues and filter paper observed during germination tests indicates potential for nanoparticle uptake or physical interference with plant-soil interactions [6]. Whether such adherence leads to internalization, translocation, or accumulation in edible plant parts is an open question requiring further investigation [114]. In addition, methodological and regulatory challenges hinder the safe deployment of chitosan-based nanomaterials [114]. The absence of standardized testing protocols specific to bio-polymeric nanocomposites limits comparability and reproducibility across studies. Moreover, regulatory frameworks in most regions do not yet differentiate between conventional chitosan and its nano-enabled counterparts, despite their distinct environmental behaviors [114]. As a result, there is an urgent need for the development of tailored risk assessment tools, ecotoxicity benchmarks, and policy guidelines that account for the unique characteristics of CNPs. While chitosan nanoparticles hold significant promise for environmental remediation and agricultural innovation, their environmental safety profile must be rigorously assessed [34]. Acute microbial toxicity, structure-dependent effects, potential phytotoxicity, environmental persistence, and the current lack of regulatory oversight highlight the importance of adopting a precautionary and research-driven approach to their use [60]. Continued interdisciplinary research is needed to bridge the knowledge gap and guide the responsible application of CNPs in sustainable environmental technologies.

5 Current and Future Prospects of Chitosan-Based Nanomaterials

Modern agriculture faces escalating challenges due to climate change, reduced arable land, and increasing global food demands. Traditional farming techniques and conventional inputs have become insufficient to sustainably enhance crop yields, particularly under biotic and abiotic stress conditions [7, 58]. This has prompted a shift toward more sustainable and innovative approaches. Among these, the development and deployment of chitosan-based nanomaterials represent a promising frontier in plant science, offering unique advantages for plant growth promotion, stress mitigation, and disease control [117]. Chitosan, a natural amino-polysaccharide derived from chitin, has long been recognized for its biocompatibility, biodegradability, and bioactivity, including antimicrobial, antifungal, and plant defense-eliciting properties. Recent advancements in nanotechnology have enabled the transformation of bulk chitosan into nanochitosan, thereby significantly improving its physicochemical and biological performance. Nanochitosan exhibits superior solubility, surface area, and reactivity compared to its bulk counterpart, enabling enhanced penetration into plant tissues, systemic action, and more controlled interaction with plant systems [74].

In the current context, CNPs have demonstrated significant potential in improving seed germination, shoot and root elongation, nutrient uptake, and tolerance to stressors [82]. Empirical studies have shown that nanochitosan treatments enhance enzymatic activities related to plant defense, such as catalase, peroxidase, polyphenol oxidase, and phenylalanine ammonia lyase, thereby strengthening the plant's innate immune response [23, 96]. Moreover, nanochitosan has been effectively used to mitigate the impact of various pathogens and improve resistance to fungal infections in crops like rice, tomato, cucumber, and chilli [27, 89]. Particularly noteworthy is the development of metal-chitosan nanocomposites, such as zinc-chitosan and copper-chitosan nanoparticles. These composites have shown not only enhanced plant growth-promoting effects but also significant antimicrobial activities [15]. For instance, chitosan-Zn nanocomposites have been shown to improve seed germination, seedling vigor, and enzymatic activity in maize, while also exhibiting antibacterial effects against *E. coli* and *Corynebacterium* [10]. Similarly, Cu-chitosan nanoparticles have demonstrated strong antifungal efficacy, notably against *A. solani* and *F. oxysporum*, and have been associated with increased seed germination and vigor in tomato fruits [109]. Importantly, these nanoparticles combine the micronutrient benefits of metals with the elicitor properties of chitosan, offering a dual mechanism for enhancing plant productivity and health [15]. In addition to direct plant benefits, chitosan-based nanomaterials have promising applications in post-harvest management. Chitosan coatings and nano-films, especially when combined with bioactive compounds like propolis or essential oils, have been shown to reduce microbial spoilage and prolong the shelf life of fruits and vegetables [113]. Such applications can play a crucial role in minimizing post-harvest losses and enhancing food safety. Looking forward, the future of chitosan-based nanotechnology in agriculture lies in the development of smart delivery systems [114]. Owing to the pH-sensitive nature of chitosan, nanoformulations can be designed to release agrochemicals, such as pesticides, fertilizers, or micronutrients, in response to specific environmental cues, like the acidification of tissues during pathogen infection [76]. This targeted release mechanism can significantly reduce the amount of agrochemicals required, lowering costs, minimizing environmental contamination, and avoiding negative plant feedback responses associated with over-application [91]. Moreover, there is great potential in designing nano-bioconjugates (NBCs) using chitosan as a base [21]. These NBCs can encapsulate multiple biologically active agents, such as micronutrients (e.g., Zn, Cu, Se), phytohormones (salicylic acid, jasmonic acid), or secondary metabolites (e.g., neem oil, thymol), to simultaneously promote growth, enhance stress resistance, and suppress pathogens [86]. This multifunctional approach aligns well with the principles of sustainable agriculture. However, while early studies have yielded promising results, the full-scale agricultural application of chitosan-based nanomaterials remains limited [114]. Challenges such as a lack of molecular-level understanding, potential ecotoxicity, regulatory ambiguity, and limited field validation need to be addressed. There is a pressing need for further research to elucidate the precise mechanisms of action, optimal dosages, and long-term effects of these nanomaterials on plant physiology and soil ecosystems. Chitosan-based nanomaterials hold immense potential to revolutionize crop production and protection strategies sustainably. Through further innovation, field testing, and integration with precision agriculture, they could become key components in meeting the global food security challenges of the twenty-first century.

6 Conclusion

CNPs stand at the intersection of nanotechnology and sustainable agriculture, offering promising avenues for addressing some of the most pressing challenges facing modern food systems. Their unique physicochemical properties, enhanced surface area, biocompatibility, and capacity for controlled release enable a wide range of applications, from targeted delivery of agrochemicals to elicitation of plant defense responses. These capabilities not only promise to increase crop productivity and resilience but also reduce the environmental footprint of conventional agricultural inputs. The adaptability of chitosan to serve as a nanocarrier, biostimulant, and antimicrobial agent further underscores its potential as a multifunctional tool for integrated crop management. However, the translation of CNPs from laboratory research to practical field applications remains constrained by several critical challenges. These include inconsistencies in nanoparticle synthesis and formulation, uncertainties surrounding environmental fate and toxicity, and a lack of standardized regulatory frameworks to guide safe usage. The complex interactions of nanoparticles with biological systems and ecosystems, while offering new opportunities, also demand caution and rigorous assessment. Without a deeper understanding of these interactions, and in the absence of harmonized regulatory oversight, the broad-scale deployment of CNPs in agriculture risks unintended consequences. To unlock the full potential of chitosan nanotechnology, a coordinated effort involving materials scientists, agronomists, environmental researchers, and policymakers is essential. Future research should prioritize scalable green synthesis methods, comprehensive risk assessments, and field-based validations across diverse agroecological contexts. Additionally, the development of international standards and regulatory guidelines will be vital to ensure both environmental safety and market access. As agriculture confronts the twin pressures of climate change and population growth, chitosan-based nanomaterials may play a pivotal role in building more resilient, productive, and environmentally sound farming systems. The insights gathered in this chapter aim to catalyze further inquiry and innovation, ultimately contributing to the responsible integration of nanotechnology in sustainable agriculture.

References

1. Abbas Q, Yousaf B, Ullah H, Ali MU, Ok YS, Rinklebe J (2020) Environmental transformation and nano-toxicity of engineered nano-particles (ENPs) in aquatic and terrestrial organisms. Crit Rev Environ Sci Technol 50(23):2523–2581
2. Abdel Latef AAH, Zayed EM, Omar AA (2025) Bioremediation and Nanotechnology for Climate Change Mitigation. Springer Nature, Singapore Pte Ltd. pp 1–652. https://doi.org/10.1007/978-981-96-3069-1
3. Alshammari BH, Lashin MM, Mahmood MA, Al-Mubaddel FS, Ilyas N, Rahman N, Sohail M, Khan A, Abdullaev SS, Khan R (2023) Organic and inorganic nanomaterials: fabrication, properties and applications. RSC Adv 13(20):13735–13785

4. Altammar KA (2023) A review on nanoparticles: characteristics, synthesis, applications, and challenges. Front Microbiol 14:1155622
5. Amenta V, Aschberger K, Arena M, Bouwmeester H, Moniz FB, Brandhoff P, Gottardo S, Marvin HJ, Mech A, Pesudo LQ (2015) Regulatory aspects of nanotechnology in the agri/feed/food sector in EU and non-EU countries. Regul Toxicol Pharmacol 73(1):463–476
6. Amiripour H, Iranbakhsh A, Saadatmand S, Mousavi F, Ardebili ZO (2025) Exogenous application of melatonin and chitosan mitigate simulated microgravity stress in the Rocket (Eruca sativa L.) plant. Plant Physiol Biochem 218:109294
7. Anyibama BJ, Orjinta KK, Omisogbon TO, Atalor SI, Daniels EO, Fadipe E, Galadima DA (2025) Modern agricultural technologies for sustainable food production: a comprehensive review of technological innovations and their impact on global food systems. Int J Innov Sci Res Technol 10(2)
8. Aranaz I, Alcántara AR, Civera MC, Arias C, Elorza B, Heras Caballero A, Acosta N (2021) Chitosan: An overview of its properties and applications. Polymers 13(19):3256
9. Arif Y, Siddiqui H, Hayat S (2022) Role of chitosan nanoparticles in regulation of plant physiology under abiotic stress. In: Sustainable agriculture reviews 53: Nanoparticles: A new tool to enhance stress tolerance. Springer, pp 399–413
10. Asghar MA, Ahmed F, Qamar AR, Khan K, Anwar A (2025) Synthesis and characterization of chitosan/zinc oxide nanocomposite for enhanced applications as antibacterial, antifungal and aflatoxin B1 adsorption. J Cluster Sci 36(2):47
11. Attia MS, Osman MS, Mohamed AS, Mahgoub HA, Garada MO, Abdelmouty ES, Abdel Latef AAH (2021) Impact of foliar application of chitosan dissolved in different organic acids on isozymes, protein patterns and physio-biochemical characteristics of tomato grown under salinity stress. Plants 10:388
12. Barkalina N, Charalambous C, Jones C, Coward K (2014) Nanotechnology in reproductive medicine: emerging applications of nanomaterials. Nanomedicine: Nanotechnol, Biol Med 10(5):e921–e938
13. Beillouin D, Cardinael R, Berre D, Boyer A, Corbeels M, Fallot A, Feder F, Demenois J (2022) A global overview of studies about land management, land-use change, and climate change effects on soil organic carbon. Glob Change Biol 28(4):1690–1702
14. Bhardwaj AK, Chejara S, Malik K, Kumar R, Kumar A, Yadav RK (2022) Agronomic biofortification of food crops: An emerging opportunity for global food and nutritional security. Front Plant Sci 13:1055278
15. Bhati VS, Singh PK, Parmar YJ, Rokins PG (2025) Foliar uptake and translocation of nanoparticles in the plants: a review. J Plant Nutr:1–38
16. Bhushan B (2007) Introduction to nanotechnology. In: Springer handbook of nanotechnology. Springer, pp 1–12
17. Bouis HE, Welch RM (2010) Biofortification—a sustainable agricultural strategy for reducing micronutrient malnutrition in the global south. Crop science 50:S-20-S-32
18. Bowman K, Leong KW (2006) Chitosan nanoparticles for oral drug and gene delivery. Int J Nanomed 1(2):117–128
19. Brunel F, Véron L, David L, Domard A, Delair T (2008) A novel synthesis of chitosan nanoparticles in reverse emulsion. Langmuir 24(20):11370–11377
20. Bundschuh M, Filser J, Lüderwald S, McKee MS, Metreveli G, Schaumann GE, Schulz R, Wagner S (2018) Nanoparticles in the environment: where do we come from, where do we go to? Environ Sci Eur 30:1–17
21. Carvalho SM, Mansur AA, Lobato ZI, Leite MF, Mansur HS (2025) Bioengineering chitosan–antibody/fluorescent quantum dot nanoconjugates for targeted immunotheranostics of non-hodgkin B-cell lymphomas. Int J Biol Macromol 294:139515
22. Castelló J, Gallardo M, Busquets MA, Estelrich J (2015) Chitosan (or alginate)-coated iron oxide nanoparticles: a comparative study. Colloids Surf, A 468:151–158
23. Chandra S, Chakraborty N, Dasgupta A, Sarkar J, Panda K, Acharya K (2015) Chitosan nanoparticles: a positive modulator of innate immune responses in plants. Sci Rep 5(1):15195

24. Chen H, Yada R (2011) Nanotechnologies in agriculture: new tools for sustainable development. Trends Food Sci Technol 22(11):585–594
25. Chhipa H, Joshi P (2016) Nanofertilisers, nanopesticides and nanosensors in agriculture. Nanosci Food Agric 1:247–282
26. Choudhary RC, Kumari S, Kumaraswamy RV, Sharma G, Kumar A, Budhwar S, Pal A, Raliya R, Biswas P, Saharan V (2019) Chitosan nanomaterials for smart delivery of bioactive compounds in agriculture. In: Nanoscale engineering in agricultural management. CRC Press, pp 124–139
27. Chun S-C, Chandrasekaran M (2019) Chitosan and chitosan nanoparticles induced expression of pathogenesis-related proteins genes enhances biotic stress tolerance in tomato. Int J Biol Macromol 125:948–954
28. Coelho FP, Rabelo TS, da Cruz Felix L, Bila DM, Ritter E (2020) Characterization and ecotoxicological evaluation of nanostructured chitosan particles. In: Sustainable environmental geotechnics: Proceedings of EGRWSE 2019. Springer, pp 173–183
29. Dadashi H, Vandghanooni S, Karamnejad-Faragheh S, Karimian-Shaddel A, Eskandani M, Jahanban-Esfahlan R (2024) A rapid protocol for synthesis of chitosan nanoparticles with ideal physicochemical features. Heliyon 10(11)
30. Darwish MA, Abd-Elaziem W, Elsheikh A, Zayed AA (2024) Advancements in nanomaterials for nanosensors: a comprehensive review. Nanoscale Adv
31. de Oliveira ALB, Cavalcante FTT, da Silva Moreira K, Lima PJM, de Castro Monteiro RR, Pinheiro BB, dos Santos KP, dos Santos JCS (2021) Chitosan nanoparticle: alternative for sustainable agriculture. Nanomater Nanotechnol: Biomed, Environ, Ind Appl:95–132
32. Desai KG (2016) Chitosan nanoparticles prepared by ionotropic gelation: An overview of recent advances. Crit Reviews^{TM} Ther Drug Carr Syst 33(2)
33. Desai N, Rana D, Salave S, Gupta R, Patel P, Karunakaran B, Sharma A, Giri J, Benival D, Kommineni N (2023) Chitosan: a potential biopolymer in drug delivery and biomedical applications. Pharmaceutics 15(4):1313
34. Divya K, Jisha MS (2018) Chitosan nanoparticles preparation and applications. Environ Chem Lett 16:101–112
35. Djajadi DT, Müller S, Fiutowski J, Rubahn H-G, Thygesen LG, Posth NR (2024) Interaction of chitosan with nanoplastic in water: the effect of environmental conditions, particle properties, and potential for in situ remediation. Sci Total Environ 907:167918
36. Duceppe N, Tabrizian M (2010) Advances in using chitosan-based nanoparticles for in vitro and in vivo drug and gene delivery. Expert Opin Drug Deliv 7(10):1191–1207
37. El Hadrami A, Adam LR, El Hadrami I, Daayf F (2010) Chitosan in plant protection. Mar Drugs 8(4):968–987
38. Elemike EE, Uzoh IM, Onwudiwe DC, Babalola OO (2019) The role of nanotechnology in the fortification of plant nutrients and improvement of crop production. Appl Sci 9(3):499
39. El-Saadony MT, Saad AM, Soliman SM, Salem HM, Desoky E-SM, Babalghith AO, El-Tahan AM, Ibrahim OM, Ebrahim AA, Abd El-Mageed TA (2022) Role of nanoparticles in enhancing crop tolerance to abiotic stress: A comprehensive review. Front Plant Sci 13:946717
40. Espinasse BP, Geitner NK, Schierz A, Therezien M, Richardson CJ, Lowry GV, Ferguson L, Wiesner MR (2018) Comparative persistence of engineered nanoparticles in a complex aquatic ecosystem. Environ Sci Technol 52(7):4072–4078
41. Esquivel R, Juárez J, Almada M, Ibarra J, Valdez MA (2015) Synthesis and characterization of new thiolated chitosan nanoparticles obtained by ionic gelation method. Int J Polym Sci 2015(1):502058
42. Fundo JF, Galvis-Sanchez AC, Delgadillo I, Silva CL, Quintas MA (2015) The effect of polymer/plasticiser ratio in film forming solutions on the properties of chitosan films. Food Biophys 10:324–333
43. Furbank RT, White R, Palta JA, Turner NC (2004) Internal recycling of respiratory CO_2 in pods of chickpea (Cicer arietinum L.): the role of pod wall, seed coat, and embryo. J Exp Bot 55(403):1687–1696

44. Gedanken A (2004) Using sonochemistry for the fabrication of nanomaterials. Ultrason Sonochem 11(2):47–55
45. Grenha A (2012) Chitosan nanoparticles: a survey of preparation methods. J Drug Target 20(4):291–300
46. Grillo R, Abhilash PC, Fraceto LF (2016) Nanotechnology applied to bio-encapsulation of pesticides. J Nanosci Nanotechnol 16(1):1231–1234
47. Guan Y, Hu J, Wang X, Shao C (2009) Seed priming with chitosan improves maize germination and seedling growth in relation to physiological changes under low temperature stress. J Zhejiang Univ Sci B 10:427–433
48. Harish V, Tewari D, Gaur M, Yadav AB, Swaroop S, Bechelany M, Barhoum A (2022) Review on nanoparticles and nanostructured materials: Bioimaging, biosensing, drug delivery, tissue engineering, antimicrobial, and agro-food applications. Nanomaterials 12(3):457
49. Hazell P, Wood S (2008) Drivers of change in global agriculture. Philos Trans R Soc B: Biol Sci 363(1491):495–515
50. Hochella Jr MF, Mogk DW, Ranville J, Allen IC, Luther GW, Marr LC, McGrail BP, Murayama M, Qafoku NP, Rosso KM (2019) Natural, incidental, and engineered nanomaterials and their impacts on the Earth system. Science 363(6434):eaau8299
51. Hosseinnejad M, Jafari SM (2016) Evaluation of different factors affecting antimicrobial properties of chitosan. Int J Biol Macromol 85:467–475
52. Iber BT, Kasan NA, Torsabo D, Omuwa JW (2022) A review of various sources of chitin and chitosan in nature. J Renew Mater 10(4):1097
53. Isibor PO, Kayode-Edwards II, Onwaeze OO (2024) Prospects and challenges of nanochitosan application in aquaculture. Nanochitosan-based enhancement of fisheries and aquaculture: Aligning with sustainable development goal 14–life below water, pp 301–320
54. Jaworska M, Sakurai K, Gaudon P, Guibal E (2003) Influence of chitosan characteristics on polymer properties. I: Crystallographic properties. Polym Int 52(2):198–205
55. Jonassen H, Kjøniksen A-L, Hiorth M (2012) Stability of chitosan nanoparticles cross-linked with tripolyphosphate. Biomacromol 13(11):3747–3756
56. Kafshgari MH, Khorram M, Khodadoost M, Khavari S (2011) Reinforcement of chitosan nanoparticles obtained by an ionic cross-linking process. Iran Polym J 20(5):445–456
57. Kashyap PL, Xiang X, Heiden P (2015) Chitosan nanoparticle based delivery systems for sustainable agriculture. Int J Biol Macromol 77:36–51
58. Kassout J, El Issaoui K, Azenzem R, Chokrane B, Chakkour S, Kadaoui K, Sahli A, Oulbi S, Chraka A, Mazri MA (2025) Phytohormone-mediated protection in oilseed crops: a key strategy against temperature stress. In: Oilseed crops under abiotic stress: Mitigation strategies and future perspectives. Springer, pp 391–413
59. Katiyar D, Hemantaranjan A, Singh B (2015) Chitosan as a promising natural compound to enhance potential physiological responses in plant: a review. Indian J Plant Physiol 20:1–9
60. Khan Z, Maqsood Q, Baradoke A, Ferreira LFR, Franco M, Schmidt JE, Hussain N (2024) Environmental and toxicological implications of chitosan nanostructures. In: Advances in chemical pollution, environmental management and protection. Elsevier, pp 137–172
61. Khot LR, Sankaran S, Maja JM, Ehsani R, Schuster EW (2012) Applications of nanomaterials in agricultural production and crop protection: a review. Crop Prot 35:64–70
62. Kumar M, Shamsi TN, Parveen R, Fatima S (2017) Application of nanotechnology in enhancement of crop productivity and integrated pest management. Nanotechnol: Agric Parad:361–371
63. Leceta I, Guerrero P, Cabezudo S, de la Caba K (2013) Environmental assessment of chitosan-based films. J Clean Prod 41:312–318
64. Li R, He J, Xie H, Wang W, Bose SK, Sun Y, Hu J, Yin H (2019) Effects of chitosan nanoparticles on seed germination and seedling growth of wheat (Triticum aestivum L.). Int J Biol Macromol 126:91–100
65. Lines MG (2008) Nanomaterials for practical functional uses. J Alloy Compd 449(1–2):242–245

66. Liu R, Lal R (2015) Potentials of engineered nanoparticles as fertilizers for increasing agronomic productions. Sci Total Environ 514:131–139
67. Lowry GV, Gregory KB, Apte SC, Lead JR (2012) Transformations of nanomaterials in the environment. ACS Publications
68. Magnuson BA, Jonaitis TS, Card JW (2011) A brief review of the occurrence, use, and safety of food-related nanomaterials. J Food Sci 76(6):R126–R133
69. Malerba M, Cerana R (2016) Chitosan effects on plant systems. Int J Mol Sci 17(7):996
70. Maluin FN, Hussein MZ (2020) Chitosan-based agronanochemicals as a sustainable alternative in crop protection. Molecules 25(7):1611
71. Marques C, Som C, Schmutz M, Borges O, Borchard G (2020) How the lack of chitosan characterization precludes implementation of the safe-by-design concept. Front Bioeng Biotechnol 8:165
72. Mirjani M, Soleimani M, Salari V (2021) Toxicity assessment of total petroleum hydrocarbons in aquatic environments using the bioluminescent bacterium Aliivibrio fischeri. Ecotoxicol Environ Saf 207:111554
73. Mu Q, Jiang G, Chen L, Zhou H, Fourches D, Tropsha A, Yan B (2014) Chemical basis of interactions between engineered nanoparticles and biological systems. Chem Rev 114(15):7740–7781
74. Nandini T, Sudhalakshmi C, Sivakumar K, Parameswari E, Thangamani C (2025) A review-chitosan nanoparticles towards enhancing nutrient use efficiency in crops. Int J Biol Macromol:141433
75. Nestel P, Bouis HE, Meenakshi JV, Pfeiffer W (2006) Biofortification of staple food crops. J Nutr 136(4):1064–1067
76. Pal P, Singh AK, Singh MP, Prakash O, Prakash A, Pandey B, Agarwal V, Sharma L, Gupta R (2025) Role of chitin-based nanoparticles for crop nutrition and fertilization. In: Chitin-based nanoparticles for the agriculture sectors. Springer, pp 103–131
77. Pan C, Qian J, Fan J, Guo H, Gou L, Yang H, Liang C (2019) Preparation nanoparticle by ionic cross-linked emulsified chitosan and its antibacterial activity. Colloids Surf, A 568:362–370
78. Patil SS, Shedbalkar UU, Truskewycz A, Chopade BA, Ball AS (2016) Nanoparticles for environmental clean-up: a review of potential risks and emerging solutions. Environ Technol Innov 5:10–21
79. Perera U, Rajapakse N (2013) Chitosan nanoparticles: preparation, characterization, and applications. In: Seafood processing by-products: trends and applications. Springer, pp 371–387
80. Pomerantseva E, Bonaccorso F, Feng X, Cui Y, Gogotsi Y (2019) Energy storage: The future enabled by nanomaterials. Science 366(6468):eaan8285
81. Poznanski P, Hameed A, Orczyk W (2023) Chitosan and chitosan nanoparticles: parameters enhancing antifungal activity. Molecules 28(7):2996
82. Prakash O, Kumari B, Pal P, Singh MP, Prakash A, Pandey B, Agarwal V, Singh AK (2025) Chitin-based nanoparticles for crop plant growth promotion. Chitin-Based Nanoparticles Agric Sect:133–154
83. Putzbach W, Ronkainen NJ (2013) Immobilization techniques in the fabrication of nanomaterial-based electrochemical biosensors: A review. Sensors 13(4):4811–4840
84. Rampino A, Borgogna M, Blasi P, Bellich B, Cesàro A (2013) Chitosan nanoparticles: Preparation, size evolution and stability. Int J Pharm 455(1–2):219–228
85. Riegger BR, Bäurer B, Mirzayeva A, Tovar GE, Bach M (2018) A systematic approach of chitosan nanoparticle preparation via emulsion crosslinking as potential adsorbent in wastewater treatment. Carbohyd Polym 180:46–54
86. Riseh RS, Vazvani MG, Vatankhah M, Kennedy JF (2024) Chitosan coating of seeds improves the germination and growth performance of plants: A review. Int J Biol Macromol:134750
87. Rizeq BR, Younes NN, Rasool K, Nasrallah GK (2019) Synthesis, bioapplications, and toxicity evaluation of chitosan-based nanoparticles. Int J Mol Sci 20(22):5776
88. Romanazzi G, Feliziani E, Sivakumar D (2018) Chitosan, a biopolymer with triple action on postharvest decay of fruit and vegetables: Eliciting, antimicrobial and film-forming properties. Front Microbiol 9:2745

89. Román-Doval R, Torres-Arellanes SP, Tenorio-Barajas AY, Gómez-Sánchez A, Valencia-Lazcano AA (2023) Chitosan: Properties and its application in agriculture in context of molecular weight. Polymers 15(13):2867
90. Ruszczyńska M, Pączkowski C, Szakiel A (2024) The effect of foliar application of chitosan on growth and biochemical characteristics of Eruca sativa. IV International Organic Fruit Symposium and II International Organic Vegetable Symposium 1427:109–114
91. Sanghi AK, Kumar R, Chaubey KK, Pal AK, Sharma S, Dayal D, Kumar A, Verma DK (2025) Application of chitin/chitosan based agro nano chemicals in crop protection. In: Chitin-based nanoparticles for the agriculture sectors. Springer, pp 155–177
92. Sangwan S, Sharma P, Wati L, Mehta S (2023) Effect of chitosan nanoparticles on growth and physiology of crop plants. In: Engineered nanomaterials for sustainable agricultural production, soil improvement and stress management. Elsevier, pp 99–123
93. Sathiyabama M, Manikandan A (2021) Foliar application of chitosan nanoparticle improves yield, mineral content and boost innate immunity in finger millet plants. Carbohyd Polym 258:117691
94. Scott N, Chen H (2012) Nanoscale science and engineering for agriculture and food systems. Ind Biotechnol 8(6):340–343
95. Sekhon BS (2014) Nanotechnology in agri-food production: an overview. Nanotechnol, Sci Appl:31–53
96. Selvaraj M, Ramasami N, Jagan EG (2022) The effect of chitosan nanoparticles on immune responses in plants. In: Role of chitosan and chitosan-based nanomaterials in plant sciences. Elsevier, pp 185–196
97. Shahrajabian MH, Chaski C, Polyzos N, Tzortzakis N, Petropoulos SA (2021) Sustainable agriculture systems in vegetable production using chitin and chitosan as plant biostimulants. Biomolecules 11(6):819
98. Sharma B, Tiwari S, Kumawat KC, Cardinale M (2023) Nano-biofertilizers as bio-emerging strategies for sustainable agriculture development: Potentiality and their limitations. Sci Total Environ 860:160476
99. Sharon M, Choudhary AK, Kumar R (2010) Nanotechnology in agricultural diseases and food safety. J Phytol 2(4)
100. Singh A, Singh K, Kumari B, Kumar P-F-St, Planet H (2025) Regulatory frameworks and biosafety protocols of nanomaterials in agro-based applications. Eco-Friendly Nanotechnology: Harnessing small-scale technologies for a cleaner and healthier planet, pp 170–188
101. Singh TG, Sharma N (2016) Nanobiomaterials in cosmetics: current status and future prospects. Nanobiomaterials Galen Formul Cosmet:149–174
102. Sogias IA, Khutoryanskiy VV, Williams AC (2010) Exploring the factors affecting the solubility of chitosan in water. Macromol Chem Phys 211(4):426–433
103. Soni SK, Dogra S, Sharma A, Thakur B, Yadav J, Kapil A, Soni R (2024) Nanotechnology in agriculture: Enhancing crop productivity with sustainable nano-fertilizers and nano-biofertilizers. J Soil Sci Plant Nutr:1–34
104. Stark WJ (2011) Nanoparticles in biological systems. Angew Chem Int Ed 50(6):1242–1258
105. Su W, Xu M, Radani Y, Yang L (2023) Technological development and application of plant genetic transformation. Int J Mol Sci 24(13):10646
106. Syberg K, Hansen SF (2016) Environmental risk assessment of chemicals and nanomaterials—the best foundation for regulatory decision-making? Sci Total Environ 541:784–794
107. Tang Z-X, Qian J-Q, Shi L-E (2007) Preparation of chitosan nanoparticles as carrier for immobilized enzyme. Appl Biochem Biotechnol 136:77–96
108. Thirugnanasambandan T, Gopinath SC (2023) Laboratory to industrial scale synthesis of chitosan-based nanomaterials: A review. Process Biochem 130:147–155
109. Tong M, Zhu W, Ren Y, Li K, Yang B, Zhang R, He J (2025) Preparation of chitosan-based coating enriched with zinc oxide nanoparticles for inhibiting Alternaria alternata on tomato fruits. J Food Meas Charact:1–16
110. Vaezifar S, Razavi S, Golozar MA, Karbasi S, Morshed M, Kamali M (2013) Effects of some parameters on particle size distribution of chitosan nanoparticles prepared by ionic gelation method. J Cluster Sci 24(3):891–903

111. Wu W (2017) Inorganic nanomaterials for printed electronics: a review. Nanoscale 9(22):7342–7372
112. Xu ZP, Zeng QH, Lu GQ, Yu AB (2006) Inorganic nanoparticles as carriers for efficient cellular delivery. Chem Eng Sci 61(3):1027–1040
113. Yan T, Ren Y, Zhang R, Li K, Yang B, Tong M, He J (2025) Biodegradable chitosan-based films decorated with biosynthetic copper oxide nanoparticle for post-harvest tomato preservation. Int J Biol Macromol 295:139595
114. Yu J, Wang D, Geetha N, Khawar KM, Jogaiah S, Mujtaba M (2021) Current trends and challenges in the synthesis and applications of chitosan-based nanocomposites for plants: A review. Carbohyd Polym 261:117904
115. Zhang S, Kawakami K (2010) One-step preparation of chitosan solid nanoparticles by electrospray deposition. Int J Pharm 397(1–2):211–217
116. Zheng Q, Qin D, Wang R, Yan W, Zhao W, Shen S, Huang S, Cheng D, Zhao C, Zhang Z (2022) Novel application of biodegradable chitosan in agriculture: Using green nanopesticides to control Solenopsis invicta. Int J Biol Macromol 220:193–203
117. Zia KM (2025) Applications of modified chitosan nanoparticles in water, food, and agriculture. In: Chitosan: Green derivatization and applications. Springer, pp 239–258

www.ingramcontent.com/pod-product-compliance
Ingram Content Group UK Ltd.
Pitfield, Milton Keynes, MK11 3LW, UK
UKHW020436270926
475829UK00013B/29
* 9 7 8 9 8 1 9 5 6 4 2 1 7 *